LUCIUS JUNIUS MODERATUS
COLUMELLA

ON AGRICULTURE X-XII
ON TREES

EDITED AND TRANSLATED BY
E. S. FORSTER
AND
EDWARD H. HEFFNER

HARVARD UNIVERSITY PRESS
CAMBRIDGE, MASSACHUSETTS
LONDON, ENGLAND

First published 1955
Revised & reprinted 1968
Reprinted 1979, 1993

ISBN 0-674-99449-3

Printed in Great Britain by St Edmundsbury Press Ltd,
Bury St Edmunds, Suffolk, on acid-free paper
Bound by Hunter & Foulis Ltd, Edinburgh, Scotland.

CONTENTS

SIGLA vii

ON AGRICULTURE

Book X 2
The Layout of the Garden. Plants To Be Grown.

Book XI 47
The Bailiff. Distribution of the Farmer's Work throughout the Year. Cultivation of Vegetables and Herbs.

Book XII 173
The Bailiff's Wife and Her Responsibilities. Preparation and Storage of Provisions: Vegetables, Fruits, Cheese, Wine, Olive Oil.

ON TREES 341
The Selecting of Land and Plants for the Vineyard. Cultivation and Pruning of the Vine. Grafting. Protection against Insects. Planting of Specific Types of Trees. Grafting. Willow, Broom, Violet, Rose.

INDEXES 413

SIGLA

S = Cod. Sangermanensis Petropolitanus 207 (9th cent.).
A = Cod. Ambrosianus L 85 sup. (9th–10th cents.).
R = all or consensus of the 15th cent. MSS.
a = Cod. Laurentianus plut. 53. 32 (15th cent.).
c = Cod. Caesenas Malatestianus plut. 24. 2 (15th cent.).
ed. pr. = editio princeps (Jensoniana), Venice, 1472.
Ald. = the first Aldine edition, Venice, 1514.
Gesn. = J. M. Gesner, *Scriptores Rei Rusticae Veteres Latini*, Leipzig, 1735.
Schneider = J. G. Schneider, *Scriptores Rei Rusticae Veteres Latini*, Leipzig, 1794.
Lundström = V. Lundström, *L. Iun. Mod. Columella Lib. I–II, VI–VII, X–XI, de Arboribus*, Upsala-Göteborg, 1897–1940.

NOTE.—Where the *apparatus criticus* is based on Lundström's recension, his siglum R is used as representing the reading of all or the majority of the twenty-five 15th-century MSS. collated by him. A new collation has been made of only the two best 15th-century MSS., for which the sigla a and c are used.

LUCIUS JUNIUS MODERATUS COLUMELLA

ON AGRICULTURE

L. IUNI MODERATI COLUMELLAE

REI RUSTICAE

LIBER X

PRAEFATIO

I. Faenoris tui, Silvine, quod stipulanti spoponderam tibi, reliquam pensiunculam percipe. Nam superioribus novem libris hac minus parte debitum, quod nunc persolvo, reddideram. Superest ergo cultus hortorum segnis[1] ac neglectus quondam veteribus agricolis, nunc vel celeberrimus. Siquidem cum parcior apud priscos esset frugalitas, largior tamen pauperibus fuit usus epularum lactis copia[2] ferinaque[3] ac domesticarum pecudum carne, velut aqua frumentoque, summis atque humillimis victum 2 tolerantibus. Mox cum sequens et praecipue nostra aetas dapibus libidinosa pretia constituerit, cenaeque non naturalibus[4] desideriis,[5] sed censibus aestimentur, plebeia paupertas submota a pretiosi-

---

[1] signis *A*. [2] copa *S* : capa *A*.
[3] ferinaque *R* : ferineque *SAa*.
[4] naturalius *SA*. [5] desieris *S*.

LUCIUS JUNIUS MODERATUS COLUMELLA

ON AGRICULTURE

BOOK X

PREFACE

I. Accept, Silvinus, the small remaining payment of interest which I had promised when you demanded it; for in my nine previous books I had discharged my debt except for this portion, and I am now going to pay you in full. So the subject which has still to be dealt with is horticulture, which the husbandman of old carried out in a half-hearted and negligent fashion but which is now quite a popular pursuit. Though, indeed, among the ancients there was a stricter parsimony, the poor had a more generous diet, since highest and lowest alike sustained life on an abundance of milk and the flesh of wild and domestic animals as though on water and corn. Very soon, when subsequent ages, and particularly our own, set up an extravagant scale of expenditure on the pleasure of the table, and meals were regarded as occasions not for satisfying men's natural desires but for the display of wealth, the poverty of the common people, forced to abstain

LUCIUS JUNIUS MODERATUS COLUMELLA

3 oribus cibis ad vulgares compellitur. Quare cultus hortorum, quoniam fructus magis in usu est, diligentius nobis, quam tradiderunt maiores, praecipiendus est: isque, sicut institueram, prosa[1] oratione prioribus subnecteretur exordiis, nisi propositum meum expugnasset frequens postulatio tua, quae praecepit,[2] ut poeticis numeris explerem Georgici carminis omissas partes, quas tamen et ipse Vergilius significaverat, posteris se memorandas relinquere. Neque enim aliter istud nobis fuerat audendum, quam ex voluntate vatis maxime vene-
4 randi: cuius quasi numine instigante[3] pigre sine dubio propter difficultatem operis, verumtamen non sine spe prosperi successus aggressi sumus tenuem admodum et paene viduatam corpore materiam, quae tam exilis est, ut in consummatione quidem totius operis annumerari veluti particula possit laboris nostri, per se vero et quasi suis finibus terminata nullo modo conspici. Nam etsi multa[4] sunt eius quasi membra, de quibus aliquid possumus effari, tamen eadem tam exigua sunt, ut, quod aiunt Graeci, ex incomprehensibili parvitate arenae funis effici non
5 possit. Quare quidquid est istud, quod elucubravimus, adeo propriam sibi laudem non vindicat, ut boni consulat, si non sit dedecori prius editis a me scriptorum monumentis. Sed iam praefari desinamus.

[1] prosa *R*: prorsus (*om.* oratione prioribus) *SA*: prorsa *Lundström*.
[2] precepit *SA*: pervicit *R*.
[3] instigante *R*: castigante *SA*.
[4] et si multa *R*: et simul *SA*.

---

[a] *Georg.* IV. 147–8.
[b] *I.e.* to deal with the subject of horticulture in verse.

from the more costly foods, is reduced to an ordinary fare. The cultivation, therefore, of gardens, since 3 their produce is now in greater demand, calls for more careful instruction from us than our forefathers have handed down; and I should be adding it in prose to my earlier books, as I had intended to do, had not your repeated appeals overruled my resolve and charged me to complete in poetic numbers those parts of the *Georgics* which were omitted by Vergil and which, as he himself had intimated,[a] he left to be dealt with by later writers. For indeed I ought not to have ventured on the task,[b] were it not in compliance with the wish of that greatly revered poet, at whose instigation, which almost seemed a divine 4 summons—tardily, no doubt, owing to the difficulty of the task, but all the same not without hope of a prosperous result—I have undertaken to deal with material which is very meagre and almost devoid of substance and so inconsiderable that in my work taken as a whole it can only be accounted as a small fraction of my task, and taken by itself and confined within its own limits it cannot by any means make much of a show. For, although there are many branches of the subject, so to speak, about which we can find something to say, they are, nevertheless, as unimportant as the imperceptible grains of sand out of which, according to the Greek saying, it is impossible to make a rope.[c] Whatever, therefore, 5 has resulted from burning the midnight oil, is so far from claiming for itself any special commendation as to be satisfied if it brings no disgrace on my previously published writings. But enough by way of preface.

[c] The proverb τὸ ἐκ τῆς ψάμμου σχοινίον πλέκειν (Aristeides 2.309 f.) is used of the undertaking of an impossible task.

LUCIUS JUNIUS MODERATUS COLUMELLA

Hortorum quoque te cultus, Silvine, docebo,
Atque ea, quae quondam spatiis exclusus iniquis,
Cum caneret laetas segetes et munera Bacchi,
Et te, magna Pales, necnon caelestia mella,
Vergilius nobis post se memoranda reliquit. 5
 Principio sedem numeroso praebeat horto
Pinguis ager, putres glebas resolutaque terga [1]
Qui gerit, et fossus graciles imitatur arenas,
Atque habilis [2] natura soli, quae gramine laeto
Parturit, et rutilas ebuli creat uvida bacas. 10
Nam neque sicca placet, nec quae stagnata [3] palude
Perpetitur querulae semper convicia ranae.
Tum quae sponte sua frondosas educat ulmos,
Palmitibusque feris laetatur, et aspera silvis
Achradis,[4] aut pruni [5] lapidosis obruta pomis 15
Gaudet, et iniussi [6] consternitur ubere mali:
Sed negat [7] helleboros, et noxia carpasa succo,
Nec patitur taxos, nec strenua toxica sudat,
Quamvis semihominis vesano gramine feta [8]
Mandragorae pariat flores; maestamque cicutam, 20
Nec manibus mitis ferulas, nec cruribus aequa
Terga rubi, spinisque [9] ferat paliouron acutis.
Vicini quoque sint amnes, quos incola durus
Attrahat auxilio semper sitientibus hortis:
Aut fons illacrimet putei non sede profunda, 25

[1] terra *SA*.
[2] atque habilis *om. A*.
[3] sed nec stagnata *S*: nec stagnata *A*.
[4] Achradis *Sa*: Aclaradis *A* (*in marg.* arcadis).
[5] primi *A*.
[6] iniussa *SA*: iniussu *R*.
[7] negat *S*: necat *A*.
[8] foetu *A*: fetu *R*.
[9] rubis pinisque *S*: rubi spinis quae *A*: rubis spinisque *R*.

BOOK X.

The cult of garden-plots I now will teach,
Silvinus, and those themes which Vergil left
For future bards, when, closed in narrow bounds,
He sang of joyous crops and Bacchus' gifts
And thee great Pales [a] and heaven's boon of honey.

First for the varied garden let rich soil
A place provide, which shows a crumbling glebe
And loosened clods, and, dug, is like thin sand.
Fit is the nature of the soil which grass
Abundant grows and moistened brings to birth   10
Elder's red berries.   Never joys afford
Dry soil and stagnant marsh which aye endures
The frog's complaint and curses; choose the ground
Which of itself the leafy elm-tree rears,
Joys in wild vines and bristles thick with groves
Of the wild pear, strewn with the stony fruit
Of plums and apple-trees' unbidden wealth,
That bears no hellebore [b] with noxious juice,
Nor suffers yews to grow, nor poisons strong
Exudes, though it may bear the maddening flower
Of the half-human mandrake,[c] hemlock drear [d]   20
And fennel cruel to the hands [e] and bramble-bush
To legs unkind and prickles sharp of thorn.[f]
Let rivers flow adjacent to your plot,
Whose streams the hardy gardener may lead
As aid to quench the garden's ceaseless thirst,
Or else a fountain should distil its tears

---

[a] The deity who protects shepherds and their flocks.

[b] *Helleborus* and *carpasa* are names of the same species, the latter being white hellebore (*Veratrum album*).

[c] The root of the mandrake was supposed to resemble the lower half of the human body (Pliny, *N.H.* XXV. § 3).

[d] Because used for the poisoning of condemned criminals.

[e] Because used for caning schoolboys.

[f] *Paliurus australis* (Christ's thorn.)

LUCIUS JUNIUS MODERATUS COLUMELLA

Ne gravis hausturis tendentibus ilia vellat.
Talis humus vel parietibus, vel saepibus hirtis [1]
Claudatur, ne sit pecori, neu pervia furi.
Neu tibi Daedaliae quaerantur munera dextrae,
Nec Polyclitea nec Phradmonis,[2] aut Ageladae   30
Arte laboretur: sed truncum forte dolatum
Arboris antiquae numen venerare Priapi
Terribilis membri, medio qui semper in horto
Inguinibus puero, praedoni falce minetur.   34

    Ergo age nunc cultus et tempora quaeque serendis
Seminibus, quae cura satis, quo sidere primum
Nascantur flores, Paestique [3] rosaria gemment,
Quo Bacchi genus, aut aliena stirpe gravata
Mitis adoptatis curvetur frugibus arbos,
Pierides tenui deducite carmine Musae.   40

    Oceani sitiens cum iam canis hauserit undas,
Et paribus Titan orbem libraverit horis,
Cum satur autumnus quassans sua tempora pomis
Sordidus et musto spumantes exprimet uvas;
Tum mihi ferrato versetur robore palae   45
Dulcis humus, si iam pluviis defessa madebit.
At si cruda manet caelo durata sereno,
Tum iussi veniant declivi tramite rivi,
Terra bibat fontes, et hiantia compleat ora.

[1] hitis *SA*.
[2] fragmonis *S* : phragmonis *Ac* : phraginonis *a*.
[3] festaque *S* : festi *A* : festi quoque *a*.

---

[a] The mythical sculptor and architect of the days of Theseus and Minos, for whom he made the labyrinth at Cnossos.
[b] The well-known athletic sculptor of the 5th century.
[c] An Argive sculptor of the 5th century of whom little is known; he competed with Polyclitus and Phidias in making a statue of an Amazon for the Ephesians.
[d] An Argive sculptor of the second half of the 6th century.

BOOK X.

Into a basin, not too deeply sunk,
Lest it should strain the drawers' panting sides.
This plot let walls or thick-set hedge enclose,
Impervious both to cattle and to thieves.
Seek not a statue wrought by Daedalus [a]
Or Polyclitus [b] or by Phradmon [c] carved   30
Or Ageladas,[d] but the rough-hewn trunk
Of some old tree which you may venerate
As god Priapus in your garden's midst,
Who with his mighty member scares the boys
And with his reaping-hook the plunderer.

Come now, ye Muses, and in slender verse
Recount the culture and the seasons due
For sowing seeds; the care the seedlings need;
Under what star the flowers first come to birth;
And Paestum's [e] rose-beds bud, and Bacchus' vines,
Or kindly trees are bent beneath the weight
Of borrowed fruits grafted on alien stock.   40

Soon as the thirsty Dog-star has drunk deep
Of Ocean and with equal day and night
Titan has poised his orb,[f] and Autumn rich,
Tossing his head and stained with fruit and must,
Presses the liquor from the foaming grapes,
Then let the sweet soil with the might of spades,
With iron shod, be turned, if now it lies
Weary and sodden by the rain of heaven;
But, if untouched, hardened 'neath sky serene,
It lies, at your command let streams descend
By sloping channels, and let thirsty earth
Drink of the founts and fill the gaping mouths.

[e] An Italian town near the Gulf of Salerno. Its rose-bushes bloomed twice in the year (*biferi rosaria Paesti*. Verg. *Georg.* IV. 119).

I.e. at the autumn equinox.

LUCIUS JUNIUS MODERATUS COLUMELLA

Quod si nec caeli nec campi competit humor, 50
Ingeniumque loci vel Iupiter abnegat imbrem,
Expectetur hiems, dum Bacchi Cnosius [1] ardor
Aequore caeruleo celetur [2] vertice mundi,
Solis et adversos metuant Atlantides ortus.
Atque ubi iam tuto necdum confisus Olympo 55
Sed trepidus profugit chelas et spicula Phoebus
Dira Nepae,[3] tergoque Croti [4] festinat equino,
Nescia plebs generis matri ne parcite falsae
Ista Promethei [5] genetrix fuit altera cretae:
Altera nos enixa [6] parens, quo tempore saevus 60
Tellurem ponto mersit Neptunus, et imum
Concutiens barathrum lethaeas terruit undas.
Tumque semel Stygium regem videre trementem
Tartara, cum pelagi streperent sub pondere manes.
Nos fecunda manus viduo mortalibus orbe 65
Progenerat, nos abruptae tum montibus altis
Deucalioneae cautes pepere. Sed ecce
Durior aeternusque vocat labor: eia age segnes
Pellite nunc somnos, et curvi vomere dentis
Iam virides lacerate comas, iam scindite amictus. 70
Tu gravibus rastris cunctantia perfode terga,

[1] noxius *SAR*.
[2] caeruletur *S*: ceruletur *A*: celeretur *R*.
[3] negat *SA*.
[4] eroti *AR*.
[5] promethei *SA*.
[6] altera nos enixa *R*: altera noxe nixa *S*: alter axae nixa *A*.

---

[a] Ariadne, daughter of Minos, King of Cnossos, was wedded to Bacchus, who presented her with a crown which was changed into a constellation under the name of Corona Borealis or Corona Cnossia.

[b] The Hyades and Pleiades are afraid when the sun rises immediately opposite to them.

BOOK X.

If sky and land no moisture can supply 50
The nature of the place and Jove himself
Withholds his rain, then wait till winter comes
And Bacchus' Cnossian love [a] in azure main
Is hidden at the topmost pole of heaven,
And Atlas' daughters [b] fear the rising sun's
Opposing rays. When Phoebus now distrusts
Heaven's safety, and, alarmed, flees the dread claws
And stings of Nepa, hastening on the back
Of horselike Crotus,[c] then, ye common herd,
Who of your race know nought, spare not the earth
Ye falsely call your mother; she a race
Mothered from clay Promethean, but *us*
Another parent bore, when 'neath the waves 60
Harsh Neptune whelmed the earth, the lowest pit
Shaking, and Lethe's streams with terror filled; [d]
Then once Hell saw the Stygian king afraid
And ghosts cried out by Ocean's weight oppressed.
Us, when the earth lacked men, a fruitful hand
Created; *us* rocks by Deucalion torn
From mountain heights brought forth. But lo! a task
Harder and endless calls. Come! drive away
Dull sleep, and let the ploughshare's curving tooth
Tear earth's green hair, and rend the robe she wears; 70
With heavy rakes cleave her unyielding back;

[c] Columella here refers to the sun's transit from the sign of the Scorpion (Nepa) to that of Sagittarius (Crotus), who was half man and half horse.

[d] Columella here addresses husbandmen telling them not to be afraid of digging the earth, from the idea that they are attacking their own mother, since they are not of the race of mankind which Prometheus formed of clay but are descended from the men who sprang from the stones cast by Deucalion and Pyrrha after the Flood.

LUCIUS JUNIUS MODERATUS COLUMELLA

Tu penitus latis eradere viscera marris [1]
Ne dubita, et summo ferventia [2] caespite mixta
Ponere, quae canis iaceant [3] urenda pruinis,
Verberibus gelidis iraeque obnoxia Cauri,[4]   75
Alliget ut saevus Boreas, Eurusque resolvat.
Post ubi Riphaeae torpentia frigora brumae
Candidus aprica Zephyrus regelaverit aura,
Sidereoque polo cedet Lyra mersa profundo,
Veris et adventum nidis cantabit hirundo,   80
Rudere tum pingui, solido vel stercore aselli,
Armentive fimo saturet ieiunia terrae,
Ipse ferens olitor diruptos [5] pondere qualos: [6]
Pabula nec pudeat fisso praebere novali,
Immundis quaecunque vomit latrina cloacis.   85
Densaque iam pluviis,[7] durataque summa pruinis
Aequora dulcis humi repetat mucrone bidentis.
Mox bene cum glebis vivacem [8] cespitis herbam
Contundat marrae vel fracti dente ligonis,
Putria maturi solvantur ut ubera campi.   90
Tunc quoque trita solo splendentia sarcula sumat,[9]
Angustosque foros adverso limite ducens,
Rursus in obliquum distinguat tramite parvo.
　Verum ubi iam puro discrimine pectita tellus,
Deposito squalore nitens sua semina poscet,   95

[1] matris *SAR*.
[2] ferventia *Gesner* : frementia *S* : frequentia *Aa*.
[3] iaceantur *S¹* : iacent *AR*.
[4] chauri *SR* : hauri *A*.
[5] diruptos *SA* : deductos *R*.
[6] qualos *R* : quales *S* : quata *A*.
[7] pluviis *S* : pulvis *AR*.
[8] vivacū *S* : viva cum *A*.
[9] sumant *SR* : summant *A*.

---

[a] The North-west wind.　　[b] The North wind.
[c] The East wind.　　[d] The West wind.

BOOK X.

Spare not with mattocks broad her inmost parts
To scrape and mingle with the topmost turf
Yet warm, that they may lie for frosts to sear
Exposed to Caurus'[a] wrath and chilling scourge,
That savage Boreas[b] may bind and Eurus[c] loose them.
But when bright Zephyr[d] with his sun-warmed breeze
Thaws the Riphaean[e] winter's numbing frost,
And Orpheus' Lyre[f] deserts the starry pole
And dives into the deep, and swallows hail
Spring's advent at their nests, the gardener   80
Should with rich mould or asses' solid dung
Or other ordure glut the starving earth
Bearing full baskets straining with the weight,
Nor should he hesitate to bring as food
For new-ploughed fallow-ground whatever stuff
The privy vomits from its filthy sewers.
Now let him with the hoe's well-sharpened edge
Again attack earth's surface packed with rain
And hard with frost; then with the tooth of rake
Or broken mattock mix the living turf
With clods of earth and all the crumbling wealth   90
Of the ripe field set free; then let him take
The shining hoe, worn by the soil, and trace
Straight, narrow ridges from the opposing bounds
And these across with narrow paths divide.

Now when the earth, its clear divisions marked
As with a comb,[g] shining, from squalor free,

---

[e] The Riphaean mountains were deemed to be in the far North.
[f] A constellation of the northern hemisphere.
[g] For this use of *discrimen* in the sense of a "parting" in the hair, *cf.* Claudian, *de nupt. Honor.*, 103.

LUCIUS JUNIUS MODERATUS COLUMELLA

Pingite tunc varios, terrestria sidera, flores,
Candida leucoia, et flaventia lumina calthae,
Narcissique comas, et hianti saeva leonis
Ora feri, calathisque[1] virentia lilia canis,
Nec non vel niveos vel caeruleos hyacinthos. 100
Tum quae pallet humi, quae frondens[2] purpurat auro,[3]
Ponatur viola, et nimium rosa plena pudoris.
Nunc medica panacem lacrima, succoque salubri
Glaucea, et profugos vinctura papavera somnos 104
Spargite: quaeque viros acuunt, armantque puellis,
Iam Megaris veniant genitalia semina bulbi,
Et quae Sicca legit Getulis obruta glebis:
Et quae frugifero seritur vicina Priapo,
Excitet ut Veneri tardos eruca maritos. 109
Iam breve chaerophylum, et torpenti grata palato
Intuba, iam teneris frondens lactucula[4] fibris,
Alliaque infractis spicis, et olentia late
Ulpica, quaeque fabis habilis fabrilia miscet.
Iam siser, Assyrioque venit quae[5] semine radix,
Sectaque praebetur madido sociata lupino, 115
Ut Pelusiaci proritet pocula zythi.

---

[1] chalatisque *SA*.
[2] frondes *SA*.
[3] purpuratabo *SA*.
[4] et lactula *SAa*.
[5] qua *SAa*: quo *c*.

---

[a] Probably *Antirrhinum majus*, snap-dragon, which in German is called "Löwenmaul."
[b] Probably *Gladiolus segetum*, never our hyacinth.
[c] The term *viola* includes both violets and purple and yellow pansies.
[d] *Chelidonium glaucium*.
[e] The tassel hyacinth, *Leopoldia comosa*.
[f] A town of Numidia.
[g] Here used as a general term for African.
[h] *Brassica eruca*.   [i] Ulpica = Cyprian garlic.

BOOK X.

Shall claim her seeds, 'tis time to paint the earth
With varied flowers, like stars brought down from
    heaven,
White snow-drops and the yellow-shining eyes
Of marigolds and fair narcissus-blooms,
Fierce lions' gaping mouths [a] and the white cups
Of blooming lilies and the corn flag bloom,[b]   100
Snow-white or blue. Then let the violet [c]
Be planted, which lies pale upon the ground
Or blooms with gold and purple blossoms crowned,
Likewise the rose too full of maiden blush.
Next scatter all-heal with its saving tear,
And celandines [d] with their health-giving juice,
And poppies which will bind elusive sleep;
Let hyacinths' [e] fruitful seed from Megara come,
Which sharpen men's desires and fit them for the girls,
And those which Sicca [f] gathers, hidden deep
Beneath Gaetulian [g] clods and rocket,[h] too,
Which, sown beside Priapus rich in fruits,
May rouse up sluggish husbands to make love.
Next lowly chervil plant and succory   110
Welcome to jaded palates, lettuce too
With fibres soft, garlic with much-cleft heads,
Or Cyprian [i] scented strong, and all that a skilled
    cook
Mixes with beans to make a labourer's meal.
'Tis time for skirwort [j] and the root which, sprung
From seeds Assyrian,[k] is sliced and served
With well-soaked lupines to provoke the thirst
For foaming beakers of Pelusian [l] beer.

  [j] *Sium sisarum*, or perhaps parsnip, *Daucus carota*.
  [k] Apparently the radish, which, however, is generally called *radix Syria* (*cf.* Book XI. 3. 59).
  [l] Pelusium is an Egyptian port on the easternmost mouth of the Nile.

LUCIUS JUNIUS MODERATUS COLUMELLA

Tempore non alio vili quoque salgama merce
Capparis, et tristes inulae, ferulaeque minaces,
Plantantur: necnon serpentia gramina mentae,
Et bene odorati flores sparguntur anethi, 120
Rutaque Palladiae bacae iutura saporem,
Seque lacessenti fletum factura sinapis,
Atque oleris pulli radix, lacrimosaque cepa
Ponitur, et lactis gustus quae condiat herba,
Deletura quidem fronti data signa fugarum,[1] 125
Vimque suam idcirco profitetur nomine Graio.
Tum quoque conseritur,[2] toto quae plurima terrae
Orbe virens pariter plebi regique superbo
Frigoribus caules, et veri cymata mittit:
Quae pariunt veteres caesposo litore Cumae, 130
Quae Marrucini, quae Signia monte Lepino,[3]
Pinguis item Capua, et Caudinis faucibus horti,[4]
Fontibus et Stabiae celebres, et Vesuia[5] rura,
Doctaque Parthenope Sebethide roscida lympha,
Quae dulcis Pompeia[6] palus vicina salinis 135
Herculeis, vitreoque Siler qui defluit amni,

[1] fucarum *SAR*.
[2] consertur *SA*.
[3] lepuno *SAR*.
[4] hortis *S* : horris *AR*.
[5] vesbia *S* : vespia *Aa*.
[6] pompeia *R* : rompheia *SAa*.

---

[a] Because used for caning schoolboys.
[b] The olive which was sacred to Pallas Athena. *Smyrnium olusatrum*.
[d] See Book XII. 8. 3, where *Lepidium silvestre* is said to be used for curdling milk.
[e] Columella connects the name of this herb ($\lambda\epsilon\pi i\delta\iota o\nu$) with the verb $\lambda\epsilon\pi i\zeta\epsilon\iota\nu$, which means to peal off the husk or skin.
[f] I.e. the cabbage.

BOOK X.

Now is the time, if pickles cheap you seek,
To plant the caper and harsh elecampane
And threatening *a* fennel; creeping roots of mint
And fragrant flowers of dill are spaced now   120
And rue, which the Palladian berry's *b* taste
Improves, and mustard which will make him weep,
Whoe'er provokes it; now the roots are set
Of alexanders *c* dark, the weepy onion,
Likewise the herb which seasons draughts of milk *d*
And will remove the brand, signal of flight,
From brows of runagates, and, by its name
In the Greek tongue, its virtue demonstrates.*e*
Then too is sown the plant,*f* which o'er the world
To common folk alike and haughty King
Its stalks in winter and in spring its sprouts
In plenty yields, grown on the turf-clad shore
Of ancient Cumae,*g* in Marrucine lands,*h*   130
At Signia *i* on the Lepine mountain's side,
On Capua's *j* rich plains, in garden plots
Nigh to the Caudine Forks,*k* at Stabiae *l*
Famed for its fountains, in Vesuvian fields,
At learned Naples *m* by Sebethis' *n* stream
Bedewed, and where the sweet Pompeian marsh
To Herculean salt-pits neighbour lies,*o*

g A former Greek colony on the coast of Calabria.
h The district round Pescara in the Abruzzi.
i The modern Sagni about forty miles S.E. of Rome.
j In Campania.
k Sixteen miles E. of Capua.
l The modern Castellamare, 17 miles S.E. of Naples. The ancient town was destroyed in A.D. 79 at the same time as Pompeii.
m Parthenope, a poetical name for Naples.
n A small river which flows into the Gulf of Naples.
o Pompeii and Herculaneum on the Gulf of Naples.

LUCIUS JUNIUS MODERATUS COLUMELLA

Quae duri praebent cymosa stirpe Sabelli,
Et Turnis lacus, et pomosi Tyburis arva,
Bruttia quae tellus, et mater Aricia porri.
    Haec ubi credidimus resolutae [1] semina terrae,   140
Assiduo gravidam cultu curaque fovemus,
Ut redeant nobis cumulato faenore messes.
Et primum moneo largos inducere fontes,
Ne sitis exurat concepto semine partum.
At cum feta suos nexus adaperta resolvit,   145
Florida cum soboles materno pullulat arvo,
Primitiis plantae modicos tum praebeat imbres
Sedulus irrorans olitor, ferroque bicorni
Pectat, et angentem sulcis exterminet herbam.
At si dumosis positi sunt collibus horti,   150
Nec summo nemoris labuntur vertice [2] rivi,
Aggere praeposito cumulatis area glebis
Emineat, sicco ut consuescat pulvere planta,
Nec mutata loco sitiens exhorreat aestus.
Mox ubi nubigenae Phrixi, nec portitor Helles,   155
Signorum et pecorum princeps caput efferet undis,
Alma sinum [3] tellus iam pandet, adultaque poscens
Semina depositis cupiet se nubere plantis:

---

[1] resolutae *Sa* : -a *AR*.
[2] vertice *R* : cortice *SA*.
[3] alma sinum *R* : masinū *SA*.

---

[a] The modern Sele, a river flowing into the Gulf of Salerno.
[b] In Latium, also known as the *lacus Iuturnae*.
[c] The modern Tivoli, 15 miles east of Rome.
[d] In the " toe " of Italy.
[e] S.E. of Rome.
[f] *I.e.* Aries, the Ram, " prince of flocks " as having a golden fleece, and prince of constellations because it rose early in March when the Roman year began. The ram bore across the straits Phrixus son of Athamas and Nephele (hence called " cloud-born ") and his sister Helle, who were pursued by

BOOK X.

Where Siler *a* rolls his crystal waters down,
'Mid the hardy Sabines, who a stem produce
With many shoots, and nigh to Turnus' lake,*b*
In fields of Tibur *c* where the fruit trees grow,
In Bruttian lands *d* and in Aricia,*e*
Mother of leeks.
          When to the loosened soil   140
We have these seeds entrusted, with fond care
And culture we must tend the pregnant earth,
That crops with interest may our toil repay.
First I would warn you water to provide
Abundantly, lest parching thirst destroy
The life conceived within the fruitful seed.
Then when the earth in travail bursts her bonds
And flowery offsprings sprout from mother soil,
Then let the careful gardener bedew
With moderate showers the infant plant and comb
The ground with two-pronged fork, and choking weeds
Cast from the furrows. If your garden lies   150
On bush-clad hills and from their wooded tops
No streams roll down, then let a bed be raised
On high-heaped mound of clods together thrown,
That the young plant may be inured to grow
In the dry dust nor, when transplanted, thirst
And dread the heat. Soon when the Prince of stars
And flocks,*f* who ferried safely o'er the straits
The cloud-born Phrixus but let Helle drown,
Lifts head above the waves, the bounteous Earth
Will ope her arms, claiming the adult seeds
Wishing to wed the plants that have been set.

their step-mother Ino. Helle was drowned in the Hellespont, to which she gave her name, but Phrixus reached Colchis, where he sacrificed the ram and dedicated its fleece to Ares.

LUCIUS JUNIUS MODERATUS COLUMELLA

Invigilate, viri: tacito nam tempora gressu
Diffugiunt, nulloque sono convertitur annus. 160
Flagitat ecce suos genetrix mitissima fetus,
Et quos enixa est partus iam quaerit alendos,
Privignasque rogat proles. Date nunc sua matri
Pignora, tempus adest; viridi redimite parentem
Progenie, tu cinge comam, tu dissere crines. 165
Nunc apio viridi crispetur florida tellus,
Nunc capitis porri longo resoluta capillo
Laetetur, mollemque sinum staphylinus inumbret.
Nunc et odoratae peregrino munere plantae
Sicaniis croceae descendant montibus Hyblae, 170
Nataque iam veniant hilari samsuca Canopo,
Et lacrimas imitata tuas, Cinyreia virgo,
Sed melior stactis [1] ponatur Achaica myrrha:
Et male damnati maesto qui sanguine surgunt
Aeacii flores, immortalesque amaranti, 175
Et quos mille parit dives natura colores,
Disponat plantis olitor, quos semine sevit.
Nunc veniat quamvis oculis inimica corambe,[2]
Iamque salutari properet lactuca sapore,
Tristia quae relevat longi fastidia morbi. 180
Altera crebra viret, fusco nitet altera crine,[3]

[1] stactis *a*: tactis *SAR*.
[2] coramve *SAR*. corambe (crambe) *Lundström*. For its effect on the eyes, *cf.* Diosc, 3.33 ii.
[3] crine *S*: criem *A*.

---

[a] *I.e.* both the plants which have grown where they were sown and those transplanted from elsewhere.
[b] Apium is more generally celery, but may also be parsley as perhaps here.
[c] *Porrum capitatum*, a leek left to grow till the bulb developed; *porrum sectile*, leek whose stalk was cut earlier.
[d] A town on the most westerly mouth of the Nile.
[e] *Origanum marjoranum*.
[f] *Myrrhis odorata*.

BOOK X.

Now, men, beware; for seasons with silent steps
Fly past, and noiselessly the year revolves; 160
Lo! gentlest mother, Earth demands her young
And longs to nurse the offspring she has borne
And her step-children.[a] To the mother give
—The time is come—the pledges of her love;
With her green progeny the parent crown,
Bedeck her hair, in order set her locks;
Now let the flowery earth with parsley[b] green
Be curly, let her joyfully behold
Herself dishevelled with the leeks'[c] long hair
And let the carrot shade her tender breast.
Let scented crocus-plants, of foreign lands
The gift, descend from the Sicilian hills 170
Of Hybla and from gay Canopus[d] come
The marjoram,[e] and let sweet cicely,[f]
Brought from Achaia, be in order set,
Which better juice than any myrrh-tree yields,
And imitates thy tears, Cinyrian maid,[g]
The flowers that sprung from Ajax' mournful blood,[h]
Wrongly condemned, and deathless amaranths[i]
And all the thousand-coloured flowers brought forth
By bounteous nature, let the gardener set out
Plants from his own sowing; and, though it harms the eyes,
Let the sea-cabbage come; with healthful juice
Let lettuce haste to come which can assuage
Sad distaste[j] caused by lingering disease; 180
One kind grows thick and green, another shines

[g] The daughter of Cinyras was changed into a myrrh-tree (Ovid, *Met.* X. 298–502).
[h] The hyacinths which sprang from the blood of Ajax, grandson of Aeacus.
[i] *Celosia cristata*.
[j] Really want of appetite.

21

LUCIUS JUNIUS MODERATUS COLUMELLA

Utraque Caecilii de nomine dicta Metelli.
Tertia, quae spisso sed [1] puro vertice pallet,
Haec sua Cappadocae servat cognomina gentis.
Et mea, quam generant Tartesi litore Gades,   185
Candida vibrato discrimine,[2] candida thyrso est.
Cypros item Paphio quam pingui nutrit in arvo,
Punicea depexa coma, sed [3] lactea crure [4] est.
Quot [5] facies, totidem sunt tempora quamque serendi.
Caeciliam primo deponit Aquarius anno,   190
Cappadocamque premit ferali mense Lupercus.
Tuque tuis, Mavors, Tartesida pange calendis,
Tuque tuis,[6] Paphie, Paphien [7] iam pange calendis ;
Dum cupit, et cupidae quaerit se iungere matri,
Et mater facili mollissima subiacet arvo,   195
Ingenera; nunc sunt genitalia tempora mundi:
Nunc amor ad coitus properat, nunc spiritus orbis
Bacchatur Veneri, stimulisque cupidinis actus [8]
Ipse suos adamat partus, et fetibus implet.
Nunc pater aequoreus, nunc et regnator aquarum,   200
Ille suam Tethyn,[9] hic pellicit Amphitriten,

[1] spissos et *SA*.
[2] discrimine *S* : -a *AR*.
[3] comas et *SA*.
[4] crura *SAa* : cura *R*.
[5] quod *SA*.
[6] suis *SAR*.
[7] Paphie Paphien *Pontedera* : paphien iterum *SAR*.
[8] aptus *SA*.
[9] thetyn *S* : tethin *A*.

---

[a] L. Caecilius Metellus, consul during the First Punic War.
[b] A province of Asia Minor, north of Cilicia: this was a Cos lettuce as was Cyprian lettuce, line 187.
[c] Columella was a citizen of Gades, the modern Cadiz, on the coast of Spain some sixty miles N.W. of Gibraltar in the district known as Tartessus.
[d] *I.e.* with lanate leaf.

BOOK X.

With tawny foliage, both Caecilian called
After Metellus;*a* a third kind is pale
With dense, smooth top and still retains the name
Of Cappadocian *b* from its place of birth;
Then there's my own, which on Tartessus' shore
Gades *c* brings forth (pale is its curled *d* leaf,
And white its stalk); and that which Cyprus rears
In Paphos' fertile field, with its purple locks
Well-combed, but its stem full of milky juice.
As many as its kinds the seasons are
For planting each: when the new year begins,
Aquarius *e* the Caecilian lettuce sets;   190
In the funereal month *f* Lupercus *g* plants
The Cappadocian; on thy Calends, Mars,
Plant the Tartessian lettuce; Paphian Queen,
On thine *h* the Paphian. While the plant desires
Its mother-earth's embrace, who longs for it,
And she most soft, beneath the yielding Earth
Lies waiting, grant her increase. Now's the time
When all the world is mating, now when love
To union hastes; the spirit of the world
In Venus' revels joins and, headlong urged
By Cupid's goads, itself its progeny
Embraces and with teeming offspring fills.
The Father of the Sea *i* his Tethys now   200
Allures, and now the Lord of all the Waves *j*

---

e The sun is in Aquarius in January.
f February, when sacrifices were offered to the infernal gods and the shades of the dead.
g Pan, in whose honour the Lupercalia was held in February.
h April was sacred to Venus, whose chief seat of worship was Paphos in Cyprus. Many editors have deemed this line corrupt.
i Oceanus.   *j* Neptune.

LUCIUS JUNIUS MODERATUS COLUMELLA

Et iam caeruleo partus enixa marito
Utraque nunc reserat pontumque [1] natantibus implet.
Maximus ipse deum posito iam fulmine fallax
Acrisioneos veteres imitatur amores,   205
Inque sinus matris violento defluit imbre.
Nec genetrix nati nunc aspernatur amorem,
Sed patitur nexus flammata cupidine tellus.
Hinc maria, hinc montes, hinc totus denique mundus
Ver agit: hinc hominum pecudum volucrumque cupido,
Atque amor ignescit menti, saevitque medullis,   211
Dum satiata Venus fecundos [2] compleat artus,[3]
Et generet varias soboles, semperque frequentet
Prole nova mundum, vacuo ne torpeat aevo.
    Sed quid ego infreno volitare per aethera cursu   215
Passus equos audax sublimi tramite raptor?
Ista canit, maiore deo quem Delphica laurus
Impulit ad rerum causas et sacra moventem
Orgia naturae, secretaque foedera caeli,
Extimulat vatem per Dindyma casta Cybeles,   220
Perque Cithaeronem, Nysaeaque per iuga Bacchi,
Per sua Parnassi, per amica silentia Musis
Pierii nemoris, Bacchea voce frementem

[1] portumque *SAa*.   [2] fecundus *SA*.
[3] arctus *SA*.

[a] Jupiter descended upon Danae in a shower of gold, after her father, Acrisius, had enclosed her in an inviolable tower.
[b] A mountain in Phrygia where Cybele was worshipped.
[c] The home of the Muses in Boeotia.
[d] Nysa was the legendary birthplace of Bacchus, sometimes placed in Arabia and sometimes in Caria, and in various other sites.

BOOK X.

His Amphitrite; each anon displays
To her caerulean lord a new-born breed,
And fills the sea with swimmers; King of gods
Himself lays down his thunder and repeats,
As once by craft with the Acrisian maid,[a]
His ancient loves and in impetuous rain
Descends into the lap of Mother Earth;
Nor does the mother her son's love refuse,
But his embrace, inflamed by love, permits.
Hence seas, hence hills, hence e'en the whole wide world
Is celebrating spring; hence comes desire
To man and beast and bird, and flames of love   210
Burn in the heart and in the marrow rage,
Till Venus, satiated, impregnates
Their fruitful members and a varied brood
Brings forth, and ever fills the world with new
Offspring, lest it grow tired with childless age.
  But why so boldly do I let my steeds
With loosened reins fly through the air and waft
Their master on the path of heaven above?
This is a theme for him whom Delphic bay
Drove on with fire more godlike to seek out
Causes of things and sacred rites explore
Of nature and the secret laws of heaven
And spurs her bard o'er the unsullied heights
Of Dindyma,[b] the home of Cybele,   220
Over Cithaeron[c] and the Nysian ridge[d]
Of Bacchus and his own Parnassian mount,[e]
And Muse-loved silence of Pierian woods[f]
With frantic cries of triumph hailing thee,

[e] Parnassus near Delphi was sacred to Apollo, Bacchus, and the Muses.
[f] A district at the foot of Mount Olympus.

25

LUCIUS JUNIUS MODERATUS COLUMELLA

Delie te Paean, et te euhie euhie Paean.
Me mea Calliope cura leviore [1] vagantem   225
Iam revocat, parvoque iubet decurrere gyro,
Et secum gracili connectere carmina filo,
Quae canat inter opus musa modulante putator
Pendulus [2] arbustis, olitor viridantibus hortis.   229
　Quare age, quod sequitur, parvo discrimine sulci
Spargantur caecis nasturcia dira colubris,
Indomito male sana cibo quas educat alvus,
Et satureia thymi referens thymbraeque saporem,
Et tenero cucumis, fragilique cucurbita collo.
Hispida ponatur cinara, quae dulcis Iaccho   235
Potanti veniat,[3] nec Phoebo grata canenti.
Haec modo purpureo surgit glomerata corymbo,
Murteolo modo crine viret, deflexaque collo,
Nunc adaperta manet, nunc pinea vertice pungit,
Nunc similis calatho, spinisque [4] minantibus [5] horret,
Pallida nonnunquam tortos imitatur acanthos.   241
Mox ubi sanguineis se floribus induit arbos
Punica, quae rutilo mitescit tegmine grani,
Tempus aris satio, famosaque tunc coriandra

[1] cura leviore *R* : curaverit ore *SA*.
[2] penulus *SA*.
[3] potati veiat ne *SA*.
[4] calathos pinis (*om.* -que *SA*).
[5] imitantibus *SA*.

---

[a] *I.e.* Apollo.
[b] A name for Bacchus, derived from "Evoe," the cry with which he was hailed.
[c] Usually the Muse of Epic poetry, here used in a wider sense.
[d] *I.e.* the cardoon goes well with wine, but is not good for the voice.
[e] *Acanthus mollis* and *A. spinosus.*
[f] The pomegranate.
[g] In this context probably *Arum maculatum*, the cuckoo-

BOOK X.

Delian,[a] and thee, Evian[b] god of wine.
Me, Calliope,[c] on a humbler quest
Roaming, recalls and bids me to confine
My course in narrow bounds and with her weave
Verse of a slender thread, which tunefully
The pruner perched amid the trees may sing
Or gardener working in his verdant plot.
    So to our next task come. In furrows close   230
Let cress be scattered, fatal to the worms
Which form unseen from ill-digested food
In sickly stomachs, and the savory,
Whose flavour thyme and marjoram recalls,
And cucumber with tender neck and gourd
With brittle stem. Let prickly cardoon
Be planted, which to Bacchus when he drinks
Are pleasant, not to Phoebus when he sings;[d]
It sometimes rises to a purple crown
Close-set and sometimes blooms with foliage
Of myrtle hue and downwards bends its neck,
And opens wide or tapers to a point
Like pine-tree's cone, or, basket's form recalls,   239
Bristling with threatening thorns, and sometimes pale
It imitates the twisted brankursine.[e]
Next, when the Punic tree,[f] whose ripened seeds
With ruddy sheath are clad, bears blood-red flowers,
Then is the time to sow the dragon-root;[g]
'Tis then that ill-famed corianders[h] sprout
And fennel-flowers which so well agree[i]

pint, though elsewhere it = the edderwort, *Dracunculus vulgaris* (*Arum dracunculus*) or the Italian arum, *Arum italicum*.

[h] *Coriandrum sativum: famosa* probably refers to the odour it imparts to the breath.

[i] "Grata: *quae cumino iungi solent in condimentis*" (Schneider). Fennel-flowers = *Nigella sativa*.

27

LUCIUS JUNIUS MODERATUS COLUMELLA

Nascuntur gracilique melanthia grata cumino,[1] 245
Et baca asparagi [2] spinosa prosilit herba,
Et moloche,[3] prono sequitur quae vertice solem: [4]
Quaeque tuas audax imitatur, Nysie, vitis,
Nec metuit sentes: nam vepribus improba surgens,
Achradas [5] indomitasque bryonias alligat [6] alnos. 250
Nomine tum Graio, ceu littera proxima primae
Pangitur in cera docti mucrone magistri:
Sic et humo pingui ferratae cuspidis ictu
Deprimitur folio viridis, pede candida beta.

Quin et odoratis messis iam floribus instat, 255
Iam ver purpureum, iam versicoloribus anni [7]
Fetibus alma parens pingi [8] sua tempora gaudet.
Iam Phrygiae loti [9] gemmantia [10] lumina promunt,
Et conniventes oculos violaria [11] solvunt,
Oscitat et leo, et ingenuo confusa rubore 260
Virgineas adaperta genas rosa praebet honores
Caelitibus templisque Sabaeum miscet odorem.[12]
Nunc vos Pegasidum comites Acheloidas oro,

[1] camino *SAR*.
[2] et ambo spargi *A*.
[3] moloche *R*: molo he *S*: molo hec *A*.
[4] solem *a*: molē *SAR*.
[5] Archradas *Ursinus*: Achrados *S*: Archados *AR*.
[6] alligant *AS*: alligit *a*.
[7] anni *R*: annus *SAa*.
[8] pingi *S²*: pium gi *A*: cingi *R*.
[9] loti *Ald.*: lotę *S*: lotae *A*.
[10] geminantia *Aac*.
[11] vivaria *SA*.
[12] odorem *R*: honorem *SA*.

---

a Black bryony (*Tamus communis*).

BOOK X.

With slender cumin, and with prickly leaves
The berry of asparagus shoots forth,
And mallows grow, whose bended heads pursue
The sun, and, Bacchus, that presumptuous flower
Which imitates thy vine nor fears the thorns;
For shameless bryony [a] in brambles rises,
And binds wild pears and untamed alder-trees.   250
And next the plant, whose name in Grecian script
Eke learned master's pen imprints on wax
Next to the first,[b] the beet, too, is with blow
Of iron-shod point thrust into the rich soil—
The beet, with its white stalk and foliage green.

But now the harvest, rich in fragrant flowers,
Draws nigh and purple spring; kind mother-earth
Joys with the season's many-coloured blooms
To deck her brow, and now its gem-like eyes
The Phrygian lotus [c] opes, and violet-beds
Unclose their half-shut eyes; the lion gapes;[d]   260
Roses, with modest blush suffused, reveal
Their maiden eyes and offer homage due
In temples of the gods, their odours sweet
Commingling with Sabaean [e] incense-smoke.
Daughters of Achelous,[f] you I call,
Companions of the Muses [g] and your choirs,

[b] *I.e.* the beet (in Latin *beta*) has the name of the second letter in the alphabet.
[c] Probably a trefoil such as *Trifolium fragiferum*.
[d] See note on line 98.
[e] Saba (or Sabaea) is the Sheba of the Bible, a district of Arabia Felix.
[f] A river of N.W. Greece, which divides Acarnania and Aetolia.
[g] The Muses are called Pegasides because they dwelt on Mt. Helicon, on the top of which was the fountain of Hippocrene which sprang from the blow of a hoof from the winged horse Pegasus.

LUCIUS JUNIUS MODERATUS COLUMELLA

Maenaliosque choros Dryadum, nymphasque Napaeas, 264
Quae colitis nemus Amphrysi, quae Thessala Tempe,
Quae iuga Cyllenes, et opaci rura Lycaei,
Antraque Castaliis[1] semper rorantia guttis,
Et quae Sicanii flores legistis Halesi,
Cum Cereris proles[2] vestris intenta[3] choreis
Aequoris Hennaei vernantia lilia carpsit, 270
Raptaque, Lethaei coniunx mox facta tyranni,
Sideribus tristes umbras,[4] et Tartara caelo
Praeposuit, Ditemque Iovi, letumque[5] saluti,
Et nunc inferno potitur Proserpina regno:
Vos quoque iam posito luctu maestoque timore 275
Huc facili gressu[6] teneras advertite[7] plantas,
Tellurisque comas sacris aptate canistris.
Hic nullae insidiae nymphis, non ulla rapina,
Casta Fides nobis colitur sanctique Penates.
Omnia plena iocis, securo plena cachinno, 280
Plena mero, laetisque vigent convivia pratis.
Nunc ver egelidum, nunc est mollissimus annus,
Dum Phoebus tener, ac tenera[8] decumbere[9] in herba
Suadet, et arguto fugientes gramine fontes
Nec rigidos potare iuvat, nec sole tepentes. 285

[1] castaliis *R* : casis *SA*.
[2] proles *R* : flores *SA*.
[3] intecta *SAR*.
[4] umbras *SR* : imbres *Aa*.
[5] iovilo etumque *SA*.
[6] gressus *SA*.   [7] avertite *SAR*.
[8] tenerans tenera *SA*.
[9] decumbere *a* : decumbit *SA*.

---

[a] A mountain of Arcadia.
[b] The Amphrysus was a river of south Thessaly.

BOOK X.

Dryads of Maenalus,[a] and woodland nymphs
Who haunt the Amphrysican [b] forests, or who range
Thessalian Tempe [c] or Cyllene's [d] heights,
Or dark Lycaeus' [e] fields, or caves bedewed
With constant drops from the Castalian spring,[f]
And ye in Sicily who plucked the flowers
Upon Halaesus' [g] banks, when Ceres' child
Gazed on your dances and the lilies plucked
That bloomed on Henna's [h] plain and, borne away   270
By Lethe's [i] tyrant-king, became his bride,
Preferring gloomy shades to stars, and Dis [j] to Jove,
And Tartarus to heaven and death's abode
To life, and in realms below she reigns
The Queen Proserpina. Come, lay aside
Your mourning and sad fears and hither turn
With gentle steps your tender feet and fill
Your sacred baskets with earth's blossoming.
Here are no snares for nymphs, no rapine here;
Pure faith we worship here and household gods
Inviolate. Everywhere is fun and wine
And care-free laughter; feasts are at their height   280
In joyous meads. Cool spring's mild hour is here,
So, too, the fairest turn in year's whole course,
When Phoebus' rays are gentle and invite
To lie on gentle grass. What joy to quaff
Fountains of water through the rustling grass
Fleeing, nor chilled by cold nor warmed by sun!

[c] A valley of northern Thessaly.
[d] A mountain of Arcadia.
[e] Another Arcadian mountain sacred to Pan.
[f] At Delphi.
[g] A river which rises in the region of Mt. Etna.
[h] In the centre of Sicily.
[i] The fountain of forgetfulness in Hades.
[j] Another name for Pluto.

LUCIUS JUNIUS MODERATUS COLUMELLA

Iamque Dionaeis redimitur floribus hortus,[1]
Iam rosa mitescit Sarrano clarior ostro.
Nec tam nubifugo Borea Latonia Phoebe [2]
Purpureo radiat vultu, nec Sirius ardor
Sic micat, aut [3] rutilus [4] Pyrois, aut ore corusco   290
Hesperus, Eoo remeat cum Lucifer ortu;
Nec tam siderео fulget Thaumantias arcu:
Quam nitidis hilares collucent fetibus horti.
Quare age vel iubare exorto iam nocte suprema,   294
Vel dum Phoebus equos in gurgite mersat Hibero,
Sicubi odoratas praetexit amaracus umbras,
Carpite narcissique comas, sterilisque balausti.[5]
Et tu,[6] ne Corydonis opes despernat Alexis,
Formoso Nais puero formosior ipsa
Fer calathis violam et nigro permixta ligustro   300
Balsama cum casia nectens croceosque corymbos,
Sparge mero Bacchi: nam Bacchus condit odores.
Et vos agrestes, duro qui pollice molles
Demetitis flores, cano iam [7] vimine textum
Sirpiculum ferrugineis cumulate [8] hyacinthis.   305

[1] hortus *vel* ortus *R* : herbis *SA*.
[2] phoobe *R* : pobe *SA*.
[3] mitigata ut *A*.
[4] rutilius *SA* : rutilis *a*.
[5] balausio *SA* : balausi *R*.
[6] tunc *SAR*.
[7] cano iam *R* : canolā *SA*.
[8] cumulate *a* : comulate *R* : tumulat *SA*.

---

[a] Venus.
[b] Sarra is another name for Tyre, a city renowned for its purple dye.
[c] Phoebe is Diana, here used for the moon as Phoebus is used for the Sun.

BOOK X.

Lo! now Dione's daughter [a] with her flowers
The garden decks, the rose begins to bloom
Brighter than Tyrian [b] purple; not so bright
Grow Phoebe's [c] radiant cheeks, Latona's child,
When Boreas blows and puts the clouds to flight,   290
Nor yet so brightly shines the ardent heat
Of Sirius, [d] or the Fire-star's [e] ruddy glow,
Or Hesperus' [f] flashing face, when Lucifer [g]
Returns in eastern rising; not so bright
Gleams Thaumas' daughter [h] with her heavenly bow,
As gay with shining flowers the gardens gleam.
Come, then, as night is ending and the ray
Of dawn appears, or when the setting sun
Plunges his steeds in the Iberian main, [i]
Where marjoram outspread its perfumed shade,
Come, pluck it and the flowers of daffodils
And wild-pomegranate trees which bear no fruit;
And that Alexis [j] may not scorn the wealth
Of Corydon, come, Naiad, fairer still
Than that fair boy, baskets of violets bring
And balsam, with black privet mixed, entwined   300
With marjoram and clustering saffron-flowers.
Sprinkle these blossoms with the unmixed wine
Of Bacchus, for with wine are perfumes seasoned.
And you, ye rustics, who with callous thumb
Cull the soft flowers, with dark-red corn-flag blooms

[d] The Dog-star.
[e] The planet Mars.
[f] The Evening Star (the planet Venus).
[g] The Morning Star (again the planet Venus).
[h] Iris, the rainbow.
[i] The Spanish Sea, *i.e.* the Atlantic.
[j] Alexis, the boy beloved by Corydon (Vergil, *Ecl.* II). The next line is an imitation of Vergil's *Formosi pecoris custos formosior ipse* (*Ecl.* V. 44).

LUCIUS JUNIUS MODERATUS COLUMELLA

Iam rosa distendat contorti [1] stamina iunci,
Pressaque flammeola rumpatur fiscina caltha,
Mercibus ut vernis dives Vortumnus [2] abundet,
Et titubante gradu multo madefactus Iaccho
Aere sinus gerulus plenos gravis urbe reportet.   310
　Sed cum maturis flavebit messis aristis,
Atque diem gemino Titan extenderit [3] astro,
Hauserit et flammis Lernaei bracchia Cancri,
Allia tunc cepis, cereale papaver anetho
Iungite, dumque virent, nexos deferte [4] maniplos,   315
Et celebres Fortis Fortunae dicite laudes,
Mercibus exactis, hilaresque recurrite in hortos.
Tum quoque proscisso riguoque inspersa novali
Ocima comprimite, et gravibus densate [5] cylindris,
Exurat sata ne [6] resoluti pulveris aestus,   320
Parvulus aut pulex irrepens dente lacessat,
Neu formica rapax populari semina possit.
Nec solum teneras audent erodere frondes
Implicitus conchae limax, hirsutaque campe:
Sed cum iam valido pinguescit [7] lurida caule   325
Brassica, cumque tument pallentia robora [8] betae,
Mercibus atque olitor gaudet securus adultis,
Et iam maturis quaerit supponere falcem,

[1] contorti *R* : -o *SA*.
[2] portunus *SAR*.
[3] expenderit *S*.
[4] deferte *Sc* : deforte *A*.
[5] densate *SR* : densatque *Ac*.
[6] satana *S*.
[7] pinguescit *SA* : turgescit *R*.
[8] robora *R* : robore *SA*.

---

[a] The god of the changing year, also of trade and exchange.
[b] *I.e.* when the sun is passing through the constellations of Gemini (the Twins) and Cancer (the Crab), and the days begin to lengthen.

BOOK X.

Pile high your baskets of hoar willow-twigs,
Let roses strain the threads of twisted rush
And let the throngs of flaming marigolds
Their panniers burst, that rich in vernal wares
Vertumnus [a] may abound and that from town
The carrier may return well soaked with wine,
With staggering gait, and pockets full of cash.   310

But when the harvest with ripe ears of corn
Grows yellow and when, passing the Twin Stars,
Titan extends the day [b] and with his flames
Consumes the claws of the Lernaean Crab,[c]
Garlic with onions join, and with the dill
Ceres' blue poppy, and to market bring
Still fresh the close-packed bunches and, with wares
All sold, to Fortune [d] solemn praises sing,
And to your garden home rejoicing go.
Now plant the basil too in fallow ground,
Well-trenched and watered; tightly press it down
With heavy rollers, lest the burning heat
Of earth dissolved in dust the seedlings scorch,   320
Or tooth of tiny ground flea creeping in
Attack, or greedy ant the seeds can waste.
For not alone the snail, wrapped in its shell,
And hairy caterpillar dare to gnaw
The tender leaves, but, when on solid stalk
The yellow cabbage now begins to swell,
And beet's pale stems increase and, free from care,
The gardener in his rich, ripe merchandise
Rejoices, ready to put in the knife,

[c] The Crab is called Lernaean because it tried to bite the heel of Hercules while he was fighting with the Hydra, a many-headed snake at Lerna near Argos.

[d] Fors Fortuna had a temple on the bank of the Tiber near Rome.

LUCIUS JUNIUS MODERATUS COLUMELLA

Saepe ferus duros iaculatur Iuppiter imbres, 329
Grandine dilapidans hominumque boumque labores:
Saepe etiam gravidis irrorat pestifer undis,
Ex quibus infestae Baccho glaucisque salictis
Nascuntur volucres, serpitque eruca per hortos;
Quos[1] super ingrediens exurit semina morsu,
Quae capitis viduata comas[2] spoliataque nudo 335
Vertice, trunca iacent tristi consumpta veneno.
Haec ne ruricolae paterentur monstra, salutis
Ipsa novas artes varia experientia rerum
Et labor ostendit miseris, ususque magister
Tradidit agricolis, ventos sedare furentes, 340
Et tempestatem Tuscis avertere sacris.
Hinc mala rubigo virides ne torreat herbas,
Sanguine lactentis catuli[3] placatur et extis.
Hinc caput Arcadici nudum cute fertur aselli
Tyrrhenus fixisse Tages in limite ruris.[4] 345
Utque Iovis magni prohiberet fulmina Tarchon,
Saepe suas sedes praecinxit vitibus albis.
Hinc Amythaonius,[5] docuit quem plurima Chiron,
Nocturnas crucibus volucres suspendit, et altis
Culminibus vetuit feralia carmina flere. 350
Sed ne dira novas segetes animalia carpant,
Profuit interdum medicantem semina pingui

[1] quos *a* : quo *SAc*.
[2] comas *Sa* : comis *A* : coma *R*.
[3] lactantis catulis *SA*.
[4] ruris *R* : rursus *SA*.
[5] aminthanus *S* : amynthanus *A*.

---

[a] The Romans derived many of their religious practices from the Etruscans.

[b] The goddess who sent mildew; her festival was held on April 25th (see Aulus Gellius, iv. 6 and Pliny, *N.H.* XVII. § 10).

BOOK X.

Oft-times fierce Jove, launching his cruel showers,
Lays waste with hail the toils of man and beast.   330
Oft too his rain, pregnant with pestilence,
Bedews the earth, whence flying creatures breed,
Foes of the vine and the grey thickets,
And, creeping through the garden, canker-worms
Bite and dry up the seedlings, as they go,
Which, of their leaves bereft, with naked tops,
Consumed by baneful wastage, ruined lie.
Lest rustics suffer from these monstrous pests,
Varied experience of herself and toil
And use, their teacher novel arts have shown
To wretched husbandmen, how to appease   340
Fierce winds and to avert by Tuscan [a] rites
The tempest. Hence, lest fell Rubigo [b] parch
The fresh, green plants, her anger is appeased
With blood and entrails of a suckling whelp;
Hence Tages,[c] Tuscan seer, they say, set up
The skinless head of an Arcadian ass
At the field's edge; hence Tarchon,[d] to avert
The bolts of mighty Jove, oft hedged his domain
With bryony; [e] and Amythaon's son,[f]
Whom Chiron [g] taught much wisdom, hung aloft
Night-flying birds on crosses and forbade
Their sad funereal cries on housetops high.   350
But that creatures horrible may not eat
The tender crop, it sometimes has availed

[c] The inventor of divination (Cicero, *de Div.* II. 23. 50).
[d] The Etruscan general who helped Aeneas against Turnus (Verg., *Aen.* VIII. 122 ff.).
[e] *Bryonia dioica*, common bryony; to be distinguished from black bryony, *Tamus communis*.
[f] Melampus, diviner and physician.
[g] The centaur, who was also tutor of Achilles.

LUCIUS JUNIUS MODERATUS COLUMELLA

Palladia sine fruge salis conspergere amurca,
Innatave laris nigra satiare favilla.
Profuit et plantis latices infundere amaros   355
Marrubii, multoque sedi [1] contingere succo.
At si nulla valet [2] medicina repellere pestem,
Dardaniae veniant artes, nudataque plantas
Femina, quae, iustis tum demum operata iuvencae [3]
Legibus, obscaeno manat pudibunda cruore,   360
Sed resoluta sinus,[4] resoluto maesta capillo,
Ter circum areolas et sepem ducitur horti.
Quae [5] cum lustravit gradiens, mirabile [6] visu!
Non aliter quam decussa pluit arbore nimbus
Vel teretis mali, vel tectae cortice glandis,   365
Volvitur in terram distorto corpore campe.
Sic quondam magicis sopitum [7] cantibus anguem
Vellere Phrixeo delapsum vidit Iolcos.

Sed iam prototomos tempus decidere caules,
Et Tartesiacos, Paphiosque revellere thyrsos,   370
Atque apio [8] fasces et secto cingere porro.
Iamque eruca salax fecundo provenit horto.
Lubrica iam lapathos, iam thamni sponte virescunt,
Et scilla,[9] hirsuto sepes nunc horrida rusco

[1] sedi *Aldus.*: seri *SAR*.
[2] valet *S²*: vacet *SA*.
[3] iuvencae *SA*: iuvente *R*.
[4] sinius *SA*. [5] quem *SA*.
[6] miserabile *SA*.
[7] sopitum *R*: soticū *SA*.
[8] apios *SA*.
[9] scila *c*: est illa *S*: stila *R*.

---

[a] Dardanus was a famous magician (Pliny, *N.H.* XIII. § i).

[b] *Quae primum menstruorum fluxum patitur* (Schneider), *iuvenca* is here used of a young woman: *Cf.* Ovid, *Her.* X. 117, where Helen is called *Graia iuvenca*.

[c] Iolcos in Thessaly was the home of Jason, who went in the ship *Argo* to fetch the golden fleece, which Phrixus had

BOOK X.

The seed to sprinkle with unsalted lees
Of oil, rich gift of Pallas, or to steep
In the black soot that in the hearth collects;
And it has also profited to drench
The plants in horehound's bitter sap or pour
On them abundance of the house-leek's juice.
But if no medicine can the pest repel,
Let the Dardanian [a] arts be called to aid,
A maiden then, who the first time obeys
Her youth's fixed laws,[b] bare-footed and ashamed   360
Of the foul blood which flows, with bosom bare
And hair dishevelled, thrice about the beds
And garden-hedge is led. What wondrous sight,
When she with gentle pace her course has run!
E'en as when from a shaken tree rains down
A shower of shapely apples or of mast
Sheathed in soft shells, so roll in twisted shapes
To earth the caterpillars. Thus of old
Iolcos saw the serpent, lulled to sleep
By magic spells, fallen from Phrixus' fleece.[c]
But now 'tis time to cut the early stalks,
Both the Tartessian and the Paphian [d] stems,   370
With leeks [e] and parsley girdling the bunch;
Now in the fruitful garden springs apace
The lustful rocket;[f] now spontaneous shoots
The slippery sorrel and bushes [g] and the squill;
Prickly with bristling butcher's broom the hedge

carried off to Colchis. Aided by the magic arts of Medea he put to sleep the serpent which guarded the fleece.

[d] See lines 185-7 and notes.

[e] See note c, p. 20.
See lines 108-9.

[g] *Thamni* should mean simply " bushes "; but as a specific plant name seems required, the original reading may have been *tamni* i.e. black bryory, *Tamus communis*.

LUCIUS JUNIUS MODERATUS COLUMELLA

Prodit, et asparagi corruda simillima filo, 375
Humidaque andrachne [1] sitientes protegit antes,
Et gravis atriplici consurgit longa phaselus,
Tum modo dependens [2] trichilis,[3] modo more chelydri
Sole sub aestivo gelidas per graminis umbras
Intortus [4] cucumis praegnansque cucurbita serpit. 380
Una neque est illis facies.   Nam si tibi cordi
Longior est, gracili capitis quae vertice pendet,
E tenui collo semen lege: sive [5] globosi
Corporis, atque utero nimium quae vasta tumescit,
Ventre leges [6] medio; sobolem dabit illa capacem 385
Naryciae picis, aut Actaei [7] mellis Hymetti,
Aut habilem lymphis hamulam, Bacchove lagoenam.
Tum pueros eadem fluviis innare docebit.
Lividus at cucumis, gravida qui [8] nascitur alvo,
Hirtus, et ut coluber nodoso gramine tectus 390
Ventre cubat flexo, semper collectus in orbem,
Noxius exacuit morbos aestatis iniquae.
Fetidus hic succo, pingui quoque [9] semine fartus.
At qui sub trichila manantem repit [10] ad undam,[11]
Labentemque sequens nimium tenuatur amore, 395
Candidus, effetae tremebundior ubere porcae,
Mollior infuso calathis modo lacte gelato,

[1] andrachie *SAR*. [2] dependes *SAR*.
[3] triplicis *SA*. [4] intortos *AR*.
[5] lege sive *R* : leges iube *SA*.
[6] ventreles *SA*.
[7] aut hactei *R* : autaei *SA*.
[8] qui *R* : que *SA*.
[9] suco pingui quoque *R* : sucoque et pingui *SA*.
[10] trichila manantem repit *R* : triclea anterepit *SA*.
[11] ad undam *R* : anundam *SA*.

---

[a] A town of Locris in southern Italy.

BOOK X.

Begins to sprout, and wild asparagus
In stalk most like to those which gardens bear;
Moist purslane now protects the thirsty rows,
And calavance grows tall, which ill consorts
With orach; and the twisted cucumber
And swelling gourd, sometimes from arbours hang,
Sometimes, like snakes beneath the summer sun,
Through the cool shadow of the grass do creep.   380
Nor have they all one form: now, if you desire
The longer shape which hangs from slender top,
Then from the narrow neck select your seed;
But if a gourd of globelike form you seek,
Which vastly swells with ample maw, then choose
A seed from the mid-belly, bearing fruit
Which makes a vessel for Narycian [a] pitch
Or Attic honey from Hymettus' [b] mount,
Or handy water-pail or flask for wine;
'Twill also teach the boys in pools to swim.[c]
But bluish cucumber [d] with swollen womb,
Hairy and like a snake with knotted grass   390
Covered, which on its curving belly lies
Forever coiled, is dangerous and makes
Still worse the cruel summer's maladies;
Foul is its juice and with fat seeds 'tis stuffed.
The white cucumber [e] 'neath the arbour's shade
Creeps towards running water and pursues
Its course—by such devotion worn and thin—
More quivering than the udder of a sow
Lately delivered, softer than the milk
Just thickened and into the cheese-vat poured;

[b] A mountain near Athens.
[c] *I.e.* gourds can be used to support those who are learning to swim.
[d] Perhaps *Cucumis flexuosus* L.
[e] The true cucumber, *Cucumis sativus* L.

LUCIUS JUNIUS MODERATUS COLUMELLA

Dulcis erit, riguoque madescit luteus arvo,
Et feret auxilium quondam mortalibus aegris.
   Cum canis Erigones flagrans Hyperionis aestu   400
Arboreos aperit fetus, cumulataque moris
Candida sanguineo manat fiscella cruore,
Tunc praecox bifera [1] descendit ab arbore ficus
Armeniisque, et cereolis, prunisque Damasci
Stipantur calathi, et pomis, quae barbara Persis   405
Miserat, ut fama est, patriis armata venenis.
At nunc expositi parvo discrimine leti
Ambrosios praebent succos, oblita nocendi.
Quin etiam eiusdem gentis de nomine dicta
Exiguo properant mitescere Persica malo.   410
Tempestiva madent, quae maxima Gallia donat;
Frigoribus pigro [2] veniunt Asiatica fetu.
At gravis Arcturi sub sidere parturit arbos
Livia Chalcidicis et Caunis aemula Chiis,   414
Purpureaeque [3] Chelidoniae, pinguesque Mariscae,
Et Callistruthis roseo quae semine ridet,
Albaque, quae servat flavae cognomina cerae,
Scissa Libyssa simul, picto quoque Lydia tergo.

[1] bifere *SA*.   [2] pigro *R*: pigri *SAa*.
[3] purpure quae *SA*.

---

[a] The Dog-star.
[b] Hyperion was father of Helios, the sun, but is often used of the sun itself.
[c] Daughter of Icarius, who when he was slain was led to his dead body by her dog and in her grief killed herself. She became the constellation Virgo and the Dog became Sirius.
[d] Armenia = apricots; cereoli a variety of plum = *Prunus cereola* and pruna Damasci = damsons.
[e] Lines 405-8 refer to the citron *Citrus medica*, often known as the "Persian fruit." For long it was believed to be inedible.
[f] *Persicum* denoted the peach; not to be confused with *persea*, an Egyptian fruit tree *Mimusops Schimperi*.
[g] The Livian, Chalcidian, Callistruthian, Libyan and Lydian fig-trees have already been mentioned in Book V. 10. and 11.

BOOK X.

Sweet will it be, and when it ripened is
And yellow grows upon well-watered ground,
Oft to sick mortals sure relief will bring.
    When Sirius,[a] kindled by Hyperion's[b] heat,   400
Dog of Erigone,[c] upon the trees
The fruit discloses and with blood-red juice
Of mulberries the white osier-basket drips,
Then from twice-bearing trees the early fig
Falls earthwards; and the panniers are piled high
With apricots and plums and damsons[d] too,
And fruits once sent by Persis barbarous
—So runs the tale—with native poisons charged;
But now with little risk of harm set forth
They yield a juice divine, forgetting quite
How to work harm.[e] Peaches in Persia grown,
Bearing that country's name,[f] with tiny fruit   410
Are quick to ripen; huge ones by Gaul supplied,
Mature in season due; those Asia yields
Are slow to grow and wait for winter's cold.
But 'tis beneath baleful Arcturus' star
That Livian[g] figs, which with Chalcidian vie,
And Caunian,[h] rivals of the Chian,[i] come
To birth, and purple Chelidonian[j]
And fat Mariscan,[k] Callistruthian
Gay with its rosy seeds, and that pale kind
Which ever keeps the name of tawny wax,[l]
And the cleft Libyan and the Lydian fig

  [h] Caunus is a town in Asia Minor on the coast of Caria.
  [i] Chios, an island off the W. coast of Asia Minor; noted for its figs.
  [j] The Chelidonian Islands lay off the coast of Lycia in Asia Minor.
  [k] The meaning of this word is uncertain: Martial (7. 25. 7) says that this type of fig was insipid.
  [l] Pliny calls this kind *albiceratus* (*N.H.* XV. 18).

LUCIUS JUNIUS MODERATUS COLUMELLA

Quin et Tardipedi sacris iam rite solutis
Nube nova seritur, caeli pendentibus undis, 420
Gongylis,[1] illustri mittit quam Nursia campo,
Quaeque Amiterninis defertur bunias arvis.
Sed iam maturis nos flagitat anxius uvis
Evius excultosque iubet claudamus ut hortos.
Claudimus, imperioque tuo paremus agrestes, 425
Ac metimus laeti[2] tua munera, dulcis Iacche,
Inter lascivos Satyros Panasque[3] biformes
Bracchia iactantes, vetulo marcentia vino.
Et te Maenalium, te Bacchum, teque Lyaeum,
Lenaeumque patrem canimus sub tecta vocantes, 430
Ferveat ut lacus, et multo[4] completa Falerno
Exundent pingui spumantia dolia musto.

Hactenus hortorum[5] cultus, Silvine, docebam
Siderei vatis referens praecepta Maronis,
Qui primus veteres ausus recludere fontes 435
Ascraeum cecinit Romana per oppida carmen.

[1] congilis *R* : congyli *SA*.
[2] leti tua *R* : letitia *A* : letita *S*.
[3] panasque *SR* : pansasque *A*.
[4] musto *SA* : multo *R*.
[5] hortorum *S²* : agrorum *Ac* : arvorum *a*.

---

[a] Vulcan, whose festival fell in August.
[b] A town of the Sabines, renowned next to Amiternum for its turnips (Pliny, *N.H.* XIX. 77).
[c] A town of Campania.
[d] Bacchus.

BOOK X.

With spotted skin. But when the sacred rites
Of the slow-footed god *a* have been performed,
When fresh clouds come and rain hangs in the sky,   420
The turnip should be sown, which Nursia *b* sends
From her famed plain, and navew from the fields
Of Amiternum *c* sent. But anxious now
For his ripe grapes the wine-god *d* summons us
And bids us shut our well-tilled garden-plots.
We farmers close them, Bacchus dear, and obey
Thine order, and with joyful hearts thy gifts
We harvest and our arms on high we raise,
By stale old wine enfeebled, mid the throng
Of wanton satyrs and of two-formed *e* Pans.
Thee god of Maenalus,*f* who loosest cares,
Lord of the wine-press, thee we celebrate,   430
Bacchus, and summon thee beneath our roofs,
That in our vats the grape-juice may ferment,
And that our jars with much Falernian filled
Foaming with rich new wine may overflow.

 Thus far, Silvinus, I have sought to teach
The cult of gardens and to call to mind
The precepts taught by Maro, seer divine,
Who first dared to unseal the ancient founts
And sang through Roman towns the Ascraean lay.*g*

---

e Half men, half goats.
f A mountain of Arcadia.
g A quotation from Vergil, *Georg.* II. 175–6. Ascra was a town of Boeotia, the home of Hesiod, the earliest writer of a didactic poem on farming.

BOOK XI

LIBER XI

I. Claudius Augustalis tam ingenuae naturae, quam eruditionis adolescens complurium studiosorum [1] et praecipue agricolarum sermonibus instigatus extudit mihi, cultus hortorum prosa ut oratione componerem. Nec me tamen fallebat hic eventus rei, cum praedictam materiam carminis legibus implicarem. 2 Sed tibi, Publi Silvine, pertinaciter expetenti versificationis nostrae gustum, negare non sustinebam, facturus mox, si collibuisset,[2] quod nunc aggredior, ut holitoris curam subtexerem [3] villici [4] officiis. Quae quamvis primo rei rusticae libro videbar aliquatenus executus; quoniam tamen ea simili [5] desiderio noster Augustalis saepius flagitabat, numerum, quem iam quasi consummaveram, voluminum excessi, et hoc undecimum praeceptum rusticationis memoriae tradidi.

[1] studiosorum *Gesner* : studiorum *SAR*.
[2] conlibisset *SA* : collibuisse *a*.
[3] subtexerem *S* : subtexeram *A* : subtexam *R*.
[4] villici *R* : vilicis *SA*.
[5] simili *AR* : similis *S*.

---

[a] The title Augustalis has two meanings : (1) a member of the college of priests instituted at Rome by Tiberius after the death of Augustus to supervise the worship of Augustus (Tac. *Ann.* I. 54); it consisted of twenty-one members, among whom was Claudius, afterwards Emperor, but if he were the person referred to here, he would hardly have been described

BOOK XI

I. Claudius, a priest of Augustus [a] and a young man equally distinguished for his ingenuous disposition and for his erudition, encouraged by the discourses of several men of learning and especially of those interested in agriculture, extorted from me a promise to write a treatise in prose on the cultivation of gardens. Nor did this circumstance escape my memory when I was seeking to confine the said subject within the rules of poetry. But when you, Publius Silvinus, persistently demanded a taste of my versification, I could not bring myself to refuse, intending presently, if it should be your pleasure, to undertake the task on which I am now entering, namely, to append the duties of a gardener to the functions of a bailiff. Although I seemed already to some extent to have accomplished this in my first book on Agriculture, yet since my friend the priest of Augustus rather often demanded it of me with an eagerness which matches your own, I have exceeded the number of books which I had already practically completed, and have published this eleventh book of the principles of husbandry.

The duties of a bailiff.

2

so simply as Claudius Augustalis. (2) An order of priests who supervised the rites of Caesar-worship and other cults in the *municipia*; they were often selected from the *libertini*. No doubt the Claudius mentioned here belonged to the latter class, but, without fuller details of his name, we have no means of identifying him.

LUCIUS JUNIUS MODERATUS COLUMELLA

3 Villicum fundo familiaeque praeponi convenit aetatis nec primae nec ultimae. Nam servitia sic tirunculum contemnunt, ut senem: quoniam alter nondum novit opera ruris, alter exequi iam non potest; atque hunc adolescentia negligentem, senectus illum facit pigrum. Media igitur aetas huic officio est aptissima: poteritque ab annis quinque et triginta[1] usque in sexagesimum et quintum[2] si non interveniant fortuita corporis vitia, satis validi
4 fungi muneribus agricolae. Quisquis autem destinabitur huic negotio, sit oportet idem scientissimus robustissimusque, ut et doceat[3] subiectos, et ipse commode faciat quae praecipit. Siquidem nihil recte sine exemplo docetur, aut discitur[4] praestatque villicum magistrum esse operariorum, non discipulum, cum etiam de patrefamiliae prisci moris exemplum Cato dixerit: " Male agitur cum domino,
5 quem villicus docet." Itaque in Oeconomico Xenophontis, quem M. Cicero Latino sermoni[5] tradidit, egregius ille Ischomachus[6] Atheniensis rogatus a Socrate utrumne, si res familiaris desiderasset, mercari villicum tamquam fabrum, an a se instituere consueverit: " Ego vero, inquit, ipse instituo. Etenim qui me absente in meum locum substituitur, et vicarius meae diligentiae succedit, is[7] ea, quae ego, scire debet." Sed haec nimium prisca, et eius quidem temporis sunt, quo idem Ischomachus
6 negabat quemquam rusticari nescire. Nos autem

---

[1] ab annis quinque et triginta S : et triginta (*om.* ab annis quinque) A : ab anno xxx R.

[2] sexagensimum et quintum SA : sexagesimum R.

[3] ut et doceat S : ut doceat A : ut edoceat R.

[4] discitur R : dicitur S : dr̄ A.

[5] latine sermone S.

[6] scomachus SA.   [7] is *om. SA*.

BOOK XI. i. 3-6

It is fitting that a bailiff should be set over your 3 farm and household who is neither in the first or in the last stage of his life. For slaves despise a novice as much as they despise an old man, since the former has not yet learned the operations of agriculture and the latter cannot any longer carry them out, and his youth makes the novice careless while his age makes the old man slow. Middle age, therefore, is best suited to this function and, if no accidental bodily defects intervene, a man will be able to perform the duties of an active enough farmer from thirty-five years until his sixty-fifth year. But whoever is 4 destined for this business must be very learned in it and very robust, that he may both teach those under his orders and himself adequately carry out the instructions which he gives; for indeed nothing can be taught or learned correctly without an example, and it is better that a bailiff should be the master, not the pupil, of his labourers. Cato, a model of old-time morals, speaking of the head of a family, said: "Things go ill with the master when his bailiff has to teach him." Therefore, in the *Economicus* of 5 Xenophon,[a] which Marcus Cicero translated into Latin, that excellent man Ischomachus the Athenian, when asked by Socrates whether, if his domestic affairs required it, he was in the habit of buying a bailiff, as he would a craftsman, or of training him up himself, answered, " I train him up myself; for he who stands in my place in my absence and acts as a deputy in my activities, ought to know what I know." But this state of affairs dates from too long ago and indeed belongs to a time when the same Ischomachus asserted that everyone knew how to farm. But let 6

[a] XII. §§ 3, 4.

LUCIUS JUNIUS MODERATUS COLUMELLA

memores ignorantiae nostrae vigentis sensus adolescentulos corporisque robusti peritissimis agricolis commendemus. Quorum monitionibus vel unus [1] ex multis, (nam est difficile erudire) non solum rusticationis, sed imperandi consequatur scientiam. Quidam enim quamvis operum probatissimi artifices, imperitandi parum prudentes aut saevius aut etiam lenius agendo rem dominorum corrumpunt. Quare, 7 sicut dixi, docendus, et a pueritia rusticis operibus edurandus, multisque prius experimentis inspiciendus erit futurus villicus, nec solum an perdidicerit disciplinam ruris, sed an etiam domino fidem ac benevolentiam exhibeat, sine quibus nihil prodest villici summa scientia. Potissimum autem est in eo magisterio scire et aestimare, quale officium et qualis labor sit cuique iniungendus. Nam nec valentissimus possit [2] exequi quod imperatur, si nesciat quid agat; nec peritissimus, si sit invalidus. 8 Qualitas [3] itaque cuiusque rei consideranda est. Quippe aliqua sunt opera tantummodo virium, tanquam promovendi onera portandique: aliqua sociata viribus et arti, ut fodiendi [4] arandique, ut segetes et prata desecandi; nonnullis minus virium, plus artis adhibetur, sicut putationibus insitionibusque vineti: plurimum etiam scientia pollet in aliquibus, ut in pastione pecoris atque eiusdem 9 medicina. Quorum omnium officiorum villicus, quod

---

[1] unus *ed. pr.*: unum *R*: *om. SA.*
[2] possit *vet. ed.*: posset *R*: *om. SA.*
[3] nostrae (§ 6)—qualitas *om. SA.*
[4] foviendi *A.*

BOOK XI. i. 6-9

us, mindful of our ignorance, place young men who are mentally active and physically strong in charge of our most skilful husbandmen, by whose advice one at least out of many (for education is difficult) may attain to a knowledge not only of farming but also of commanding others; for there are some who, although they are highly approved for the manner in which they carry out their crafts, have very little skill in commanding others and ruin their masters by handling the matter with either too much severity or even too much leniency. Therefore, as I have said, your future bailiff must be taught and must be hardened from boyhood to the operations of husbandry and must be first tested by many trials to see not only whether he has thoroughly learned the science of farming but also whether he shows fidelity and attachment to his master, for, without these qualities, the most perfect knowledge possessed by a bailiff is of no use. The most important thing in this kind of superintendence is to know and estimate what duties and what tasks should be enjoined on each person; for the strongest man could not carry out an order, unless he knows what he has to do, nor the most skilful, if he lacks the strength. The nature therefore of each operation must be taken into consideration; for some tasks require strength only, such as the moving and carrying of heavy loads, others require a combination of strength and skill, as in digging and ploughing and mowing field crops and meadows; to some less strength and more skill is applied as in the pruning and grafting of a vineyard, and for some knowledge is of the chief importance, for instance, in the feeding and doctoring of cattle. Of all these tasks a bailiff cannot,

LUCIUS JUNIUS MODERATUS COLUMELLA

iam dixi prius, aestimator bonus esse non potest, nisi fuerit etiam peritus, ut in unoquoque corrigere queat perperam factum. Neque enim satis est reprehendisse [1] peccantem, si non doceat recti viam. Libenter igitur eadem loquor [2] tam docendus est futurus villicus, quam futurus figulus aut faber. Et haud facile dixerim, num illa tanto expeditiora sint discentibus artificia, quanto minus ampla sunt.
10 Rusticationis autem magna et diffusa materia est, partesque si velimus eius percensere, vix numero comprehendamus. Quare satis admirari nequeo, quod primo scriptorum meorum exordio iure conquestus sum, ceterarum artium minus vitae necessariarum repertos antistites, agriculturae neque discipulos neque praeceptores inventos; nisi magnitudo rei fecerit [3] reverentiam vel discendi vel profitendi paene immensam scientiam, cum tamen non ideo turpi desperatione oportuerit eam neglegi.[4]
11 Nam nec oratoria disciplina deseritur, quia perfectus orator nusquam repertus est; nec philosophia, quia nullus consummatae sapientiae:[5] sed e contrario plurimi semetipsos exhortantur vel aliquas partes earum addiscere, quamvis universas percipere non possint. Etenim quae probabilis ratio est obmutescendi, quia nequeas orator esse perfectus, aut in socordiam compelli, quia desponderis sapientiam?
12 Magnae rei quantulumcunque possederis, fuisse participem, non minima gloria est. Quis ergo,

[1] reprehendisse R : reprehendi SA.
[2] loquor R : loquatur A : loquitur S.
[3] fecerit AR : ceperit S.
[4] negligi R.
[5] sapientiae R : scientiae SA.

---

[a] Book I, Preface, § 5.

as I have said, be a good judge unless he has also become skilled in them, so that he can correct what has been done amiss in any one of them; for it is not enough to have reproved one who errs, if he fail to show him the way to do the task aright. Therefore I wish to say what I said before, namely, that the future bailiff must be taught his job just like the future potter or the mechanic. I could not readily state whether these trades are more quickly learnt because they have a narrower scope; but certainly the subject-matter of agriculture is extensive and widespread and, if we wished to reckon up its various parts, we should have difficulty in enumerating them. I cannot, therefore, sufficiently express my surprise, as I justly complained at the beginning of my treatise,[a] at the fact that, while instructors can be found in the other arts which are less necessary for life, for agriculture neither pupils nor teachers have been discovered. Perhaps the magnitude of the subject has produced a feeling of awe about the learning or professing a science which has practically no limits, though it ought not for that reason be neglected in a base spirit of despair. For neither is the art of oratory abandoned because no perfect orator has anywhere been found, nor philosophy, because no man of consummate wisdom has been discovered; but, on the contrary, very many people encourage themselves to acquire a knowledge of at least some parts of them, though they cannot grasp them as a whole. For what justifiable motive is there for keeping silent just because you cannot be a perfect orator, or for being driven into idleness because you despair of attaining wisdom? However little is the portion which you have attained of something great, it is no

LUCIUS JUNIUS MODERATUS COLUMELLA

inquis, docebit futurum villicum, si nullus professor est? Et ego intelligo difficillimum esse ab uno velut auctore cuncta rusticationis consequi praecepta. Verumtamen ut universae disciplinae vix aliquem consultum, sic plurimos partium eius invenies [1] magistros, per quos efficere queas perfectum villicum. Nam et arator [2] reperiatur aliquis bonus, et optimus fossor, aut faeni sector, nec minus arborator et vinitor, tum etiam veterinarius et probus pastor, qui singuli rationem scientiae suae desideranti non subtrahant.

13 Igitur complurium agrestium formatus artibus, qui susceperit officium villicationis, in primis convictum domestici, multoque etiam magis exteri vitet.[3] Somni et vini sit abstinentissimus, quae utraque sunt inimicissima diligentiae. Nam et ebrioso cura officii pariter cum memoria subtrahitur; et somniculosum plurima effugiunt.[4] Quid enim possit aut ipse agere aut cuiquam dormiens impe-
14 rare? Tum etiam sit a venereis amoribus aversus: quibus si se dediderit, non aliud quidquam possit cogitare quam illud quod diligit. Nam vitiis eiusmodi pellectus animus nec praemium iucundius quam fructum libidinis nec supplicium gravius quam frustrationem cupiditatis existimat.

[1] invenies *scripsi*: invenias *codd.*
[2] arator *SA*: orator *R.*
[3] exteri vitet *AR*: exteribit *S.*
[4] officiunt *SAa.*

BOOK XI. I. 12-14

little glory to have possessed yourself of a share. Who, therefore, you say, will teach the future bailiff, if there is no professor of the art? I, too, am aware that it is very difficult to attain to a knowledge of all the precepts of agriculture from, as it were, a single authority; nevertheless, though you will find scarcely anyone acquainted with the whole art, yet you will find very many masters of parts of it, with whose help you can form the perfect bailiff. For a good ploughman can be found and an excellent digger or mower, and likewise a forester and vine-dresser, also a farrier and a good shepherd, no one of whom would refuse to impart to one desirous of learning them the principles of his art.

Therefore, when he who is to take up the duties of a bailiff has been instructed in the arts of a number of different husbandmen, let him particularly avoid intimacy with members of the household and even much more with strangers. He should be most abstemious in respect of wine and sleep, both of which are quite incompatible with diligence; for as a drunkard loses his memory, so he becomes careless in his duties, and very many things escape the notice of one who is unduly given to sleep. For what can a man himself do or order someone else to do, while he is asleep? Further, he should also have an aversion to sexual indulgence; for, if he gives himself up to it, he will not be able to think of anything else than the object of his affection; for his mind being effused by vices of this kind thinks that there is no reward more agreeable than the gratification of his lust and no punishment more heavy than the frustration of his desire.

LUCIUS JUNIUS MODERATUS COLUMELLA

Igitur primus omnium vigilet, familiamque semper ad opera cunctantem pro temporibus anni festinanter producat et strenue ipse praecedat. Plurimum enim refert colonos a primo mane opus aggredi
15 nec lentos per otium pigre procedere. Siquidem Ischomachus[1] idem ille: Malo, inquit, unius agilem atque industriam, quam decem hominum negle-
16 gentem et tardam operam. Quippe plurimum affert mali, si operario tricandi[2] potestas fiat. Nam ut in itinere conficiendo saepe dimidio maturius pervenit is, qui naviter et sine ullis concessationibus permeabit, quam is, qui cum sit una profectus, umbras arborum fonticulorumque amoenitatem vel aurae refrigerationem captavit: sic in agresti negotio dici vix potest, quid navus operarius ignavo et
17 cessatore praestet. Hoc igitur custodire oportet villicum, ne statim a prima luce familia cunctanter et languide procedat, sed velut in aliquod proelium cum vigore et alacritate animi praecedentem eum tamquam ducem strenue sequatur, variisque exhortationibus in opere ipso exhilaret laborantes: et interdum, tanquam deficienti succursurus, ferramentum auferat parumper, et ipse fungatur eius officio, moneatque sic fieri debere, ut ab ipso fortiter sit effectum.
18 Atque ubi crepusculum incesserit, neminem post se relinquat,[3] sed omnes subsequatur more optimi pastoris, qui e grege nullam pecudem patitur in

---

[1] scomachus *A* : comachus *S*.
[2] tricandi *SA* : meretricandi *R*.
[3] reliquerit *S*.

---

[a] Xenophon, *Oecon.*, XX. § 16.

BOOK XI. i. 14-18

Therefore he should be the first of all to wake up and, according to the season of the year, always march out smartly to their work the slaves who are slow and himself walk briskly at their head; for it is of the greatest importance that farm-workers should begin work at early dawn and should not proceed slowly with it through laziness. For Ischomachus, already mentioned, says: "I prefer the prompt and assiduous work of one man to the negligent and slothful work of ten."[a] It occasions a great deal of harm if a workman be given the opportunity to trifle away his time; for as in accomplishing a journey a man who goes forward energetically without any loitering arrives by one-half earlier than he who, having set out at the same time, has kept looking out for shady trees and pleasant springs and cooling breezes; so in the business of farming it is difficult to say how far superior an active workman is to one who is lazy and a loiterer. A bailiff, therefore, must watch that the slaves go out immediately at early dawn not in a dilatory and half-hearted manner, but that, like soldiers marching to some battle, with vigour and eagerness of mind they follow him energetically as he marches in front of them as their leader, and he should encourage them as they labour at their actual work with various exhortations, and from time to time, as if to aid one whose strength is failing, he should take his iron tool from him for a while and do his work for him and tell him that it ought to be carried out in the vigorous manner in which he himself has done it.

And when twilight has come on, he should leave no one behind but should walk in rear of them, like a good shepherd, who suffers no member of the flock to

LUCIUS JUNIUS MODERATUS COLUMELLA

agro relinqui. Tum vero, cum tectum subierit, idem faciat, quod ille diligens opilio: nec [1] in domicilio suo statim delitescat, sed agat cuiusque maximam curam; sive quis, quod accidit plerumque, sauciatus in opere noxam [2] ceperit, adhibeat fomenta: sive aliter languidior est,[3] in valetudinarium [4] confestim deducat, et convenientem ei

19 ceteram curationem adhiberi iubeat. Eorum vero, qui recte valebunt, non minor habenda erit ratio, ut cibus et potio sine fraude a cellariis praebeatur, consuescatque rusticos circa larem domini focumque familiarem semper epulari; atque ipse in conspectu eorum similiter epuletur, sitque frugalitatis exemplum: nec nisi sacris diebus accubans cenet, festosque sic agat, ut fortissimum quemque et frugalissimum largitionibus prosequatur, nonnumquam etiam mensae suae adhibeat, et velit aliis [5] quoque honoribus dignari.

20 Tum etiam per ferias instrumentum rusticum, sine quo nullum opus effici potest, recognoscat, et saepius inspiciat ferramenta; eaque semper duplicia comparet,[6] ac subinde refecta custodiat, ne si quod in opere vitiatum fuerit, a vicino petendum sit; quia plus in operis servorum avocandis, quam in pretio rerum eiusmodi dependitur.

[1] opilioni ec *S* : oppinio nec *A*.
[2] noxam *R* : noxiam *Sa* : noxium *A*.
[3] est *Schneider* : at *SAR*.
[4] valetudinari *SAR*.
[5] alis *SA*.
[6] comparet *SR* : comparat *A*.

be left in the field. Then, when he has come indoors, let him act like that careful herdsman and not immediately hide himself in his house but exercise the utmost care for every one of them; and if, as generally happens, any one of them has received some hurt in the course of his work and is wounded, let him apply fomentations, or, if anyone is rather ill in some other respect, let him immediately convey him to the infirmary and order that any other treatment which is suitable to his case be applied. No less considera- 19 tion should be shown for those who are in good health, that their food and drink may be provided by those in charge of the supplies without their being defrauded, and he should accustom the farm-labourers always to take their meals at their master's hearth and household fire; and he should himself eat under similar conditions in their presence and be an example to them of frugality. He should not take his meals reclining on a couch except on solemn holidays, and he should celebrate festal days by bestowing largesse on the strongest and most frugal among them, sometimes even admitting them to his own table and showing himself willing also to confer other honours upon them.

Then also during the holidays he should inspect 20 the farming implements, without which no work can be carried out, and he should rather often examine the iron tools. These he should always provide in duplicate, and, having repaired them from time to time, he should keep his eye upon them, so that, if any of them have been damaged in the course of work, it may not be necessary to borrow from a neighbour; for it costs more than the price of these things if you have to call off the slaves from their work.

LUCIUS JUNIUS MODERATUS COLUMELLA

21 Cultam vestitamque familiam utiliter magis habeat quam delicate, id est munitam diligenter a frigoribus et imbribus, quae utraque prohibentur optime pellibus manicatis, et sagatis cucullis: idque si fiat, omnis paene hiemalis dies[1] in opere tolerari possit. Quare tam vestem servitiorum, quam, ut dixi, ferramenta bis debebit omnibus[2] mensibus recensere. Nam frequens recognitio nec impuni-
22 tatis spem nec peccandi locum praebet. Itaque mancipia vincta, quae sunt ergastuli, per nomina quotidie citare debebit atque explorare, ut sint diligenter compedibus innexa: tum etiam custodiae sedes an tuta et recte munita sit: nec, si quem dominus aut ipse vinxerit, sine iussu patrisfamiliae resolvat.[3] Sacrificia nisi ex praecepto domini facere nesciat: haruspicem sagamque sua sponte non noverit, quae utraque genera vana superstitione rudes
23 animos infestant. Non urbem, non ullas nundinas nisi vendendae aut emendae rei necessariae causa, frequentaverit. Neque enim coloniae suae terminos egredi debet, nec absentia sua familiae cessandi aut delinquendi spatium dare. Semitas novosque limites in agro fieri prohibeat. Hospitem, nisi ex amicitia domini, quam rarissime recipiat. Ad ministeria sua conservos non adhibeat, nec ulli terminos egredi,

---

[1] hiemalis dies *R*: hiemalibus diebus *SA*: hiemalis diebus *a*.
[2] omnibus *SA*: singulis *R*.
[3] solvat *SA*.

He should keep his slaves turned out and clothed 21
serviceably rather than daintily, that is to say, carefully protected against cold and rain, both of which are best kept off by coats of skin with sleeves and thick hoods. If this be done, almost every winter's day can be endured while they are at work. Therefore the bailiff ought to examine the clothing in the same way as the iron tools, as I have already said, twice every month; for regular inspection allows no hope of impunity and no opportunity for wrong-doing. And so he will have to call over the names of the 22 slaves in the prison, who are in chains, every day and make sure that they are carefully fettered and also whether the place of confinement is well secured and properly fortified; and he should not release anyone whom his master or he himself has bound without an order from the lord of the house. He must not think of offering sacrifices except on an instruction from his master; and he must not on his own initiative have any acquaintance with a soothsayer or fortune-teller, both of which classes of persons disturb ignorant minds with vain superstition. He should not fre- 23 quent the town or any fairs except for the sale or purchase of something necessary; for he ought not to go beyond the limits of the estate nor by his absence give the slaves the opportunity of stopping work or committing misdemeanours. He should prevent the making of paths and new boundaries on the estate. He should receive a visit from any stranger as rarely as possible, and then only when he comes as a friend of his master. He should not make use of his fellow-slaves for any service to himself nor allow any of them to go beyond the boundaries of the estate except under stress of great

LUCIUS JUNIUS MODERATUS COLUMELLA

24 nisi magna coegerit necessitas, permittat. Pecuniam domini neque in pecore nec in aliis rebus promercalibus occupet. Haec enim res avocat villici curam, et eum negotiatorem potius facit quam agricolam, nec unquam sinit eum cum [1] rationibus domini paria facere; sed ubi nummum est [2] numeratio, res pro nummis ostenditur. Itaque tam istud vitandum habebit, quam hercule fugiendum venandi vel aucupandi studium, quibus rebus plurimae operae avocantur.

25 Illa iam, quae etiam in maioribus imperiis difficulter custodiuntur, considerare debebit, ne aut crudelius aut remissius agat cum subiectis: semperque foveat bonos et sedulos, parcat etiam minus probis, et ita temperet,[3] ut magis [4] eius vereantur severitatem, quam ut saevitiam detestentur. Poteritque id custodire, si maluerit cavere, ne peccet operarius, quam, cum peccaverit, sero punire. Nulla est autem vel nequissimi hominis amplior custodia quam quoti-
26 diana operis exactio. Nam illud verum est M. Catonis oraculum, nihil agendo homines male agere discunt. Itaque curabit villicus, ut iusta reddantur. Idque [5] non aegre consequetur, si semper se reprae-
27 sentaverit. Sic enim et magistri singulorum officiorum diligenter exequentur sua munia, et familia post operis exercitationem fatigata cibo quietique potius ac somno quam maleficiis operam dabit.

[1] eum *om. AR.*
[2] nummum est *Lundström*: non est *SAa.*
[3] temperet *R*: temperat *S*: temperate *Aa.*
[4] magis *om. S.*
[5] idque *Schneider*: itaque *R*: atque *a.*

necessity. He should not employ his master's money in purchasing cattle or anything else which is bought or sold; for doing this diverts him from his duties as a bailiff and makes him a trader rather than a farmer and makes it impossible to balance accounts with his master; but when a reckoning-up in money is being held, goods are displayed instead of cash. He will, therefore, regard it as a practice to be shunned, just as much indeed as zeal for hunting and fowling, by which many employees are distracted from their duties, is to be avoided.

He will have to observe those principles which it is difficult to maintain in larger spheres of government, namely, not to deal either too cruelly or too leniently with those set under him: he should always cherish the good and diligent and spare those who are not as good as they ought to be, and use such moderation that they may rather respect his strictness than hate his cruelty. He will be able to guard against the latter contingency, if he prefers rather to take care that a labourer commits no offence than to punish him later on, when he has done so; for there is no more efficient means of restraining even the most wicked man than by daily exacting from him his task. For the oracular saying of Marcus Cato is full of truth: "By doing nothing men learn to do evil." The bailiff will, therefore, take care that a full day's work is rendered, and he will obtain this without difficulty, if he himself always puts in an appearance; for thus the foremen of the different duties will diligently carry out their tasks, and the slaves, being tired after the performance of their work, will turn their attention to food and rest and sleep, rather than to evil doing.

LUCIUS JUNIUS MODERATUS COLUMELLA

In universa porro villicatione, sicut in cetera vita, pretiosissimum est intelligere quemque, nescire se quod nesciat, semperque cupere, quod ignoret, 28 addiscere. Nam etsi multum prodest scientia,[1] plus tamen obest imprudentia vel neglegentia, maxime in rusticatione; cuius est disciplinae caput semel fecisse quicquid exegerit ratio culturae. Nam[2] quamvis interdum emendata sit perperam facti imprudentia vel neglegentia, res tamen ipsa iam domino decoxit, nec mox in tantum exuberat, ut et iacturam capitis amissi restituat, et quaestum re- 29 sarciat. Praelabentis vero temporis fuga quam sit irreparabilis, quis dubitet? Eius igitur memor praecipue semper caveat, ne improvidus ab opere vincatur. Res est agrestis insidiosissima cunctanti, quod ipsum expressius vetustissimus auctor Hesiodus hoc versu significavit:

Αἰεὶ δ' ἀμβολιεργὸς ἀνὴρ ἄταισι παλαίει.

Quare vulgare illud de arborum positione rusticis usurpatum, serere ne dubites, id villicus ad agri totum[3] cultum referri iudicet, credatque, praetermissas non duodecim horas sed annum periisse, nisi sua quaque die quod instat effecerit. 30 Nam cum propriis paene momentis fieri quidque debeat, si unum opus tardius quam oporteat[4] peractum sit, ceterae quoque, quae sequuntur, culturae post iusta tempora serius adhibentur,

---

[1] scientia *vett. edd.*: inscientia *R*.
[2] nam *R*: num *SA*.
[3] totam *S*: totū *AR*.
[4] oporteat *R*: oportet *SA*.

---

[a] *Works and Days*, 413.

Furthermore, in everything which concerns the bailiff's profession, as in life generally, it is of the greatest value that every one should realize that he does not know that of which he is ignorant and that he should always desire to learn it. For, although knowledge is a great advantage, ignorance or carelessness does more harm than knowledge does good, especially in agriculture, of which art the chief point is to have done, once and for all, whatever the method of cultivation shall have required; for, though ignorance and carelessness, which have caused something to be done amiss, can sometimes be remedied, yet the master's property has already been impaired and cannot afterwards yield a great enough increase to make up for the loss of capital and restore the lost profit. Who can doubt how irreparable is the flight of time as it slips away? The bailiff, mindful of this, should always beware above all things, lest, through want of forethought, he be overcome by his work; for agriculture is very apt to deceive the dilatory man, a fact which the very ancient author Hesiod has expressed rather forcibly in this line:

He who delays must aye with ruin strive.[a]

Wherefore let the bailiff hold that the opinion about the planting of trees common on the lips of husbandmen, "never hesitate to plant," is applicable to the whole of agriculture, and let him be sure that not merely twelve hours but a whole year has been lost, if pressing work is not carried out on its own proper day. For since everything has, one might say, its own proper moment when it should be done, if one piece of work is carried out later than it should be, all the other tasks of agriculture also which follow it

LUCIUS JUNIUS MODERATUS COLUMELLA

omnisque turbatus operis ordo spem totius anni frustratur. Quare necessaria est menstrui cuiusque officii [1] monitio ea, quae pendet ex ratione siderum et
31 caeli. Nam ut ait Vergilius,

> Tam sunt Arcturi sidera nobis
> Haedorumque dies servandi et lucidus anguis,
> Quam quibus in patriam ventosa per aequora vectis
> Pontus et ostriferi fauces tentantur Abydi.

Contra quam observationem multis argumentationibus disseruisse me non infitior [2] in iis libris, quos adversus astrologos composueram. Sed illis disputationibus exigebatur id, quod improbissime Chaldaei pollicentur, ut certis quasi terminis, ita
32 diebus statis [3] aëris mutationes respondeant: in hac autem ruris disciplina non desideratur eiusmodi scrupulositas; sed, quod dicitur, pingui Minerva quamvis utile [4] continget [5] villico tempestatis futurae praesagium, si persuasum habuerit, modo ante, modo post, interdum etiam stato die orientis vel occidentis competere vim sideris. Nam satis providus erit, cui licebit ante multos dies cavere suspecta [6] tempora.

II. Itaque [7] praecipiemus, quid quoque mense faciendum sit, sic temporibus accommodantes opera ruris, ut permiserit status caeli: cuius varietatem

---

[1] officium *SA*.
[2] infitior *A* : inficior *SR*.
[3] statis *S* : satis *AR*.
[4] quamvis utile *S* : quamvis inutile *A* : quantumvis utile *R*.
[5] continget *R* : -it *SA*.
[6] suspecta *R* : suscepta *SA*.
[7] itemque *SA*.

---

[a] *Georg.* I. 204 ff.
[b] A work of Columella which has not survived.

are put in train too late after the proper dates, and all the order of work, being disturbed, causes the hopes of the whole year to be disappointed. Therefore warning about the duties of each month, dependent on a consideration of the stars and sky, is necessary: for as Vergil says:

> Arcturus' star, the Kids and gleaming Snake
> We must observe as carefully as men
> Who, sailing homewards o'er the wind-swept sea
> Through Pontus and Abydos' narrow jaws,
> The breeding-ground of oysters, seek to pass.[a]

Against this observation I do not deny that I have disputed with many arguments in the books which I wrote *Against the Astronomers*.[b] But in those discussions the point which was being examined was the impudent assertion of the Chaldaeans [c] that changes in the air coincide with fixed dates, as if they were confined within certain bounds; but in our science of agriculture scrupulous exactitude of that kind is not required, but the prognostication of future weather by homely mother-wit, as they say, will prove as useful as you can desire to a bailiff, if he has persuaded himself that the influence of a star makes itself felt sometimes before, sometimes after, and sometimes on the actual day fixed for its rising or setting. For he will exercise sufficient foresight if he shall be in a position to take measures against suspected weather many days beforehand.

II. We will, therefore, prescribe what work must be done each month, accommodating the operations of agriculture to the seasons of the year, as the state

[c] An early people of Assyria, pioneers in astronomy and astrology.

LUCIUS JUNIUS MODERATUS COLUMELLA

mutationemque si ex hoc commentario fuerit praemonitus villicus, aut nunquam [1] decipietur, aut certe non frequenter. Et ne desciscamus [2] ab optimo vate, quod ait ille,

Vere novo terram proscindere incipiat.

2 Novi autem veris principium non sic observare rusticus debet, quemadmodum astrologus, ut expectet certum illum diem, qui veris initium facere dicitur: sed aliquid etiam sumat de parte hiemis, quoniam consumpta bruma, iam intepescit annus, permittitque clementior dies opera moliri.

3 Possit igitur ab idibus Ianuariis (ut principem mensem Romani anni observet) auspicari [3] culturarum officia; quorum alia ex pristinis residua consummabit, atque alia futuri temporis incohabit. Satis autem erit per dimidios menses exsequi quodque [4] negotium, quia neque praefestinatum opus nimium immature videri possit ante quindecim dies factum, nec rursus post totidem nimium tarde.

4 Idibus Ianuariis ventosa tempestas et incertus status. XVIII cal. Feb. tempestas incerta. XVII cal. Feb. sol in Aquarium transit; Leo mane incipit occidere; Africus, interdum Auster cum pluvia. XVI cal. Feb. Cancer desinit occidere; hiemat. XV cal. Feb. Aquarius incipit oriri; ventus Africus

---

[1] nonnumquam *SA*.
[2] desciscamus *R*: desistamus *S*: desinamus *A*.
[3] auspicari *R*: rusticari *SA*.
[4] quoque *SA*.

---

a Vergil, *Georg.* I. 43.

BOOK XI. ii. 1-4

of the weather shall permit, and if the bailiff has been warned by our brief explanation about the variety and changing of the weather, he will never, or, at any rate, very seldom be deceived, and, not to depart from what our excellent poet says:

In the new spring let him begin to cleave
The soil of earth.[a]

But the husbandman ought not to observe the beginning of spring, in the same way as the astronomer, by waiting for the fixed day which is said to mark the entry of spring, but let him even take in something also from the part of the year which belongs to winter, since, when the shortest day is passed, the year is already beginning to grow warmer and the more clement weather allows him to put work in hand.

Therefore from January 13th (so that he may pay regard to the first month of the Roman year) he would be able to enter upon tasks of cultivation, some of which, left over from the past, he will complete, and others, belonging to the future, he will begin. But it will be enough to finish each piece of work by half-months, because work finished before the time fifteen days too early cannot be regarded as too hurriedly performed, nor on the other hand can work finished fifteen days late be regarded as too slowly carried out.

On January 13th the weather is windy and conditions are uncertain. On January 15th the weather is uncertain. On January 16th the sun passes into Aquarius: the Lion begins to set in the morning: there is a south-west or sometimes a south wind with rain. On January 17th the Crab finishes setting: the weather is wintry. On January 18th Aquarius

tempestatem significat. XI cal. Feb. Fidicula vespere
5 occidit. Dies pluvius. IX[1] calen. Febr. ex occasu
pristini sideris significat tempestatem: interdum
etiam tempestas. VI calend. Feb. Leonis quae est
in pectore clara stella occidit, nonnumquam significatur: hiems bipertitur.[2] V calend. Febr. Auster,
aut Africus, hiemat, pluvius dies. III calend. Febr.
Delphinus incipit occidere. Item Fidicula occidit.
Pridie calen. Februar. eorum, quae[3] supra sunt,
siderum occasus tempestatem facit: interdum
tantummodo significat.

6 Hoc igitur semestrium et deinceps sequentia
tempestatibus annotatis percensuimus, quo cautior
villicus (ut iam dixi) vel abstinere possit operibus,
vel festinationem adhibere. Itaque ab idibus Ianuariis, quod habetur tempus inter brumam et adventum Favonii, si maior est vineae vel arbusti modus,
quicquid ex autumno putationis superfuit, repetendum est, sed ita ne matutinis temporibus vitis
saucietur, quoniam pruinis et gelicidiis nocturnis
7 adhuc rigentes materiae ferrum reformidant. Itaque dum hae regelatae secentur[4] usque in horam
secundam vel tertiam poterunt vepres attenuari,
ne[5] incremento suo agrum occupent, segetes emundari, acervi virgarum[6] fieri, ligna denique confici,

[1] VIIII *AR* : VIII *S*.
[2] bipertitur *R* : biperitur *SA*.
[3] qui *SAR*.
[4] secentur *Lundström* : secuntur *S* : sectuntur *A* : sequuntur *R*.
[5] attenuari ne *R* : attenurine *SA*.
[6] occupent—virgarum *om. A*.

begins to rise: the south-west wind is the sign of a storm. On January 22nd the Lyre sets in the evening: it is a rainy day. On January 24th there are forebodings of a storm from the setting of the constellation of the Whale: sometimes the storm actually occurs. On January 27th the bright star on the breast of the Lion sets: sometimes it is a sign: the winter is divided into two parts at this point. On January 28th there is either a south or a south-west wind: the weather is wintry and the day rainy. On January 30th the Dolphin begins to set: also the Lyre sets. On January 31st the setting of the constellations mentioned above causes a storm, but sometimes is merely a sign of it.

We have now gone through this half-month and those which follow afterwards noting the storms which occur, in order that, as I have already said, the bailiff who exercises greater caution may be able either to abstain from his tasks or else make greater haste. Therefore from January 13th, which is regarded as the time between mid-winter and the coming of the west wind, if there is a somewhat large extent of vineyard or of plantation for supporting vines, you must take up again any pruning which was left over from the autumn, but in such a way that the vine may not be damaged in the mornings, for the hard-wood, when it is still stiff with hoar-frost and nocturnal freezing dreads the knife. Therefore, until these have thawed and can be cut, that is, until the second or third hour, the thinning of brier-hedges can be undertaken, to prevent their taking up space by their growth, and the corn-field can be weeded and heaps of twigs can be formed, and, lastly, wood may be cut for fuel, so that

LUCIUS JUNIUS MODERATUS COLUMELLA

ut tum demum tepenti iam die putatio administretur.

Apricis etiam et macris aut aridis locis prata iam purganda et a pecore sunt defendenda, ut faeni sit copia. Siccos quoque et pingues agros tempestivum est proscindere. Nam uliginosi et mediocris habitus sub aestatem vervagendi[1] sunt; macerrimi vero et aridi post aestatem primo autumno[2] arandi,[3] et subinde conserendi. Sed iugerum agri pinguis hoc tempore anni commode duabus operis proscinditur, quia hibernis pluviis adhuc madens terra facilem cultum sui praebet. Eodemque mense ante cal. Feb. sarriendae segetes autumnales, sive illae seminis adorei sunt, quod quidam for vernaculum[4] vocant, seu tritici: earumque tempestiva sartio est, cum enata frumenta quattuor fibrarum esse coeperunt. Hordeum quoque maturum, quibus superest opera, nunc demum sarrire debebunt. Sed et faba eandem culturam exigit, si iam coliculus eius in quattuor digitos altitudinis creverit.[5] Nam prius sarrivisse nimium teneram non expedit. Ervum melius quidem priore mense, nec tamen improbe hoc ipso vel proximo seremus.[6] Nam Martio nullo modo terrae committendum esse rustici praecipiunt. Vineae, quae sunt palatae et ligatae, recte iam fodiuntur. Surculi, qui primum florem afferunt, statim circa idus inserendi sunt, ut cerasorum, tuberum,[7] amygdalarum, persicorumque. Ridicis vel etiam palis con-

---

[1] vervagendi *a*: vervagiendi *SA*.   [2] autem non *R*: *om. SA*.
[3] post—arandi *om. SA*.   [4] vemaculum *SA*.
[5] crevit *S*.   [6] proximo serenus *S*: proximos seremus *AR*.
[7] tuberum *R*: tuburum *SAa*.

---

[a] According to Pliny, *N.H.* XV. 47, this fruit was brought from Africa in the reign of Augustus, probably product of *Zizyphus vulgaris*.

finally pruning can be carried out when the day is already getting warm.

In places that are exposed to the sun and are poor and arid the meadows must now be cleaned up and protected against cattle, so that there may be an abundant crop of hay. It is time also to break up dry and rich lands; for lands which are marshy and of medium quality should be turned up towards summer, while very lean and dry soil should be ploughed up after the summer at the beginning of autumn and then sown. But a *iugerum* of rich land is conveniently broken up at this season of the year by two labourers, because the soil, being still wet from the winter rains, admits of easy cultivation. In the same month, before February 1st, the autumn-sown corn-lands must be hoed, whether they be sown with naked wheat, which some people call Italian wheat, or with emmer wheat; and the proper time for hoeing them is when the corn has sprung up and begins to consist of four blades. Those who have labour to spare ought now to hoe the early barley also. The bean also requires the same treatment, if its stalk has already grown to the height of four inches; for it should not be hoed earlier when it is too tender. It is better to have sown bitter vetch the month before, but it will not be amiss if we sow it in this or the following month; for it is a precept among husbandmen that it should on no account be committed to the earth in March. Now is the proper time to trench vines which have been propped and tied up. Scions which are bearing their first blossom, should be immediately grafted about the 13th of the month, such as those of cherry-trees, tuber-apple-trees,[a] almond-trees and peach-trees. It is also a fit time

LUCIUS JUNIUS MODERATUS COLUMELLA

ficiendis, idoneum tempus est. Nec minus in aedificia succidere arborem convenit. Sed utraque melius fiunt luna decrescente ab vigesima usque in trigesimam, quoniam omnis materia sic caesa [1]
12 iudicatur carie non infestari. Palos una opera caedere et exputatos acuere centum numero potest: ridicas [2] autem querneas, sive oleagineas findere, et dedolatas utraque parte exacuere numero sexaginta. Item ad lucubrationem vespertinam palos decem vel ridicas quinque conficere; [3] totidemque per
13 antelucanam lucubrationem. Materies si roborea est, ab uno fabro dolari ad unguem per quadrata debet pedum xx: haec [4] erit vehis una. Pinus autem v et xx pedum aeque ab uno expeditur, quae et ipsa vehis dicitur: nec minus ulmus et fraxinus pedum xxx; cupressus autem pedum xl: tum etiam sexagenum pedum abies atque populus,[5] singulis operis ad unguem quadrantur, atque omnes eae
14 mensurae similiter vehes appellantur. His etiam diebus maturi agni et reliqui fetus pecudum, nec minus maiora quadrupedia charactere signari debent.

Cal. Feb. Fidis incipit occidere, ventus Eurinus, et interdum Auster cum grandine est. III nonas Feb. Fidis tota, et Leo medius occidit. Corus [6] aut Septentrio, nonnunquam Favonius. Nonis Febr. mediae partes Aquarii oriuntur, ventosa tempestas.
15 vii idus Febr. Callisto sidus [7] occidit; Favonii spirare

---

[1] sic caesa *R* : sicca eē *A* : siccae sa *S*.
[2] radicas *c* : oleagina eas *SA*.
[3] conficeret *S* : configeret *A*.
[4] hoc *SAR*.
[5] populus *R* : populis *SA*.
[6] chorus *SAR*.
[7] fidis *SA* : fidus *a*.

for making vine-props or even stakes, and it is also equally suitable for cutting down trees for buildings; but both these operations are better carried out when the moon is waning, from the 20th to the 30th of the month, since all wood so cut is considered not to be attacked by decay. One workman can cut down, strip and sharpen a hundred stakes a day; he can also split, smooth on both sides and sharpen sixty oaken or olive-wood props, and he can finish ten stakes or five props by artificial light in the evening and the same number by artificial light before dawn. If the wood is oak, twenty square feet ought to be perfectly hewn by one workman in a day; this will make a wagon's load. Twenty-five square feet of pine-wood can be finished in the same condition by one man (and this is also called a load), and likewise thirty feet of elm or ash; and forty feet of cypress and also sixty feet of fir-wood and poplar can be perfectly squared by a single workman, and all these amounts are likewise called loads. In these days, too, early lambs and the other young of cattle and the larger four-footed beasts also ought to be marked with the branding-iron.

On February 1st the Lyre begins to set; the wind is from the east, and sometimes from south accompanied by hail. On February 3rd the whole of the Lyre and half of the Lion set: the wind is in the North-west or North and sometimes in the West. On February 5th the middle portions of Aquarius rise: the weather is windy. On February 7th the constellation of Callisto[a] sets: the westerly winds begin to blow. On February 8th the

[a] *I.e.* the Great Bear into which Callisto was changed by Diana.

incipiunt. VI idus Febr. ventosa tempestas. III id. Feb. Eurus.

Per hosce dies locis maritimis et calidis ac siccis prata vel arva purgantur, et in fenum submittuntur.
16 Reliquae partes vinearum propter brumam vel frigora omissae,[1] nunc palandae et alligandae[2] sunt, ne postea tumentes gemmae laedantur et oculi atterantur. Item vinearum fossio iisdem locis peragenda, arbustorumque sive putatio sive alligatio finienda[3] est, quorum iusta certa esse non possunt. Inter nonas deinde et idus pomorum seminaria facienda sunt, et maturae plantae de seminariis in
17 scrobes transferendae. Pastinatio quoque, quae mense Decembri vel Ianuario coepta est, iam nunc includenda et vitibus conserenda est. Pastinatur autem terreni iugerum ita, ut[4] solum in altitudinem trium pedum defodiatur operis LXXX: vel in altitudinem dupondii[5] semissis[6] operis L: vel ad bipalium,[7]
18 quae[8] est altitudo duorum pedum, operis XL. Haec tamen in agro sicco surculis conserendis minima pastinationis mensura est. Nam holeribus deponendis possit vel[9] sesquipedalis altitudo satisfacere, quae plerumque in singula iugera triginta operis conficitur.

Hoc eodem tempore stercoris pars in prata digerenda,[10] pars oleis et ceteris arboribus[11] inspergenda.

---

[1] frigore omisso *A*.
[2] alligande *AR* : -a *S*.
[3] finienda *R* : fienda *S* : figenda *Aa*.
[4] ita ut *R* : itautem *S* : id at̄ *A*.
[5] dupondius *S* : dupundius *A* : dipundii *R*.
[6] semissis *S* : semis *A* : semisses *R*.
[7] vipedalium *AS* : bipalium *R*.
[8] quae *S* : que *Aa* : qui *R*.
[9] vel *om. SA*.
[10] egerenda *S* : degerenda *Aa* : egredienda *R*.

BOOK XI. II. 15-18

weather is windy. On February 11th the east wind blows.

During these days in districts near the sea which are warm and dry the meadows and cornfields are cleansed and put under hay. The remaining portions of the vineyards, which were passed by on account of the winter and the cold, must now be propped and tied up, so that later the swelling buds may not be damaged and the " eyes " rubbed off. Also in the same districts the digging up of the vineyards should be carried out and the pruning of the trees which support the vines and the attachment to them of the vines must be finished; but there cannot be any fixed time for these operations. Then between the 5th and 13th nurseries must be made for fruit-trees, and early plants transferred from the nurseries to the planting-holes. The trenching too, which was begun in December or January, must now be finished and the ground so treated planted with vines. A *iugerum* of land is trenched, in such a way as to be dug to a depth of three feet, by eighty labourers working for one day, or to the depth of two and a half feet by fifty labourers, or to the depth of a double mattock, which is two feet, by forty labourers. This last is the minimum depth for trenching when you have to plant young shoots in dry soil; for planting vegetables even a depth of a foot and a half might be enough, and this is generally carried out by thirty labourers to a *iugerum*.

At the same period part of the dung must be spread on the meadows and part sprinkled round the olive-trees and the other trees. Moreover nurseries for

---

[11] arboribus *S* : oleribus *A*.

LUCIUS JUNIUS MODERATUS COLUMELLA

Quinetiam vinearia diligenter facienda, malleolusque
19 quam curiosissime pangendus. Populos [1] et salices et
fraxinos,[2] prius quam frondeant, plantasque ulmorum
nunc ponere utile est, aut ante satas nunc exputare,
et circumfodere, ac summas earum aestivas radiculas
amputare. Sarmenta e vineis [3] nondum fossis atque
arbustis et segetibus ramos et rubos, quicquid
denique iacens facile [4] fodientem vel alio genere
terram molientem [5] potest impedire, nunc egerere [6] et
ad sepem applicare [7] oportet: rosaria nova conserere,
vel antiqua curare: arundineta nunc ponere, vel
etiam pristina colere: salicta facere, vel deputata
runcare ac fodere: genistam semine vel plantis in
20 pastinato vel etiam sulco deponere. Trimestrium
quoque satio non est aliena huic tempori, quamvis
tepidis regionibus melius administretur [8] per mensem
Ianuarium. Idibus Februariis Sagittarius vespere [9]
occidit; vehementer hiemat. xvi calend. Martii
vespere Crater oritur; venti mutatio. xv cal. Martii
sol in Pisces [10] transitum facit, nonnumquam ventosa
21 tempestas. xiii et xii cal. Martii Favonius vel Auster
cum grandine et nimbis. x cal. Martii Leo desinit
occidere; venti Septentrionales, qui vocantur ὀρνιθίαι,
per dies triginta esse solent; tum et hirundo advenit.
ix cal. Martii Arcturus prima nocte oritur, frigidus

---

[1] populus *SAR*.
[2] fraxinus *SAR*.
[3] sarmenta e vineis *S* : sarmente vineis *A*.
[4] iacens facile *Lundström* : facile *SA* : iacens *R*.
[5] mollientem *S²* : molliem *A*.
[6] egerere *Sa* : egere *A* : egeri *R*.
[7] applicari *SAR*.
[8] administretur *R* : -aretur *SA*.
[9] vesperi *R* : vesper *SA*.
[10] pisces *R* : pisce *Ac* : piscē *S*.

vines must be carefully made and mallet-shoots planted in them with the most meticulous care. It 19 is expedient at this time to put in poplars and willows and ash-trees before they put forth leaves, and elm-plants, or else to prune those already planted and to dig round them and to cut off the topmost summer rootlets. You should also now remove the twigs from the vines and supporting trees before they have been dug round, and the branches and briers from the cornfields—in a word, any which may be lying about and may easily get in the way of one who is digging or working the soil in some other way—and use them to form a hedge. Now is the time to plant new rose-beds and attend to the old ones, and to plant beds of reeds or cultivate the old ones also, and to make willow-beds or, after cutting them back, to weed and dig round them, and to sow greenweed-seed or set plants of it in trenched ground or even in a furrow. The sowing of three- 20 month corn also is not amiss at this season, although in warm regions it is better to carry out this operation during the month of January.

On February 13th Sagittarius sets in the evening: the weather is extremely wintry. On February 14th the Bowl rises in the evening: there is a change of wind. On February 15th the sun passes into the 21 Fishes: the weather is sometimes windy. On February 16th and 17th the wind is South or West with hail and rain-storms. On February 20th the Lion ceases to set: the North winds, which are called Ornithian,[a] usually continue for thirty days: it is then that the swallow arrives. On February 21st Arcturus rises early in the night: the day is

[a] *I.e.* the winds which bring back the birds of passage.

LUCIUS JUNIUS MODERATUS COLUMELLA

dies Aquilone vel Coro¹ interdum pluvia. VIII cal. Martii Sagitta crepusculo incipit oriri, variae tempestates: Halcyonei dies vocantur, in Atlantico quidem mari summa tranquillitas notata est.

22 VII cal. Martii ventosa tempestas, hirundo conspicitur. Per hos dies frigidis² locis earum rerum, quas supra scripsimus, tempestiva est administratio. Locis autem calidioribus, quamvis sera, tamen necessaria. Ceterum malleoli et viviradicis positio 23 huius esse temporis videtur optima. Nec tamen deterior etiam inter cal. et idus sequentis mensis, utique si non sit ferventissima regio: si vero etiam magis frigida, vel melior. Insitio quoque arborum atque vitium tepidis locis hoc tempore commode administrabitur.

Cal. Martii Africus interdum Auster cum grandine.
24 VI nonas Martii Vindemiator apparet, quem Graeci τρυγητῆρα dicunt: Septentrionales venti. IV Nonas Martii Favonius, interdum Auster; hiemat. Nonis Martii Equus mane oritur; flatus Aquilonis. III idus Martii Piscis aquilonius desinit oriri, Septentrionales venti. Pridie idus Martii Argo navis exoritur, Favonius aut Auster, interdum Aquilo.

25 His diebus commode instruuntur horti, de quibus suo loco dicam secretius, ne inter hanc quasi turbam³ operum neglegentius holitoris officia descripsisse videar, aut nunc ordinem reliquarum culturarum

¹ coro *c* : choro *SAR*.
² frigidus *S*.   ³ turba *A*.

---

a Called after the bird ἀλκυών, which was said to make its nest on the sea, and this was only possible when the weather was calm.

cold with wind in the North or North-west, and it sometimes rains. On February 22nd at dusk the Arrow begins to rise: the weather is variable: the days are called the Halcyon days*a*: in the Atlantic indeed the greatest calm has been observed.

On February 23rd the weather is windy: the swallow is seen. During these days in cold regions it is the proper season for carrying out those operations which we described above; but in warmer regions it is essential to carry them out, though it is late to do so. But the best time for setting mallet-shoots and quick-sets seems to fall in this season; nevertheless it is no worse between the 1st and 15th of the following month, especially if it is not a very hot district; but if it is rather cold, so much the better. The grafting of trees and vines will also be conveniently carried out at this season in warm places.

On March 1st the wind is in the South-west, or sometimes in the South accompanied by hail. On March 2nd the Vintager appears whom the Greeks call τρυγητήρ: the winds are northerly. On March 4th the wind is in the West but sometimes in the South: the weather is wintry. On March 7th the Horse (Pegasus) rises early: the wind is from the North-east. On March 13th the northern Fish ceases to rise: the winds are northerly. On March 14th the ship Argo rises: the wind is West or South, sometimes North.

During these days it is proper that the gardens should be put in order, concerning which I will speak more particularly in the proper place, lest, amid the crowded array of tasks, I should seem to have given a somewhat careless description of the gardener's duties or to have interrupted at this point the

LUCIUS JUNIUS MODERATUS COLUMELLA

26 coeptum interrupisse. Igitur a cal. Martiis eximia est vitium putatio usque in decimum calend. Apriles, si tamen se gemmae nondum movent. Surculi quoque silentes ad insitionem nunc praecipue utiliter leguntur, et ipsa insitio vitium atque arborum nunc est optima. Frigidis quoque locis et humidis vitium satio nunc praecipua est, sed et ficulnea cacumina iam tumentia utilissime [1] deponuntur.[2] Sartura quoque frumentorum iteratur egregie. Modios tres
27 una opera recte sarrit. Prata purgare et a pecore defendere iam tempestivum est: locis quidem calidis et siccis etiam a [3] mense Ianuario, ut supra diximus, id fieri debet, nam frigidis vel a Quinquatribus prata
28 recte submittuntur. Scrobes omnis generis, quos eris autumno consiturus, hoc tempore fieri oportebit: eorum quaternarii,[4] hoc est quoquoversus pedum IV, si est commodum terrenum, XIV ab uno fiunt; ternarii autem XVIII. Ceterum ad deponendas vites, vel non magni incrementi arbores, sulcus qui sit [5] pedum centum et viginti, latitudine bipedanea, in altitudinem deprimi debet dipondio semissis,[6] eumque
29 similiter una opera efficit. Rosarium serotinum perfossum et cultum habere iam tempus est. Oleis laborantibus circum radices amurcam,[7] quae salem

---

[1] utilissima *A*.
[2] deponuntur *S*: deponitur *A*: ponuntur *R*.
[3] a *om. S*.
[4] quaternũ *S*: quaternu *A*: quaternum *a*.
[5] sit *om. SA*.
[6] dupondii semissis *S*: dupundii: semisse *R*: se dupundii dimissis *A*.
[7] amurca *S*.

list which I have begun of the other branches of agriculture. Therefore from March 1st to 23rd is an excellent time for pruning vines, provided, however, that there is not any movement yet in the buds. Now also is above all the time for choosing with advantage scions which are not yet shooting, and it is the very best moment for engrafting vines and trees. Also in cold and damp localities it is the principal time for planting vines, and it is also the time when the tops of fig-trees, which are already beginning to swell, can be set most advantageously. It is an excellent time also for hoeing the cornfields a second time; one labourer can very well hoe in a day land in which three modii of seed have been sown. It is now seasonable to cleanse the meadows and prevent the cattle from entering them; indeed in warm, dry districts this ought to be done, as we have said above, from the month of January onwards; for in cold places the meadows may well be allowed to go to grass only from the Quinquatria.[a] It is the time when planting-holes of every kind will have to be made for the trees which you are going to plant in the autumn; if the ground is suitable, one man in a day can make fourteen of them measuring four feet each way or eighteen measuring three feet each way. But for planting vines or trees of no great growth a furrow a hundred and twenty feet long and two feet broad ought to be sunk to a depth of two feet and a half, and this task a labourer similarly carries out in a day. Now is the time to have finished digging and preparing your late rose bed. It will also be advantageous to pour oil-lees, which have no salt in them,

[a] The Greater Quinquatria, a feast in honour of Minerva, was held on March 19th to 23rd (Ovid, *Fasti* III. 809 ff.).

LUCIUS JUNIUS MODERATUS COLUMELLA

non habeat,[1] nunc conveniet [2] infundere: maximis sex [3] congii, mediocribus arboribus urna satisfaciunt, ceteris aestimanda erit portio. Sed tamen quae nihil vitii habuerint,[4] aliquanto laetiores fient,[5] si
30 amurca rigentur insulsa. Nonnulli hoc optimum tempus esse seminariis instituendis dixerunt. Tum etiam bacas [6] lauri et myrti ceterorumque viridium semina in areolas disserere praeceperunt. Orthocissos,[7] et ederas ab idibus Februariis vel etiam cal. Martiis poni oportere iidem censuerunt.

Idibus Mart. Nepa incipit occidere; significat
31 tempestatem. XVII calen. April. Nepa occidit; hiemat. XVI cal. April. sol in Arietem transitum facit; Favonius, vel Corus.[8] XII calen. April. Equus occidit mane; Septentrionales venti. X cal. April. Aries incipit exoriri; pluvius dies; interdum ningit.[9] IX et [10] VIII calendarum Aprilium, Aequinoctium vernum tempestatem significat.

32 Ab idibus eadem, quae supra, utique peragenda sunt: optime autem uliginosa et pinguia loca nunc demum proscinduntur: et quae [11] mense Ianuario vervacta [12] fecimus,[13] nunc ultima parte Martii sunt iteranda: et siquae pergulae vitium generosarum, vel siquae in agris aut vepribus singulares arbores

[1] habeat *R* : habeant *SA*.
[2] conveniet *R* : converi et *SA*.
[3] sex *om. SA*.
[4] habuerit *SAR*.
[5] fiet *A*.
[6] bacas *R* : bagas *S* : bachas *c* : baucas *A*.
[7] ortocissos *R* : *om. SAa*.
[8] chorus *SAR*.
[9] ningit *R* : ninguit *SAa*.
[10] IX et *om. SA*.
[11] atque *SARa*.

round the roots of olive-trees which are in poor condition; six *congii* are enough for the largest trees, and an *urna* for those of moderate size, and a proportionate amount will have to be reckoned for the rest. Even those, however, which have no defects will thrive somewhat better if they are wetted with unsalted lees of oil. Some people have pronounced this to be the best time for establishing seed-nurseries, and have advised this time for the sowing broadcast of the berries of laurel and myrtle and the other evergreens in beds. The same persons have expressed the opinion that upright and other kinds of ivy ought to be planted from February 13th onwards and even as late as March 1st.

On March 15th the Scorpion begins to set; it signifies a storm. On March 16th the Scorpion sets: the weather is wintry. On March 17th the Sun passes into Aries (the Ram): the wind is North or Northwest. On March 21st the Horse (Pegasus) sets in the early morning: the winds are northerly. On March 23rd Aries begins to rise: it is a rainy day and occasionally it snows. On March 24th and 25th the spring equinox is a sign of storm.

From the 15th onwards the operations mentioned above must certainly be carried through; but now at last it is the best time for breaking up marshy and rich lands, and fallows which we broke up in the month of January should now be similarly treated in the last part of March; and if there are any arbours [a] of good vines or any single trees mated to vines in

---

[a] See Book I. 21. 2.

---

¹² verbacta *SAa*.
¹³ fecimus *Gesner*: facimus *SAR*.

LUCIUS JUNIUS MODERATUS COLUMELLA

maritae a putatoribus[1] relictae sunt, ante calend. April. utique deputari debent; post quem diem
33 sera[2] et infructuosa fit eiusmodi rerum cura. Milii quoque et panici haec prima satio est, quae peragi debet circa idus April. utriusque seminis sextarii quini singula iugera occupant. Quinetiam pecus lanatum ceteraque quadrupedia tempus idoneum est castrandi. Locis autem tepidis ab idibus Februariis usque in idus Apriles, at[3] locis frigidis ab idibus Martiis usque in idus Maias omnia recte pecora castrantur.
34 Cal. Aprilibus Nepa occidit mane, tempestatem significat. Nonis April. Favonius aut Auster cum grandine, nonnumquam hoc idem pridie. Octavo idus Aprilis Vergiliae vespere celantur, interdum hiemat. Septimo idus Aprilis, et sexto, et quinto Austri et[4] Africi, tempestatem significat. Quarto idus Aprilis, sole oriente, Libra occidere incipit, interdum tempestatem significat.
35 Pridie id. Aprilis Suculae celantur, hiemat. His diebus locis frigidis prima vinearum fossio utique ante idus peragenda est: quaeque mense Martio post confectum aequinoctium fieri debuerunt, nunc denique quam primum exsequenda sunt. Fici vitesque adhuc recte inseruntur: seminaria, quae sunt ante facta, runcari et adhuc commode fodiri possunt. Oves Tarentinae radice lanaria lavari debent, ut tonsurae praeparentur.

[1] a putatoribus *SR*: amputatoribus *A*.
[2] sera *R*: cetera *SA*.
[3] at *Lundström*: ad *SA*: in *R*: et *a*.
[4] et *om. SA*.

---

[a] See Book VII. 2. 3.   [b] *Saponaria officinalis*.

the fields or thickets left out by the pruners, they should certainly be pruned before April 1st, after which date any care bestowed on such things is late and ineffectual. The earliest sowing of common 33 millet and Italian millet is also made at this time and should be finished about April 13th. Five *sextarii* of both these seeds cover one *iugerum*. It is also a suitable time for castrating woolly cattle and the other four-footed beasts; in warm districts, however, it is correct to castrate any cattle between February 13th and April 13th, and in cold districts between March 15th and May 15th.

On April 1st the Scorpion sets in the morning; it 34 is the sign of a storm. On April 5th the wind is in the West or South accompanied by hail: the same thing occurs sometimes the day before. On April 6th the Pleiads are hidden in the evening: the weather is sometimes wintry. On April 6th, 7th and 8th the winds are South and South-west: storm is portended. On April 10th at sunrise, Libra (the Balance) begins to set; sometimes a storm is portended.

On April 12th the Hyades are hidden; the weather 35 is wintry. During these days in cold districts the first digging of the vineyards must be finished, at any rate before the 13th, and such operations as ought to have been performed in March when the equinox was over, ought now at last to be carried out as quickly as possible. It is still the right season for engrafting fig-trees and vines; and seed-nurseries, which have been made earlier, can still advantageously be weeded and dug up. Tarentine sheep [a] ought to be washed with soapwort,[b] that they may be made ready for shearing.

LUCIUS JUNIUS MODERATUS COLUMELLA

Idibus Aprilibus, ut supra, Libra occidit, hiemat.
36 Decimooctavo calen. Maias ventosa tempestas et
imbres, nec hoc constanter. xv cal. Maias sol in
Taurum transitum facit, pluviam significat. xiv cal.
Maias Suculae se [1] vespere celant; pluviam [2] significat. xi cal. Maias ver bipertitur, pluvia, et
nonnunquam grando. Decimo cal. Maias Vergiliae cum sole oriuntur, Africus vel Auster, dies
humidus.[3] Nono cal. Maias prima nocte Fidicula
apparet, tempestatem significat. Quarto calen.
37 Maias Auster fere cum pluvia. Tertio cal. Maias
mane Capra exoritur, Austrinus dies, interdum
pluviae. Pridie cal. Maias Canis se vespere celat:
tempestatem significat.

Per hos dies eadem quae supra persequemur,
possuntque, si iam librum remittunt, inseri oleae,
vel emplastrari, ceteraeque pomiferae arbores
38 eodem emplastrationis genere inseri. Sed et prima
pampinatio recte inchoatur, dum prorepentes oculi
digito decuti possint.[4] Siqua praeterea in vineis
aut fossor disturbavit, aut neglegentia omisit,
diligens vinitor restituere debet, et fracta iuga
considerate resarcire, aut disiectos palos reponere,
ita ne teneros pampinos explantet. Eodem tempore secundi fetus pecudes signari oportet.

---

[1] se *om. SA.*   [2] pluviam *om. SA.*
[3] vel idus *S.*   [4] possẽt *A* : possunt *Sa.*

BOOK XI. II. 35-38

On April 13th Libra, mentioned above, sets: the weather is windy. On April 14th there is windy weather and there are showers of rains but not continually. On April 17th the sun passes into the Bull; rain is portended. On April 18th the Hyades hide themselves in the evening: rain is portended. April 21st marks the division of spring into two parts: there is rain and sometimes hail. On April 22nd the Pleiads rise with the sun: the wind is South-west or South: it is a wet day. On April 23rd Fidicula (the Lyre) appears early in the night: storm is portended. On April 28th the wind is generally in the South accompanied by rain. On April 29th the Goat rises in the morning: it is a day of southerly wind: there is sometimes rain. On April 30th the Dog hides himself in the evening: a storm is portended.

During these days we shall carry on the same operations as are mentioned above; also, if they are already beginning to loosen their bark, olive-trees can be engrafted or budded, and the other fruit-trees can be engrafted by the same kind of budding. It is also correct to begin the first trimming of the vines, while the "eyes" which are creeping forth can be struck off with the finger. Moreover, if a digger has disarranged anything in the vineyards or through carelessness omitted to do anything, the careful vine-dresser ought to set things right and cautiously mend any broken frame or replace poles which have been dislodged in such a way as not to pull off the young tendrils. At the same time the cattle of the second birth [a] ought to be branded.

[a] *I.e.* the young born in the second half of the previous year from dams which bear twice annually.

LUCIUS JUNIUS MODERATUS COLUMELLA

39 Cal. Maiis, hoc biduo[1] sol unam dicitur tenere particulam; Sucula cum sole exoritur. vi nonas Maias Septentrionales venti. v nonas Maias Centaurus totus apparet, tempestatem significat. iii nonas Maias idem sidus pluviam significat. Pridie nonas Maias Nepa medius occidit; tempestatem significat. Nonis Maiis Vergiliae exoriuntur mane, Favonius. vii idus Maias aestatis initium, Favonius aut Corus,[2] inter-
40 dum etiam pluvia. vi idus Vergiliae totae apparent, Favonius aut Corus, interdum et pluviae.[3] Tertio idus Maias Fidis mane oritur, significat tempestatem.

Per hos dies runcandae[4] segetes sunt, faenisciae instituendae. Bonus operarius prati iugerum desecat, nec minus mille ducentos manipulos unus alligat,[5] qui sint singuli quaternarum librarum. Arbores quoque tempus est ablaqueatas[6] circumfodere, et operire: una opera novellas circumfodiet arbores octuaginta, mediocres lxv,[7] magnas quinquaginta.
41 Hoc mense seminaria omnia crebro fodere oportet. Sed et a calendis Martiis usque in idus Septembres, omnibus mensibus non solum seminariis vineis sed etiam novellis danda fossio est. Iisdem diebus, ubi praegelidum et pluvium caelum est, oleae putantur, et emuscantur. Ceterum tepidis regionibus duobus temporibus anni facere istud oportebit, primo ab idibus Octobribus usque in idus Decembres, iterum

---

[1] viduo SA^2. [2] chorus SAR.
[3] pluviae S^2: pluvia AR.
[4] runcande A^2R: nonandae S.
[5] alligat S: obligat AR.
[6] ablaqueatas R: -is SA: -us a.
[7] lxv AR: lxx S.

On May 1st, the sun is said to keep for this and the next day in the same degree of the ecliptic: the Hyades rise with the sun. On May 2nd the winds are northerly. On May 3rd the Centaur is completely visible: a storm is portended. On May 5th the same constellation portends rain. On May 6th the middle of the Scorpion sets; a storm is portended. On May 7th the Pleiads rise in the morning: the wind is in the West. On May 9th is the beginning of summer: the wind is in the West or North-west: sometimes too there is rain. On May 10th all the Pleiads are visible: the wind is in the West or North-west: sometimes too there are rains. On May 13th the Lyre rises in the morning: a storm is portended.

During these days the cornfields must be weeded and the cutting of the hay must be begun. A good labourer cuts down a *iugerum* of meadow-land and one man can likewise bind one thousand two hundred bundles of hay weighing four pounds each in a day. This is also the time for digging round trees after the soil has been loosened about them and for covering them. One labourer will dig round eighty young trees, sixty-five trees of moderate size, and fifty large ones in a day. During this month you ought frequently to dig up all your seed-nurseries, and from March 1st to September 13th digging should be carried out every month not only in the nursery-vineyards but also in the new vineyards. During the same period, when the weather is very cold and rainy, olive-trees are pruned and cleared of moss, but in warm districts this process will have to be carried out at two seasons of the year, firstly between October 15th and December 13th and again between

LUCIUS JUNIUS MODERATUS COLUMELLA

42 ab idibus Februariis usque in idus Martias, si tamen arbor librum non remittit. Hoc eodem mense in pastinato seminario novissima positio est olearis taleae,[1] eamque oportet, cum panxeris, fimo et cinere mixtis oblinere, et superponere muscum, ne sole findatur; sed hoc opus melius fiet ultima parte mensis Martii, vel prima mensis April. et ceteris temporibus, quibus praecipimus seminaria plantis vel ramis conserere.

43 Idibus Maiis[2] Fidis[3] mane exoritur. Auster aut Eurinus, interdum dies humidus. XVII calen. Iunias idem quod supra. XVI et XV cal. Iunias Eurinus vel Auster cum pluvia. XIV cal. Iun. sol in Geminos introitum facit. XII calen. Iun. Suculae exoriuntur; Septentrionales venti, nonnumquam Auster cum pluvia. XI et X calen. Iunias Arcturus mane occidit, tempestatem significat. VIII et VII et VI cal. Iun. Capra mane exoritur; Septentrionales venti.

44 Ab idib. usque in calend. Iunias veteranam vineam priusquam florere incipiat, iterum fodere oportet, eandemque[4] et ceteras omnes vineas identidem pampinare. Quod si saepius feceris, puerilis una opera iugerum vineti pampinabit. Quibusdam regionibus oves nunc tondentur, et pecoris nati aut amissi[5] ratio accipitur. Item qui lupinum stercorandi agri causa serit, nunc demum aratro[6] subvertit.

45 Cal. Iun. et IV non. Aquila exoritur; tempestas ventosa et interdum pluvia. VII idus Iun. Arcturus

---

[1] taleae S : tale AR.
[2] Maiis R : maias SA.
[3] fidis AR : fides S.
[4] eandemque R : eandem quā SA.
[5] aut amissi R : et mulsi SA.
[6] aratro R : arat po S : arat qua A.

February 13th and March 15th, provided that the tree does not loosen its bark. In the same month is 42 the last season for planting cuttings of olive in a nursery-bed which has been trenched and, when you have set them, you should smear them with a mixture of dung and ashes and put moss over them, so that they may not be split by the sun's heat; this work, however, will be better carried out in the last part of March or in the first part of April and at the other times at which we recommend the planting of nurseries with plants or branches.

On May 15th the Lyre rises early; the wind is 43 South or East; it is sometimes a moist day. May 16th is of the same character as the day before. On May 17th and 18th the wind is East or South accompanied by rain. On May 19th the Sun makes his entry into the Twins. On May 21st the Hyades rise: the winds are northerly or sometimes there is a South wind with rain. On May 22nd and 23rd Arcturus sets in the morning: a storm is portended. On May 25th, 26th and 27th the Goat rises in the morning: the winds are northerly.

From May 15th to June 1st you should again dig 44 over an old-established vineyard before it begins to flower, and trim it and all the other vineyards. If you do this rather frequently, one boy-worker will trim a *iugerum* of vineyard in a day. In some districts sheep are sheared at this time, and a survey is taken of the stock which has been born and that which has been lost. Also he who sows lupines for manuring the land now finally turns it in with the plough.

On June 1st and 2nd Aquila rises: the weather is 45 windy and sometimes rainy. On June 7th Arcturus

occidit; Favonius aut Corus.[1] IV id. Iun. Delphinus vespere exoritur; Favonius; interdum rorat.

46 His diebus, si opere victi sumus, eadem, quae extremo mense Maio, facienda sunt: item omnes arbores frugiferae[2] circumfossae aggerari debent, ut ante solstitium id opus peractum sit. Quinetiam pro conditione regionis et caeli terra vel proscinditur vel iteratur: eaque, si est difficilis, proscinditur operis tribus, iteratur duabus, tertiatur una, lirantur[3] autem iugera duo opera una. At si facilis est terra, proscinditur iugerum duabus operis, iteratur una, lirantur una iugera quattuor, cum in subacta iam terra 47 latiores porcae sulcantur. Quae ratio colligit, ut per autumnum facile possint uno iugo tritici obseri modii centum quinquaginta, ceterorumque leguminum modii centum.

Iisdem his diebus area triturae praeparanda est: ut quaeque res desecta erit, in eam conferatur. Vinearum quoque cultus, quibus maior est modus, 48 iteratus esse debet ante solstitium. Pabulum, si facultas est, vel nunc vel etiam superioribus XV diebus, qui fuerunt ante calen. Iunii pecori praeberi oportet.[4] A cal. autem Iuniis, si iam deficit viridis herba, usque in ultimum Autumnum frondem caesam praebebimus.

[1] chorus *SAR*.
[2] frugifere *S* : fructifere *AR*.
[3] linantur *SA*.
[4] pabulum—oportet *om. SA*.

sets: the wind is West or North-west. On June 10th the Dolphin rises in the evening: the wind is in the West: there is sometimes a dew.

During these days, if we have been overwhelmed 46 with work, the same tasks must be carried out which should have been done at the end of May; also all fruit-bearing trees, after being dug round, ought to be earthed up, in order that this work may have been finished before the solstice. Moreover, according to local and climatic conditions, the earth is either broken up for the first time or the process is repeated. If the soil is difficult to work, it takes three labourers a day to break it up for the first time, two for the second time, and one for the third time, but two *iugera* can be harrowed by one labourer in a day; but if the soil is easy to work, a *iugerum* can be broken up for the first time by two labourers in a day, for the second time by one, while four *iugera* can be harrowed by one labourer, since broader ridges can be furrowed in soil which has been already worked. From this reckoning we can infer that in the autumn 47 a hundred and fifty *modii* of wheat can easily be sown with a single yoke of oxen, and a hundred *modii* of any kind of pulse.

During the same days the threshing-ground must be made ready so that whatever is cut down may be brought to it. The cultivation too of those vineyards which are on an unusually large scale ought to be repeated before the solstice. Fodder, if it is avail- 48 able, ought to be provided for cattle either now or else during the fifteen days which precede June 1st; but from June 1st if green grass is already lacking, up to the end of the autumn we shall cut and provide the foliage of trees.

LUCIUS JUNIUS MODERATUS COLUMELLA

49 Idibus Iuniis calor incipit. XIII calen. Iul. sol introitum Cancro¹ facit; tempestatem significat. XI calen. Iulii Anguifer, qui a Graecis dicitur ὀφιοῦχος, mane occidit, tempestatem significat. Octavo et VII et VI cal. Iulii Solstitium; Favonius et calor. Tertio cal. Iul. ventosa tempestas.

50 His diebus eadem, quae supra. Sed et viciam in pabulum secare oportet, priusquam siliquae² eius durentur; hordeum metere; fabam serotinam ducere; fabam maturam conterere, et paleam eius diligenter recondere, hordeum terere, paleasque omnes recondere; alveos castrare, quos³ subinde nono quoque aut decimo die a⁴ cal. Maiis considerare et curare oportet. Nunc autem si sunt pleni atque operculati favi, demetendi sunt: sin autem maiore parte vacant,⁵ aut sine operculis adaperti⁶ sunt, nondum esse maturos significatur: itaque mellatio est differenda. Quidam in provinciis transmarinis vel hoc vel sequente mense sesama serunt.

51 Calen. Iuliis Favonius, vel Auster, et calor. Quarto non. Iul. Corona occidit mane. Pridie nonas Iul. Cancer medius occidit; calor. Octavo idus Iul. Capricornus medius occidit. Septimo Idus Iul. Cepheus⁷ vespere exoritur, tempestatem significat. Sexto⁸ id. Iul. Prodromi flare incipiunt.

52 His diebus eadem quae supra. Sed et proscissum vervactum optime nunc iteratur, et silvestris ager decrescente Luna utilissime extirpatur.

---

¹ Cancro *om. S.*
² siliqua *SA.*
³ quas *SAR.*
⁴ a *SA :* ad *R.*
⁵ bachant A^1S : vachant A^2.
⁶ adaperti *R :* adoperti *A :* adopertis *S.*
⁷ Cetheus *SA :* cepheus *a.*
⁸ VI *AR :* V *S.*

ᵃ *Vicia sativa*

BOOK XI. II. 49–52

On June 13th the heat begins. On June 19th the sun makes its entrance into the Crab: a storm is portended. On June 21st Serpentarius, which the Greeks call ὀφιοῦχος, sets in the morning; a storm is portended. On June 24th, 25th and 26th is the solstice: there is a West wind and it is hot. On June 29th there is windy weather.

On these days the same tasks should be performed as I mentioned above; but you must also cut tare *a* for fodder before its pods become hard, reap barley, pull up the late beans, crush ripe beans and store up their pods carefully, thresh the barley and store up all the chaff. You should clean out the beehives, which ought to be examined and attended to at times every ninth or tenth day from May 1st. If the honey-combs are full and sealed up, they should now be harvested, but, if they are for the most part empty and left open and unsealed, it is a sign that they are not yet ripe, and so the gathering of the honey must be postponed. In the overseas provinces some people sow sesame in this or the following month.

On July 1st the wind is West or South-west: the weather is hot. On July 4th the Crown sets in the morning. On July 6th the middle of Cancer (the Crab) sets: the weather is hot. On July 8th the middle of Capricorn sets. On July 9th Cepheus rises in the evening: a storm is portended. On July 10th the Harbinger Winds begin to blow.

During these days the tasks mentioned above must be carried on; but it is also a very good thing at this time to break up for the second time fallow land which has already been broken up once, and woodland can most advantageously be cleared of stumps while the moon is waning.

LUCIUS JUNIUS MODERATUS COLUMELLA

Idibus Iuliis Procyon exoritur mane,[1] tempestatem significat. Tertiodecimo cal. Augustas[2] Sol in Leonem transitum facit; Favonius. Octavo calen. Augustas Aquarius incipit occidere clare; Favonius,[3] vel Auster. Septimo cal. Augustas Canicula apparet;
53 caligo aestuosa. Sexto cal. Augustas Aquila exoritur. Quarto calendas Augustas Leonis in pectore clara stella exoritur; interdum tempestatem significat. Tertio calen. Augustas Aquila occidit; significat tempestatem.

54 His diebus locis temperatis et maritimis messis conficitur, et intra dies triginta quam desecta est, stramenta praecisa in acervum congeruntur. Iugerum stramentorum opera una desecat, quibus remotis priusquam sol acrior exurat terram, omnes arbores, quae fuerant in segete, circumfodere et adobruere oportet. Item quibus magna sementis
55 praeparatur, nunc debent iterare. Nam de fodiendis[4] colendisve novellis vineis saepius dixi nullum esse mensem omittendum, donec autumnale aequinoctium conficiatur. Meminisse autem oportebit, ut per hos et Augusti mensis dies antelucanis et vespertinis temporibus frondem pecudibus, caedamus. Item quascumque vineas culturi sumus, ne per aestum, sed mane usque in tertiam, et a decima
56 usque in crepusculum fodiamus. Quibusdam regionibus, sicut in Cilicia et Pamphylia, hoc mense

[1] mane *om. SA.*
[2] Augustas *AR*: Augustus *S.*
[3] *post* Favonius *ordinem verborum in codicibus turbatum restituit Lundström, praeeuntibus Gesnero et aliis.*
[4] fodiendis *om. SA.*

---

[a] Columella regards Procyon as distinct from Canicula, the Lesser Dog. The two names are generally both used for the

BOOK XI. ii. 52–56

On July 15th Procyon [a] rises early; a storm is portended. On July 20th the Sun passes into the Lion: the wind is in the West. On July 25th Aquarius visibly begins to set: the wind is in the West or South. On July 26th the Lesser Dog appears: the weather is dark and sultry. On July 27th Aquila rises. On July 29th the bright star in the breast of the Lion rises: it sometimes portends bad weather. On July 30th Aquila sets: bad weather is portended.

During these days the harvest finishes in temperate places near the sea, and, within thirty days of the cutting of the corn, the straw which has been cut is gathered into heaps. One labourer can cut a *iugerum* of straw and, when it has been removed, and before the more violent heat of the sun burns the earth, you should dig round all the trees which were on the corn-field and heap earth about them. Also those who are preparing for a heavy sowing should plough the land for a second time. As regards digging and cultivating new vineyards I have rather often said that no month ought to be allowed to pass without doing these things, until the autumn equinox is finished. But it will have to be remembered that during these days and those of the month of August we should cut foliage for the cattle in the hours before daylight and in the evening. Also any vineyards which we intend to cultivate we must remember not to dig during the heat but in the early morning till the third hour of the day and from the tenth hour till dusk. In some districts, such as Cilicia and Pamphylia, sesame is sown in this month; but in

Lesser Dog: *cf.* Cicero, *N.D.* II. 44. 114. Antecanem Procyon Graio qui nomine fertur, and Pliny, XVIII. § 268.

LUCIUS JUNIUS MODERATUS COLUMELLA

sesama seruntur: Italiae autem regionibus humidis possunt ultimo mense Iunio seri. Quinetiam tempus est ficulneis arboribus caprificum suspendere; quod quidam existimant idcirco fieri debere, ne fructus decidat, et ut celerius ad maturitatem perveniat.

57 Calen. Augustis; Etesiae. Pridie non. Augusti Leo medius exoritur, tempestatem significat. VII id. Augusti Aquarius occidit medius, nebulosus aestus. Pridie idus Aug. Fidis occidit mane, et Autumnus incipit.

His diebus eadem quae supra. Nonnullis tamen locis favi demetuntur: qui si non sunt melle repleti nec operculati, differenda est in mensem Octob. mellatio.

Idib. Augustis Delphini occasus tempestatem significat. XIX calen. Septemb. eiusdem sideris
58 matutinus occasus tempestatem significat. XIII cal. Septemb. sol in Virginem transitum facit. Hoc et sequenti die tempestatem significat, interdum et tonat. Hoc eodem die Fidis occidit. Decimo cal. Septemb. ex eodem sidere tempestas plerumque oritur, et pluvia. VI cal. Septemb. Vindemiator exoritur mane, et Arcturus incipit occidere; interdum pluvia. III cal. Septemb. humeri Virginis exoriuntur: Etesiae desinunt flare, et interdum hiemat. Pridie cal. Septemb. Andromeda vespere[1] exoritur; interdum hiemat.

59 His quidem diebus arbores ficorum inoculantur; quod genus insitionis emplastratio vocatur. Idque licet vel commodius facere superiore mense post idus

---

[1] vespere *S* : vesper *A* : vesperi *R*.

---

[a] See Book V. 11.

BOOK XI. II. 56–59

the damp regions of Italy it can be done in the last part of the month of June. It is, moreover, the time to hang branches of the wild fig on the fig-trees, which some people think ought to be done to prevent the fruit from falling off and that it may come more quickly to maturity.

August 1st the Etesian winds blow. On August 4th the middle of the Lion rises: a storm is portended. On August 7th the middle of Aquarius sets: the weather is foggy and hot. On August 12th the Lyre sets in the early morning and autumn begins.

On these days the same tasks as are mentioned above are carried out; but in some places the honey-combs are gathered in, but if they are not full of honey nor sealed up, the honey-harvest must be put off till the month of October.

On August 13th the setting of the Dolphin is a sign of a storm. On August 14th the morning setting of the same constellation is a sign of a storm. On August 20th the Sun passes into Virgo: on this and the following day a storm is portended: sometimes it thunders; on the same day the Lyre sets. On August 23rd, as the effect of the same constellation, a storm generally springs up and there is rain. On August 26th the Vintager rises in the morning; and Arcturus begins to set: it sometimes rains. On August 30th the shoulders of Virgo appear: the Etesian winds cease to blow and the weather is sometimes stormy. On August 31st Andromeda rises in the evening: it is sometimes stormy.

During these days fig-trees are inoculated: this kind of grafting is called " emplastration." [a] This may be more conveniently done in the previous

LUCIUS JUNIUS MODERATUS COLUMELLA

Iul. quo tempore etiam aliarum arborum nonnulli emplastrationem faciunt. Quibusdam locis, ut in Baetica maritimis regionibus, et in Africa vindemia conficitur. Sed frigidioribus regionibus pulverationem faciunt, quam vocant rustici occationem, cum omnis gleba in vineis refringitur et solvitur[1] in pulverem. Hoc eodem tempore prius quam vineae pulverentur, si perexilis est terra, vel rara[2] ipsa vitis, lupini modii tres vel quattuor in singula iugera sparguntur, et ita inoccantur; qui, cum fruticaverint,[3] prima tum fossione conversi satis bonum stercus vineis praebent. Multi etiam, si pluvius est status caeli, sicut[4] suburbana regione Italiae, pampinis vitem spoliant, ut percoqui fructus possint, nec putrescere imbribus. At e contrario locis calidioribus, ut modo nominatis provinciis, circa vindemiam adumbrantur vel stramentis[5] vel aliis tegumentis uvae, ne ventis aut caloribus exarescant. Hoc idem tempus est aridis uvis ficisque conficiendis, de quibus quemadmodum passae fiant, suo loco dicemus, cum villicae persequemur officia. Filix quoque aut carex, ubicunque nascitur, Augusto mense recte extirpatur, melius tamen circa idus Iulias ante Caniculae exortum.

Calend. Septembribus calor. Quarto nonas Septemb. Piscis austrinus desinit occidere, calor. Non. Septemb. Arcturus exoritur, Favonius, vel Corus.[6] VII idus Septemb. Piscis aquilonius desinit

---

[1] solvitur *SAa* : resolvitur *R*.
[2] terra vel rara *Schneider* : vel terra vel *SA* : vel rara vel *R*.
[3] fructificaverunt *R*.
[4] sicut *AR* : suicui *S*.
[5] *Post* stramentis *add*. vel is *SA*.
[6] chorus *SAR*.

month after July 15th, at which time some people carry out the "emplastration" of other trees also. In some places, for instance in Baetica in the coastal regions and in Africa, the vintage is finished; but in the colder districts they carry out the pulverization of the soil, which the farmers call harrowing, when all the clods in the vineyards are broken up and reduced to powder. In the same period, before the vineyards are pulverized, if the soil is very thin or the vine itself scanty, three or four *modii* of lupine-seed are scattered on each *iugerum* and then harrowed in; these when they have sprouted, having then been turned under with the first digging, provide the vines with quite good manure. Many people also, if the climatic conditions are rainy, as in the district of Italy near the capital, strip the vines of their tendrils, so that the fruit may be able to ripen and not be rotted by the rain. But in warmer places, on the contrary, as in the provinces named just now, about the time of the vintage the grapes are shaded either with straw or some other covering, so that they may not be dried up by the winds or the heat. This is also the season for making raisins and dried figs; how they are to be dried, we will describe in the proper place, when we treat of the duties of the bailiff's wife.[a] It is also right to uproot ferns and sedge, wherever they grow, during the month of August; it had better, however, be done about July 15th before the rising of the Dogstar.

On September 1st the weather is hot. On September 2nd the Southern Fish ceases to set: the weather is hot. On September 5th Arcturus rises: the wind is West or North-west. On September

---

[a] Book XII. 15.

occidere, et Capra exoritur, tempestatem significat. Tertio idus Septembris Favonius aut Africus, Virgo media exoritur.

64 His diebus locis maritimis et calidis vindemia et cetera, quae supra scripta sunt, commode administrantur. Iteratio quoque arationis peracta esse debet, si serius terra proscissa est. Sin autem celerius, etiam tertiatum[1] solum esse convenit. Hoc etiam tempore qui consueverunt vina condire, aquam marinam praeparant, et advectam decoquunt: de qua conficienda praecipiam, cum villicae officia exsequar.[2]

65 Id. Sept. ex pristino sidere nonnumquam tempestatem significat. XV cal. Oct. Arcturus exoritur; Favonius aut Africus, interdum Eurus, quem quidam Vulturnum appellant. XIV cal. Oct. spica Virginis exoritur; Favonius, aut Corus.[3] XIII cal. Octob. sol in Libram transitum facit, Crater matutino tempore apparet. XI cal. Oct. Pisces occidunt mane, item
66 Aries occidere incipit; Favonius aut Corus,[4] interdum Auster cum imbribus. X cal. Octob. Argo navis occidit; tempestatem significat, interdum pluviam. Nono cal. Octob. Centaurus incipit mane oriri; tempestatem significat, interdum et pluviam.[5] Octavo cal. Octob. et septimo et sexto Aequinoctium autum-

---

[1] tertiatum *a*: tertia cum *SA*.
[2] exequar *R*: persequar *SAa*.
[3] chorus *SAR*.
[4] chorus *SAR*.
[5] interdum et pluviam *om. S*.

---

[a] Book XII. 19 ff.
[b] So-called from Vultur, a mountain in Apulia. Strictly speaking it is a S.E. by one-third S. wind.

BOOK XI. ii. 63-66

7th the Northern Fish ceases to set, and the She-goat rises: bad weather is portended. On September 11th the wind is West or South: the middle of Virgo rises.

During these days, in places which are near the sea and warm, the vintage and the other operations mentioned above are fittingly accomplished. Also, if the ground was broken up for the first time rather late, a second ploughing ought to be carried out; but if it was broken up too early even a third ploughing can be given to the soil with advantage. At this time too those who have been in the habit of seasoning wines provide themselves with sea-water and, when they have brought it, boil it; about the way to prepare it I will give instructions when I deal with the duties of the bailiff's wife.[a]

On September 13th a storm is sometimes portended from the effect of the constellation of the Whale. On September 17th Arcturus rises: the wind is West or South, but sometimes East, which some people call Vulturnus.[b] On September 18th the *Spica Virginis* [c] rises: the wind is West or Northwest. On September 19th the Sun passes into the Balance: the Bowl appears in the morning. On September 21st the Fishes set early in the morning, also the Ram begins to set: the wind is West or North-west, sometimes South accompanied by rain-showers. On September 22nd the ship Argo sets: a storm is portended, sometimes rain. On September 23rd the Centaur begins to rise early in the morning: a storm is portended, sometimes also rain. On September 24th, 25th, and 26th the autumnal

[c] The ear of corn in the left hand of Virgo and the brightest star in that constellation.

LUCIUS JUNIUS MODERATUS COLUMELLA

nale[1] pluviam significat. Quinto cal. Oct. Haedi exoriuntur; Favonius, nonnumquam Auster cum pluvia. Quarto cal. Octob. Virgo desinit oriri; tempestatem significat.

67 His diebus vindemiae pluribus[2] regionibus fiunt, quarum maturitatem alii aliter interpretati sunt. Quidam cum vidissent partem aliquam uvarum virescere, crediderunt tempestivam esse vindemiam: quidam cum coloratas et perlucidas uvas animadvertissent: nonnulli etiam cum pampinos ac folia decidere considerassent. Quae omnia fallacia sunt: quoniam immaturis uvis eadem omnia possunt 68 accidere propter intemperiem solis aut anni. Itaque nonnulli gustu explorare maturitatem tentaverunt, ut sive dulcis esset sapor uvae, sive acidus, proinde aestimarent. Sed et haec ipsa res habet aliquam fallaciam. Nam quaedam genera uvarum numquam dulcedinem capiunt propter austeritatem 69 nimiam. Itaque optimum est, quod nos[3] facimus[4] ipsam naturalem contemplare maturitatem. Naturalis autem maturitas est, si cum expresseris vinacea, quae acinis celantur, iam infuscata, et nonnulla propemodum nigra fuerint. Nam colorem nulla res vinaceis potest afferre, nisi naturae maturitas, praesertim cum ita in media parte acinorum sint ut[5] et a sole et a ventis protegantur, humorque ipse non patiatur ea praecoqui,[6] aut infuscari, nisi suapte 70 natura. Hoc igitur cum exploratum habuerit villi-

[1] autumnale *om. SA.*
[2] plurimis *Sa* : plurimi *A* : pluribus *R.*
[3] nos *R* : non *SA.*
[4] fecimus *SAa* : facimus *R.*
[5] ut *R* : *om. SAa.*
[6] precoqui *a* : precipi *SR* : p̄cipia *A.*

BOOK XI. ii. 66–70

equinox portends rain. On September 27th the Kids rise: the wind is West, sometimes South accompanied by rain. On September 28th Virgo ceases to rise: a storm is portended.

During these days the vintage takes place in most 67 districts, different people having inferred from different signs that the grapes are ripe. Some people have thought that the time for the vintage has come when they have seen that some part of the grapes is becoming green, others when they have noticed that the grapes are highly coloured and transparent, still others when they have observed that the tendrils and foliage are falling. All these signs are deceptive, because all the same things may happen to unripe grapes owing to the excessive heat of the sun or the inclemency of the time of year. Some people there- 68 fore have attempted to test the ripeness of grapes by tasting them, so that they might judge thereby whether their flavour is sweet or acid. But this method also has some deception about it, for some kinds of grapes never acquire any sweetness because of their excessive harshness. It is best, therefore, to 69 do what we do and to consider the natural ripeness in itself. There is natural ripeness if the grape-stones which are hidden in the berries, when you press them out, are already dusky in colour, and, in some cases, almost black; for nothing can give colour to the grape-stones except nature's own ripeness, especially as they are situated in the middle of the berries, so that they are protected from the sun and the winds, and the moisture itself does not allow them to be ripened prematurely or to turn to a dusky colour except by a natural process of their own. There- 70 fore, when the bailiff has assured himself on this point,

cus, sciet vindemiam sibi esse faciendam. Sed antequam fructum cogere incipiat, cuncta praeparanda erunt superiore si fieri possit mense : si minus, certe ut ante quindecim dies dolia partim picata, partim defricata et diligenter lauta marina, vel aqua salsa et
71 recte siccata; tum et opercula colaque et cetera, sine quibus probe confici mustum non potest; torcularia vero, et fora diligenter emundata lautaque, et si res ita exegerit, picata[1]; praeparataque habeat ligna, quibus defrutum et sapam decoquat. Tum etiam salem atque odoramenta, quibus condire vina consueverit, multo ante reposita esse oportet.

Nec tamen haec cura totum avocet eum a cetera ruris cultura. Nam et napinae itemque rapinae siccaneis locis per hos dies fiunt. Farraginaria quoque pecori futura per hiemem praesidio, itemque siliqua, quod rustici faenum Graecum vocant, nec minus in
72 pabulum vicia nunc demum conseruntur. Tum etiam lupini haec[2] erit praecipua satio, quem quidam vel ab area protinus in agrum deferri putant oportere. Milium et panicum hoc tempore demetitur, quo faseolus[3] ad escam seritur. Nam ad percipiendum semen ultima parte Octobris circa calendas Novembres melius obruitur. Quare cum haec cuncta in agris exsequi debeat, possit eorum curam, quae

[1] picta *A*.   [2] haec *vel* hec *R* : hac *SA*.
[3] faseolus *R* : passolus *SA*.

he will know that the vintage must be carried out. But before he begins to collect the fruit, all preparations must, if possible, be made in the previous month; if not, he must at any rate see that the wine-jars are some treated with pitch and others scoured and carefully washed with sea-water or salted water and properly dried fifteen days beforehand, and also the lids and strainers and all the other things, without which must cannot be properly made. He must prepare the wine-presses and tubs which have been carefully cleaned and washed and, if necessary, treated with pitch, and the bailiff should have in readiness firewood which he may use for boiling down the must to a third or half its original volume. Also salt and spices, which he has been accustomed to use in the seasoning of wine, ought to be stored up in good time beforehand.

Nevertheless his attention to this task must not call him wholly away from the other departments of agriculture; for during these days beds for navews and turnips are made in naturally dry places. Mixed fodder also, which will be a great stand-by for the cattle in the winter, and the pods which the country-folk call fenugreek and likewise tares are now sown for fodder. Also the principal sowing of lupine will take place at this time, which some people think ought to be brought straight from the threshing-floor to the field. Common and Italian millet are reaped at this time when the calavance is sown for food; for in order to obtain seed from it, it is better to cover it up in the ground at the end of October towards November 1st. Since all these tasks have to be performed in the fields, the bailiff ought to delegate the superintendence of those

LUCIUS JUNIUS MODERATUS COLUMELLA

intra villam facienda sunt, villicae delegare: ita tamen, ut ipse consideret an recte facta sint.

Cal. Octobribus, et sexto non. interdum tempestatem significat. Quarto non. Octobris Auriga occidit mane, Virgo desinit[1] occidere; significat nonnumquam tempestatem. Tertio non. Octobris Corona incipit exoriri, significat tempestatem. Pridie non. Octobris Haedi oriuntur vespere; Aries medius occidit; Aquilo. Octavo id. Octobris Coronae clara stella exoritur. Sexto id. Octobris Vergiliae exoriuntur vespere; Favonius, et interdum Africus cum pluvia. Tertio et pridie idus Octobris Corona tota mane exoritur, Auster hibernus, et nonnumquam pluvia.

Per hos dies frigidis regionibus vindemia et cetera, quae supra scripta sunt, fieri solent, iisdemque regionibus frumenta[2] matura seruntur, et praecipue far adoreum. Locis etiam opacis triticum nunc recte seritur. Et quoniam sementis mentionem fecimus, non intempestive quantum cuiusque seminis iugerum agri recipiat referemus. Iugerum agri recipit tritici modios quattuor vel quinque, farris adorei modios novem vel decem, hordei modios[3] quinque vel sex, milii vel panici[4] sextarios quattuor vel quinque, lupini modios octo vel decem, phaseli[5] modios quattuor, pisi modios tres vel quattuor, fabae modios sex, lentis modium unum[6] vel paulo amplius, lini seminis modios octo vel decem, cicerculae modios tres vel quattuor, ciceris modios tres vel quattuor, sesami sextarios quattuor vel quinque, viciae

[1] desint *S*.
[2] isdemque regionibus frumenta *R* : isdem frumenta quae regionibus *SA*.
[3] novem—modios *om. SA*.
[4] vel panici *om. SA*.

which have to be carried out within the farm buildings to his wife, but under such conditions that he can himself observe whether they are properly done.

On October 1st and 2nd bad weather is sometimes portended. On October 4th the Wagoner sets early in the morning and Virgo ceases to set: bad weather is sometimes portended. On October 5th the Crown begins to rise: bad weather is portended. On October 6th the Kids rise in the evening: the middle of Aries sets: the wind is in the North-east. On October 8th the bright star of the Crown rises. On October 10th the Pleiads rise in the evening: the wind is in the West and sometimes in the Southwest accompanied by rain. On October 13th and 14th the whole of the Crown rises in the early morning: there is a stormy South wind and sometimes rain.

During these days in cold regions the vintage usually takes place and the other operations described above, and in the same regions the early-ripe cereals are sown, particularly two-grained wheat; also in shady places it is the right time to sow emmerwheat. And since we have mentioned sowing, it will not be out of place if we state what quantity of each kind of seed a *iugerum* of land takes. A *iugerum* of land takes four or five *modii* of bread-wheat, nine or ten of emmer-wheat, five or six of barley, four or five of common millet or Italian millet, eight to ten of lupines, four of calavance, three or four of peas, six of broad-beans, one or a little more of lentils, eight to ten of linseed, three or four of chickling-vetch, four or five *sextarii* of sesame, seven or eight

---

⁵ faseli *SA*: faseoli *R*. ⁶ unum *om. SA*.

LUCIUS JUNIUS MODERATUS COLUMELLA

pabularis modios septem vel octo,[1] viciae seminalis modios quinque vel sex, ervi modios quattuor vel quinque, farraginis hordeaceae modios septem vel octo, siliquae modios sex. Medicae singulos cyathos [2] serere oportet in areolis longis pedum denum, latis pedum [3] quinum. Cannabis grana sex in pede quadrato ponuntur.

76 Idibus Octobribus et sequenti biduo interdum tempestas,[4] nonnunquam rorat, tantummodo; Iugulae exoriuntur vespere.[5] Quarto et decimo calendas
77 Novembres sol in Scorpionem transitum facit. Tertiodecimo et duodecimo calendas Novembris solis exortu Vergiliae incipiunt occidere; tempestatem significat. Undecimo calendas Novembres Tauri cauda occidit; Auster, interdum pluvia. Octavo calendas Novembres Centaurus exoriri mane desinit;
78 tempestatem significat. Septimo calendas Novembres Nepae frons exoritur; tempestatem significat. Quinto calendas Novembres Vergiliae occidunt; hiemat cum frigore et gelicidiis. Quarto calendas Novembres Arcturus vespere occidit; ventosus dies. Tertio calendas Novembres et pridie Cassiopea incipit occidere; tempestatem significat.

79 Per hos dies quaecunque semina differri debent, arbusculaeque omnis generis recte ponuntur. Ulmi quoque vitibus recte maritantur, ipsaeque vites in arbustis et vineis commode propagantur. Seminaria runcare et fodere tempus est, tum etiam arbores ablaqueare, nec minus vineas, easdemque putare, itemque in arbustis vitem deputare. Se-

---

[1] octo *om. Sa.*
[2] qui ad hos *S* : quiathos *A*.
[3] denum latis pedum *om. SA.*
[4] tempestas *R* : temperat *S* : temperatas *A*.

BOOK XI. ii. 75-79

modii of tares for fodder, five or six of seed-vetch, four or five of bitter vetch, seven or eight of mixed barley, six of fenugreek. A *cyathus* of medic should be sown in beds ten feet long and five feet broad, and six grains of hemp-seed are planted in a square foot of ground.

On October 15th and the two following days there is sometimes bad weather: occasionally dew falls: Orion's Belt rises in the evening. On October 20th the Sun crosses into the Scorpion. On October 20th and 21st at the rising of the Sun the Pleiads begin to set: a storm is portended. On October 22nd the tail of the Bull sets: the wind is in the South and there is sometimes rain. On October 25th the Centaur ceases to rise in the morning: a storm is portended. On October 26th the forehead of the Scorpion rises: a storm is portended. On October 28th the Pleiads set: the weather is wintry with cold and frosts. On October 29th Arcturus sets in the evening: it is a windy day. On October 30th and 31st Cassiopea begins to set: bad weather is portended.

During these days it is the proper thing that any seedlings which ought to be transplanted and shrubs of every kind be placed in position. It is also correct to mate the elms with the vines and it is a convenient time to propagate the vines themselves both where they are supported by trees and in vineyards. It is the time to weed and dig up the nurseries and to dig up the earth round the trees and likewise in the vineyards and also to prune them and to cut back the vines which are supported

---

⁵ tantummodo iugulae exoriuntur vespere *S* : *om. R.*

LUCIUS JUNIUS MODERATUS COLUMELLA

minaria, quae suo tempore pampinata non sunt, arbusculaeque[1] ficorum in seminariis[2] putari et ad singulos stilos redigi debent: quae tamen melius dum tenerae sunt, per germinationem pampinantur. Sed cum omnia in agricultura strenue facienda sint,
80 tum maxime sementis. Vetus est agricolarum proverbium, maturam sationem saepe decipere solere, seram nunquam, quin mala sit.[3] Itaque in totum praecipimus: ut quisque natura locus frigidus erit, is primus[4] conseratur; ut quisque calidus, novissimus.

81 Vicia et faba stercorare agrum dicuntur. Lupinum nisi in flore verteris,[5] nihil agrum stercoraveris. Sed nec ulla res magis vacuis operariis[6] aut seritur, aut conditur. Nam et primis temporibus ante ullam sementem possit[7] id obrui, et novissimis post
82 coactos fructus tolli. Sementi facta inoccare oportet, quod sparseris. Duo iugera tres operae commode occabunt, arboresque quae intererunt ablaqueabunt; quamvis antiqui singulis[8] operis singula iugera sarriri et occari velint: quod an recte fieri possit, affirmare non ausim.

Eodem tempore fossas rivosque[9] purgare, et elices
83 sulcosque aquarios facere convenit. Iisdem temporibus si sit, fraxineam; si minus, orneam; si nec haec sit, iligneam frondem bubus recte praebebimus. Glandis quoque non inutile est singulis

[1] arbusculaque *A* : arbuscula quae *S*.
[2] seminaris *SA*.   [3] sit *R* : sint *SA*.
[4] primus *S* : primis *A*.
[5] verteris *om. A*.
[6] operaris *S* : operis *A*.
[7] possis *Lundström* : possit *SAR*.
[8] singulis *SR* : similis *A*.
[9] rivosque *vett edd.* : rivos *SAR*.

by trees. The nurseries which have not been trimmed at the proper time, and small fig-trees growing in nurseries, ought to be pruned and reduced to single stems; these, however, while they are young, are better trimmed when they are budding. While every operation of agriculture ought to be performed with alacrity, this is especially true of sowing. It is an old proverb among husbandmen that early 80 sowing is often apt to deceive us, late sowing never— it is always a failure. And so our advice on the whole is that the ground which is naturally coldest should be sown first, that which is warmest last.

Tares and beans are said to manure the ground; 81 but unless you turn in lupines when they are in flower, you will not manure the ground at all. There is nothing more commonly sown or stored than lupines by labourers who have nothing to do, for they can be put into the ground at the beginning of the sowing-time before any other seed and can be taken up last of all when the fruits have been gathered in. After 82 you have sown your seed, you ought to harrow in what you have scattered. Three labourers will easily harrow two *iugera* a day and dig round the trees which grow among the crops, though the ancients would have it that labourers could each hoe and harrow a *iugerum* a day; but I would not venture to assert whether it could be properly done.

During the same period it is proper to clean out the ditches and channels and make drains and gutters for the water. At the same season we shall do well to 83 provide the oxen with ash leaves, if we have them, but, if not, with mountain-ash leaves, and, if we have not these either, with holm-oak leaves. It is also not amiss to give each yoke of oxen one *modius* of

LUCIUS JUNIUS MODERATUS COLUMELLA

iugis modios singulos dare: nec tamen amplius, ne laborent, nec minus diebus xxx praebueris. Nam si paucioribus diebus datur, ut ait Hyginus, per ver[1] scabiosi boves fiunt. Glans autem paleis immiscenda est, atque ita bubus apponenda. Tum etiam silvam si quis barbaricam, id est consemineam velit facere, recte conseret glandibus et ceteris seminibus. Tum et olea destringenda est, ex qua velis viride oleum efficere; quod fit optimum ex varia oliva, cum incipit nigrescere.[2] Nam acerbum nisi ex alba olea fieri non debet.

84 Calen. Novembribus et postridie caput Tauri occidit; pluviam significat. III non. Novembres Fidicula mane exoritur; hiemat et pluit. VIII idus Novembres idem sidus totum exoritur; Auster vel Favonius; hiemat. VII idus Novembres tempestatem significat et hiemat. Sexto idus Novembres Vergiliae mane occidunt, significat tempestatem; hiemat. Quinto idus Novembres stella clara Scorpionis exoritur; tempestatem significat; vel Vulturnus; interdum rorat. IV idus Novembres hiemis initium, Auster, aut Eurus, interdum rorat.

85 His diebus usque in idus, quae superiore mense facere non potueris,[3] adhuc tolerabiliter efficies. Sed et proprie hoc observabis, ut pridie, quam plenilunium sit; si minus, certe ipso plenilunio

[1] per ver *R*: pervi *a*: om. *SA*.
[2] incipit nigrescere *Sa*: incipit increscere *R*: incipiunt crescere *A*.
[3] poteris *SA*.

---

[a] See note on Book I. 1. 13.

mast; but do not provide them with more than this, lest they fall sick, nor for a shorter period than thirty days, for if it is given for fewer days than this (so Hyginus [a] tells us) the oxen become scabby in the spring. Now the mast should be mixed with chaff and then put before the oxen. At this time too if anyone is minded to make a wilderness, that is a wood where various trees are planted together, he will do well to plant it with acorns and seeds of the other trees. At this time too olive-trees should be stript if you wish to make green oil from them; it is best made from speckled berries when they begin to turn black, for bitter oil ought not to be made except from white olives.

On November 1st and the following day the head of Taurus sets; rain is portended. On November 3rd the Little Lyre rises in the morning; it is wintry weather and there is rain. On November 6th the whole of the same constellation rises: the wind is South or West, and the weather wintry. On November 7th a storm is portended; the weather is wintry. On November 8th the Pleiads set in the morning; a storm is portended and the weather is wintry. On November 9th the bright star of the Scorpion rises: a storm is portended, or the wind is South-east: sometimes there is a fall of dew. On November 10th is the beginning of winter, the wind is South or South-east: sometimes there is a fall of dew.

During these days up to November 13th you will still be able to carry out tolerably well the tasks which you could not do in the previous month. But you will take care in particular, that on the day before full moon, or, if not, at any rate on the day of the

LUCIUS JUNIUS MODERATUS COLUMELLA

omnem, quam saturus es, fabam uno die spargas:[1] sed postea licebit ab avibus et pecore defensam obruas: eamque, si ita competierit lunae cursus, ante idus Novembres occatam habeas quam pinguissimo et novo loco: si minus, quam stercoratissimo.

86 Satis erit in singula iugera vehes stercoris comportare numero decem octo. Vehis autem stercoris una[2] habet modios octoginta. Ex quo colligitur, oportere in denos quoquoversus pedes modios quinos[3] stercoris spargere. Quae ratio docet universo iugero

87 satisfacere modios MCCCCXL. Tum etiam convenit oleas ablaqueare, et si sunt parum fructuosae vel cacuminibus retorridae frondes, magnis arboribus quaternos modios stercoris caprini circumspergere, in ceteris autem pro magnitudine portionem servare: eodem tempore vineis ablaqueatis columbinum stercus ad singulas vites, quod sit instar unius sextarii, vel urinae hominis congios, vel alterius generis quaternos sextarios stercoris infundere. Iugerum vinearum in senos pedes positarum duae operae ablaqueant.

88 Idibus Novembribus dies incertus tamen placidus. Sextodecimo calendas Decembres Fidis[4] exoritur mane; Auster, interdum Aquilo magnus. Quintodecimo calendas Decembres Aquilo, interdum Auster cum pluvia. Quartodecimo calendas Decembres sol in Sagittarium transitum facit; Suculae mane oriuntur, tempestatem significat. Duodecimo calendas Decembres Tauri cornua vesperi occidunt;

89 Aquilo frigidus, et pluvia. Undecimo calendas

---

[1] spargas R: peragas SA.
[2] una S: uno A: om. R.
[3] quinos R: quinque SAa.
[4] fidis R: fides SA.

full moon itself, you scatter on the same day all the
beans which you intend to sow; afterwards you may
cover them up with earth as a protection against birds
and cattle, and, if the course of the moon shall fall
suitably, have them harrowed before November 13th
in ground which is fresh and as rich as possible and,
if not, in soil as well manured as possible. It will be 86
enough to cart eighteen loads of dung to each *iugerum*.
Now one load of dung consists of eighty *modii*, from
which it can be inferred that you ought to scatter
five *modii* of dung over an area measuring ten feet
each way. This calculation shows us that one
thousand four hundred and forty *modii* are enough
for a whole *iugerum*. At this time it is also proper to
dig round the olive-trees and, if they are not fruitful 87
enough, or if the foliage on their tops is shrivelled up,
you ought to sprinkle four *modii* of goat's dung around
each big tree and observe the same proportion for
the others according to their size. At the same time,
after the vineyards have been dug up, you ought to
pour what amounts to one *sextarius* of pigeon's dung on
each vine, or a *congius* of human urine or four *sextarii*
of some other kind of dung. Two labourers in one
day dig up a *iugerum* of vines planted six feet apart.

November 13th is an uncertain day: it is oftener 88
calm than not. On November 16th the Lyre rises in
the morning: the wind is South, though sometimes it
is strong North-east. On November 17th the wind is
from the North-east, sometimes from the South
accompanied by rain. On November 18th the Sun
makes its entry into Sagittarius: the Hyades rise in
the morning: a storm is portended. On November
20th the horns of the Bull set in the evening: there is
a cold North-east wind and rain. On November 89

LUCIUS JUNIUS MODERATUS COLUMELLA

Decembres Sucula mane occidit, hiemat. Decimo calend. Decembres Lepus occidit mane, tempestatem significat. Septimo calend. Decembres Canicula occidit solis ortu, hiemat. Pridie calendas Decembres totae Suculae occidunt; Favonius aut Auster, interdum pluvia.

90 His diebus, quae praeterita erunt superioribus, opera consequi oportebit.[1] Et, si non plurimum serimus, optimum est intra calendas Decembres sementem fecisse.[2] Sed etiam longis noctibus ad diurnum[3] tempus aliquid adiciendum est. Nam multa sunt, quae in lucubratione recte agantur,[4] Sive enim vineas possidemus, pali et ridicae possunt dolari exacuique:[5] sive regio ferulae vel corticis ferax est, apibus alvaria fieri debent: sive palmae spartive fecunda est, fiscinae sportaeque: seu 91 virgultorum, corbes ex vimine. Ac ne cetera nunc persequar, nulla regio non aliquid affert, quod ad lucubrationem confici possit. Nam inertis est agricolae expectare diei brevitatem, praecipue in iis regionibus, in quibus brumales dies horarum novem 92 sunt, noctesque horarum quindecim. Possit etiam salix decisa pridie ad lucubrationem expurgari, et ad vitium ligamina praeparari. Quae si natura minus lenta est, ante dies quindecim praecidenda, et purgata in stercore obruenda est, ut lentescat. Sin autem iam pridem[6] caesa exaruit, in piscina maceranda est. Tum etiam per lucubrationem ferra-

---

[1] oportebit *S* : oportet *AR*.
[2] fecisse *SAa* : confecisse *R*.
[3] ad diuturnum *R* : ad diū t̄ nū *S* : addiuntur *A*.
[4] agantur *Sa* : aguntur *AR*.
[5] exacuiq; *R* : ex ad quae *S* : exaquiq; *A*.
[6] pridem *R* : pridie *SAa*.

21st one of the Hyades sets in the morning: the weather is wintry. On November 22nd the Hare sets in the morning: a storm is portended. On November 25th the Dog-star sets at sunrise: the weather is wintry. On November 30th the Hyades set completely: the wind is from the West or the South; it sometimes rains.

During these days you will have to finish the tasks which have been omitted during the previous days, and if we do not carry out very extensive sowing, it is best to have completed it before December 1st. But also, when the nights are long, some time must be added to the period of daylight; for there are many things which can be properly done by artificial light. For if we possess vineyards, poles and props can be hewn and sharpened; or if the district is productive of fennel or bark, hives should be made for the bees; or if it is rich in palm-trees or broom, frails and baskets can be made; or if it abounds in twigs, hampers can be made from osiers. Not to go now into detail of all the other things that can be made, there is no district which does not provide something which can be made by artificial light; for he is the lazy farmer who waits for the short day to begin, especially in regions when the winter days last for only nine hours while the night goes on for fifteen hours. Willows also cut down the previous day can be cleaned by artificial light and prepared as ties for the vines; if they are not naturally tough enough, they should be cut down fifteen days beforehand and, after being cleaned, buried in dung, in order that they may be toughened; but if they have been cut a long time ago and have become dry, they should be soaked in a pond. Then too you should sharpen iron tools

LUCIUS JUNIUS MODERATUS COLUMELLA

menta acuere, et ad ea facere, vel facta manubria aptare, quorum optima sunt ilignea, deinde carpinea, post haec fraxinea.

93 Calendis Decembribus dies incertus, saepius tamen placidus. Octavo idus Decembres Sagittarius medius occidit; tempestatem significat. Septimo idus Decembres Aquila mane oritur; Africus, interdum Auster, et rorat.[1] Tertio idus Decembres Corus,[2] vel Septentrio, interdum Auster cum pluvia.

His diebus quae praeterita erunt superiore mense opera peragi debebunt, utique in locis temperatis aut calidis: nam locis frigidis recte fieri iam non possunt.

Idibus Decembribus Scorpio totus mane exoritur; 94 hiemat. Sextodecimo calendas Ianuarias sol in Capricornum transitum facit, brumale solstitium, ut Hipparcho placet: itaque tempestatem saepe significat. xv calend. Ianuarias ventorum commutationem significat. x calendas Ianuarias Capra occidit mane, tempestatem significat. Nono[3] calendas Ianuarias brumale solstitium, sic Chaldaei observant. Sexto calend. Ianuarias Delphinus incipit oriri mane; tempestatem significat. Quarto calendas Ianuarias Aquila occidit vespere; hiemat. Tertio calendas Ianuarias Canicula vespere occidit; tempestatem significat. Pridie calendas Ianuarias, tempestas ventosa.

95 His diebus qui religiosius rem rusticam colunt, nisi si vinearum causa[4] pastines, negant debere terram

[1] et rorat *SA* : et inrorat *a* : irrorat *R*.
[2] chorus *SAR*.
[3] VIIII *AR* : VIII *S*.
[4] causam *S* : causas *Aa*.

by artificial light and make handles for them or fit to them handles already made, the best material being the wood of the holm-oak, then that of the hornbeam and next that of the ash.

December 1st is an uncertain day, but more often than not it is calm. On December 6th one half of Sagittarius sets: a storm is portended. On December 7th Aquila rises early: the wind is South-east, but sometimes South, and there is dew. On December 11th the wind is North-west or North, but sometimes South accompanied by rain.

During these days tasks which have been omitted in the preceding month will have to be completed, obviously in temperate and warm regions; for in cold regions they cannot now be properly performed.

On December 13th the whole constellation of the Scorpion rises early: the weather is wintry. On December 17th the Sun makes its crossing into Capricorn: it is the winter solstice according to Hipparchus,[a] and so a storm is often portended. On December 18th a change of wind is portended. On December 23rd the Goat sets early: a storm is portended. On December 24th is the winter solstice as observed by the Chaldaeans. On December 27th the Dolphin begins to rise early in the morning: a storm is portended. On December 29th Aquila sets in the evening: the weather is wintry. On December 30th the Dog-star sets in the evening: a storm is portended. On December 31st the weather is windy.

During these days, according to those who practise husbandry with unusually scrupulous care, the soil ought not be disturbed with any iron tool, unless you

[a] See note on Book I. 1. 4.

ferro commoveri. Itaque quidquid citra id genus effici potest, id ab his comprehenditur ut olea legatur, et conficiatur, ut vitis paletur et capite tenus alligetur,[1] ut iuga vineis imponantur et capis-
96 trentur. Ceterum palmare, id est materias alligare, hoc tempore non expedit, quia plurimae propter rigorem qui fit ex frigore, franguntur. Possunt etiam his diebus cerasa[2] et tuberes et Armeniacae atque amygdalae ceteraeque arbores quae primae florent, inseri commode. Nonnulli etiam legumina serunt.

97 Calendis Ianuariis dies incertus. Tertio nonas Ianuarias Cancer occidit; tempestas varia. Pridie nonas Ianuarii media hiems; Auster multus, interdum pluvia. Non. Ianuariis Fidis[3] exoritur mane; tempestas varia. Sexto idus Ianuarias Auster, interdum Favonius. Quinto idus Ian. Auster,
98 interdum imber. Pridie idus Ian. incertus status caeli.

Per hos quoque dies abstinent terrenis operibus religiosiores agricolae, ita tamen ut ipsis calen. Ianuariis auspicandi causa omne genus operis instaurent, ceterum differant terrenam molitionem usque in proximas idus.

Sed nec ignorare debebit villicus, quid uni iugo[4] boum quoquo mense per singulos dies praestari satis sit. Quare huius quoque curae rationem subicie-

---

[1] alligetur *AR* : -entur *S*.
[2] cerasia *SA* : cerasa *R*.
[3] fidis *R* : fides *SA*.
[4] iugo *AR* : iuga *S*.

trench it for the sake of the vines. Therefore whatever can be done outside this kind of work is included in the following list: the gathering of olives and the preparing of them, the staking of vines and fastening them right up to their heads, the placing of frames in the vineyards and binding them together. But the process called "palmation," that is, the tying of hard-wood shoots to the frame, is not expedient at this time, because very many of them break off owing to the stiffness caused by the cold. Also during these days cherry-trees and tuber-apple-trees and apricot-trees and almond-trees and the other early-flowering trees can conveniently be engrafted. Some people also sow pulse.

On January 1st the weather is uncertain. On January 3rd the Crab sets: the weather is changeable. January 4th marks the middle of winter: there is much wind from the South and sometimes rain. On January 5th the Lyre rises in the morning: the weather is changeable. On January 8th the wind is from the South but sometimes from the West. On January 9th the wind is from the South and sometimes there is a rain storm. On January 12th the state of the weather is uncertain.

During these days also the more scrupulous husbandmen abstain from operations upon the soil, except that, on January 1st, for the sake of good luck, they make a beginning of work of every kind, but they put off the working of the soil until the ensuing 13th day of the month.

A bailiff will have to have a thorough knowledge of what daily amount of food is enough throughout each month to be provided for a yoke of oxen. Therefore we will add an account of this charge also.

LUCIUS JUNIUS MODERATUS COLUMELLA

99 mus. Mense Ianuario paleas cum ervi macerati [1] sextariis sex, vel paleas cum cicerculae fresae semodio, vel frondis corbem pabulatorium modiorum viginti, vel paleas quantum velint, et faeni pondo viginti, vel affatim viridem frondem ex ilice [2] vel lauro, vel, quod his omnibus praestat, farraginem hordeaceam dabit siccam.[3] Februario mense idem, Martio idem, vel, si opus facturi sunt, faeni pondo
100 quinquaginta. Aprili frondem querneam et populneam ex cal. ad idus vel paleas vel faeni pondo quadraginta. Maio pabulum affatim: Iunio ex calend. frondem affatim: Iulio idem, Augusto idem, vel paleas ex ervo [4] pondo quinquaginta. Septembri frondem affatim, Octobri frondem et ficulnea [5]
101 folia. Novemb. ad idus frondem vel folia ficulnea, quae sint corbis unius. Ex idibus glandis modium unum [6] paleis immixtum, et lupini macerati modium unum paleis immixtum, vel maturam farraginem. Decemb. frondem aridam, vel paleas cum ervi semodio macerato, vel lupini, quod ex semodio macerato exierit,[7] vel glandis modium unum, ut supra scriptum est, vel farraginem.

III. Et quoniam percensuimus opera, quae suis quibusque temporibus anni villicum exequi oporteret, memores polliciti nostri subiungemus cultus hortorum, quorum aeque curam [8] suscipere debebit, ut et quotidiani victus sui levet sumptum, et advenienti domino praebeat, quod ait poëta, inemptas ruris

[1] macerati *SR* : -is *Aa*.
[2] siliquis et *SA* : ilice vel *R*.
[3] siccam om. *SA*. [4] eruo *SA* : arvo *R*.
[5] ficulnea *R* : -eam *SA*.
[6] unum om. *SA*.
[7] exierit *SR* : exigerit *A*.
[8] curam *R* : cura *S* : cultura *A* : culturam *a*.

In the month of January he will give them chaff with 99
six *sextarii* of soaked bitter vetch, or chaff with half a
modius of bruised chickling vetch, or a basket filled
with twenty *modii* of leaves, or as much chaff as they
want and twenty pounds of hay, or green foliage of
the holm-oak or bay-tree in abundance, or, what is
better than all these, a dry mash of barley. In the 100
month of February he will give them the same diet
and again in March or, if they are going to work,
fifty pounds of hay. In April he will give them oak
and poplar leaves from the 1st to the 13th or chaff
or forty pounds of hay. In May he will provide
fodder in abundance, in June from the 1st day of the
month leaves in abundance, the same in July and
August or fifty pounds of chaff from bitter vetch; in 101
September leaves in abundance; in October leaves
and the foliage of the fig-tree; in November up to
the 13th day of the month, leaves or the foliage of the
fig-tree, one basketful; from the 13th day onwards
one *modius* of mast mixed with chaff and one *modius*
of lupines soaked and mixed with chaff or a mash
of seasonable ingredients; in December dried
leaves or chaff with half a *modius* of bitter vetch
soaked in water, or the amount produced from
half a *modius* of lupines soaked in water, or one
modius of mast, as prescribed above, or a mixed
mash.

III. Now that we have enumerated the tasks which *Of the culture of gardens and of garden herbs.*
the bailiff has to perform, each in its proper season,
mindful of our promise we will next deal with the cultivation of gardens, of which he will likewise have to
undertake the superintendence, in order to lessen the
cost of his daily sustenance and provide his master,
when he visits the farm, with what the poet calls " the

LUCIUS JUNIUS MODERATUS COLUMELLA

2 dapes.[1] Democritus in eo libro, quem Georgicon[2] appellavit, parum prudenter censet eos facere, qui hortis extruant munimenta, quod neque latere fabricata maceries[3] perennare possit, pluviis ac tempestatibus plerumque infestata, neque lapides supra rei dignitatem poscat impensa.[4] Si vero amplum modum sepire quis velit, patrimonio esse opus. Ipse igitur ostendam rationem, qua non magna opera hortum ab incursu hominum pecudumque munimus.[5]

3 Vetustissimi auctores vivam sepem structili praetulerunt, quia non solum minorem impensam desideraret, verum etiam diuturnior immensis temporibus[6] permaneret: itaque vepris efficiendi consitis 4 spinis rationem talem reddiderunt. Locus, quem sepire[7] destinaveris, ab aequinoctio autumnali simulatque terra maduerit imbribus, circumvallandus est duobus sulcis tripedaneo spatio inter se distantibus. Modum altitudinis eorum abunde est esse bipedaneum: sed eos vacuos perhiemare patiemur praeparatis seminibus quibus obserantur. Ea sint vastissimarum spinarum, maximeque rubi et paliuri et eius quam Graeci vocant κυνόσβατον, nos sentem 5 canis appellamus. Horum autem[8] ruborum semina quam maturissima legere[9] oportet, et ervi

[1] inemptas rured apes S: inemptas rure de apes A: inempta ruris dapes R.
[2] Georgicam R: om. SA.
[3] maceries Ursinus: materies R: om. SA.
[4] inpensam R: om. SA.
[5] Democritus—munimus om. SA.
[6] temporibus c: operibus SAa.
[7] sepiri SA.
[8] autem om. SA.
[9] legere Lundström: legi S: elegi c: religi AR.

BOOK XI. III. 1-5

country's unbought feasts."[a] Democritus,[b] in the book which he called the *Georgic*, expresses the opinion that those people do not act wisely who build strong defences round their gardens, because neither can an enclosing wall made of brick last for a long time, since it is usually damaged by rain and tempests, nor does the expense involved call for the use of stone, which is too sumptuous for the purpose; indeed a man needs to possess a fortune, if he wishes to enclose a large space of ground. I will, therefore, myself show you a method whereby, without much trouble, we can make a garden safe from the incursions of man and cattle.

The most ancient authors preferred a quick-set hedge to a constructed wall, on the grounds that it not only called for less expense but also lasted for a much longer time, and so they have imparted to us the following method of making a hedge by planting thorn-trees. The area which you have decided to enclose must, as soon as the ground has been moistened by rain after the autumn equinox, be surrounded by two furrows three feet distant from one another. It is quite enough if the measurement of their depth is two feet, but we shall allow them to remain empty through the winter after having got ready the seed with which they are to be sown. These should be those of the largest thorns, especially brambles and Christ's thorn[c] and what the Greeks call κυνόσβατον and we call dog's thorn.[d] The seeds of these briers must be picked as ripe as possible and

---

[a] Vergil, *Georgics*, IV. 133.
[b] That Democritus of Abdera the well-known philosopher is meant here is clear from Book I. 207.
[c] *Paliurus australis.*   [d] *Rosa sempervirens.*

LUCIUS JUNIUS MODERATUS COLUMELLA

moliti farinae immiscere: quae cum est aqua conspersa, illinitur vel nauticis veteribus funibus, vel quibuslibet aliis restibus. Siccati deinde funiculi reponuntur in tabulato: mox ubi bruma confecta est, intermissis quadraginta diebus, circa hirundinis adventum, cum iam Favonius exoritur, post idus Februarias si qua in sulcis per hiemem constitit aqua, exhauritur, resolutaque humus, quae erat autumno regesta, usque ad mediam sulcorum 6 altitudinem reponitur. Praedicti deinde funes de tabulato prompti explicantur, et in longitudinem per utrumque sulcum porrecti obruuntur, sed ita, ut non nimium supergesta terra semina spinarum, quae inhaerent toris[1] funiculorum, enasci possint.[2] Ea fere citra[3] trigesimum diem prorepunt: atque ubi coeperunt aliquod incrementum habere, sic insuesci debent, ut in id spatium, quod sulcis inter-7 iacet, inclinentur. Oportebit autem virgeam sepem interponere, quam super se pandant sentes utriusque sulci, et sit quo interdum quasi adminiculo priusquam corroborentur, acquiescant. Hunc veprem manifestum est interim non posse, nisi radicitus effodere velis. Ceterum etiam post ignis iniuriam melius renasci, nulli dubium est. Et haec quidem claudendi horti ratio maxime est antiquis probata.

8 Locum autem eligi conveniet, si permittit agri situs, iuxta villam, praecipue pinguem, quique adveniente rivo, vel si non sit fluens aqua, fonte

[1] inherent toris *R*: inhaerent oris *S*: inherento nis *A*: tortis *coni.* Anderson.
[2] possit *SA*. [3] citra *SA*: circa *R*.

mixed with meal of well-ground bitter-vetch. This mixture, after having been sprinkled with water, is smeared either on old ships' hawsers or any other kind of rope, and these are then dried and put away in a loft. Then, when mid-winter is passed, after an interval of forty days, about the time the swallow arrives, when the West wind is already rising after February 13th, if any water has stood in the furrows during the winter, it is drawn away, and the loose soil, which had been thrown out in the autumn, is replaced so as to fill half the depth of the furrows. Then the ropes already mentioned are produced from the loft and uncoiled and stretched lengthways along each furrow and covered up in such a manner that the seeds of the thorns, which adhere to the strands of the ropes, may not have too much earth heaped upon them but may be able to sprout. By about the thirtieth day the plants creep forth and, when they begin to grow somewhat, should be trained to grow in the direction of the space lying between the furrows. You will have to place a row of sticks in the middle, over which the thorns from both the furrows may spread, and which may provide a kind of support on which they may rest for a time until they grow strong. It is obvious that this thorn-hedge cannot be destroyed unless you care to dig it up by the roots; but there is also no doubt that even if it has been damaged by fire, it only grows up again all the better. This is indeed the method of enclosing a garden which was most approved of by the ancients.

It will be a good plan, if the nature of the ground allows, for a site to be chosen close to the villa, preferably where the soil is rich and it can be irrigated by a stream running into it, or, if there be no flowing

LUCIUS JUNIUS MODERATUS COLUMELLA

puteali possit rigari. Sed ut certam perennitatis puteus habeat fidem, tum demum effodiendus est,[1] cum sol ultimas partes Virginis obtinebit, id est mense Septemb. ante aequinoctium autumnale: siquidem maxime explorantur vires fontium, cum ex longa siccitate aestatis terra caret humore pluviatili.[2]

9 Providendum est autem, ne hortus areae subiaceat, neve per trituram venti possint paleas aut pulverem in eum perferre: nam utraque sunt holeribus inimica. Mox ordinandi pastinandique soli duo sunt tempora: quoniam duae quoque holerum sationes: nam et autumno et vere plurima seruntur; melius tamen vere riguis locis, quoniam et nascentis anni clementia excipit prodeuntia semina, et sitis aestatis restin-
10 guitur fontibus. At ubi loci natura neque manu[3] illatam, neque sua sponte[4] aquam ministrari patitur, nullum quidem aliud auxilium est, quam hiemales pluviae. Potest tamen etiam in siccissimis locis opus custodiri, si depressius pastinetur solum: eiusque abunde est gradum effodere tribus pedibus, ut in
11 quattuor consurgat regestum. At ubi copia est rigandi, satis erit non alto bipalio, id est, minus quam duos pedes ferramento novale converti. Sed curabimus,[5] ut ager, quem vere conseri oportet, autumno circa calend. Novembres pastinetur: quem deinde velimus autumno instituere, mense Maio converta-

---

[1] est *om. SA.*
[2] pluviatili *SA* : pluviali *R.*
[3] manu *S* : in manu *AR.*
[4] sua sponte *scripsi* : suae sponte *SA* : suae spontis *R.*
[5] curabimus *a* : curavimus *SAR.*

water, with water from a well. But that the well may provide the certainty of a continual supply, it should be dug when the sun occupies the last part of Virgo, that is, in the month of September before the autumn equinox; for that is the best time to test the capacity of springs, since then the ground is short of rain-water after the long summer drought. But you must take care that the garden is not situated below a threshing-floor, and that the winds during the time of threshing cannot carry chaff and dust into it; for both these things are harmful to vegetables. Next there are two seasons for putting the ground in order and trenching it, since there are also two seasons for sowing vegetables, most of them being sown in the autumn and in the spring. The spring, however, is better in well-watered places, since the mildness of the growing year gives a kindly welcome to the seedlings as they come forth and the drought of summer is quenched by the springs of water. But where the nature of the place does not allow a supply of water to be brought by hand nor to be supplied spontaneously, there is indeed no other resource than the winter rains. Yet even in the driest localities your labour can be safeguarded if the ground is trenched rather deeper than usual, and it is quite enough to dig up a spit of three feet, so that the earth which is thrown up may rise to four feet. But where there is an ample opportunity for irrigation, it will suffice if the fallow ground is turned over with a not very deep double-mattock, that is, with an iron tool measuring less than two feet. But we shall take care that the field which has to be planted in the spring is trenched in the autumn about November 1st; then let us turn over in the month of May the land which we intend

LUCIUS JUNIUS MODERATUS COLUMELLA

mus, ut aut hiemis frigoribus aut aestivis solibus et gleba solvatur, et radices herbarum necentur: nec multo ante stercorare debebimus; sed[1] cum sationis appropinquabit[2] tempus, ante quintum diem exherbandus erit locus, stercorandusque, et ita diligenter fossione iterandus, ut fimo terra commisceatur.

12 Optimum vero stercus est ad hunc usum asini, quia minimum herbarum creat: proximum vel armenti vel ovium, si sit anno maceratum: nam quod homines faciunt, quamvis habeatur excellentissimum, non tamen necesse est adhibere, nisi aut nudae glareae, aut sine ullo robore solutissimae arenae, cum maior scilicet vis alimenti desideratur.

13 Igitur solum, quod vere conserere destinaverimus, post autumnum patiemur effossum iacere brumae frigoribus et pruinis inurendum: quippe e contrario sicut calor aestatis, ita vis frigoris excoquit terram, fermentatamque solvit. Quare peracta bruma tum demum stercus inicietur, et circa idus Ianuarias humus refossa in areas dividitur; quae tamen sic informandae sunt, ut facile runcantium manus ad dimidiam partem latitudinis earum perveniant, ne, qui prosequuntur herbas, semina proculcare cogantur: sed potius per semitas ingrediantur, et alterna vice dimidias areas eruncent.

14 Haec, ante sationem quae facienda sunt, dixisse abunde est. Nunc[3] quid quoque tempore vel colen-

---

[1] sed *SR* : et *Aa*.
[2] appropinquavit *SAR*.
[3] nunc *S* : in uno *AR*.

to lay out in the autumn, in order that the clods of earth may be broken up by the winter cold or the summer heat, and the roots of the weeds may be killed; and we shall have to manure it not long in advance; but when the time for sowing approaches, the ground will have to be weeded and manured five days beforehand and so carefully dug up a second time that the earth mixes with the dung. The best dung for the purpose is that of asses, because it grows the fewest weeds; next is that of either cattle or sheep, if it has soaked for a year. Human ordure, although it is reckoned to be most excellent, should not necessarily be employed except for bare gravel or very loose sand which has no strength, that is, when more powerful nourishment is required. We shall, therefore, after the autumn let the soil which we have decided to sow in the spring lie, when it has been dug up, so that it may be nipped by the cold and frosts of mid-winter; for by a contrary process like the heat of summer, so the violence of the cold refines the earth and breaks it up after causing it to ferment. Therefore, when mid-winter is past, the time has at last come when the dung is to be spread, and about January 13th the ground, having been dug over again, is divided into beds, which, however, should be so contrived that the hands of those who weed them can easily reach the middle of their breadth, so that those who are going after weeds may not be forced to tread on the seedlings, but rather may make their way along paths and weed first one and then the other half of the bed.

We have now spoken at sufficient length about what has to be done before the season for sowing. Let us now give directions as to what must be cul-

LUCIUS JUNIUS MODERATUS COLUMELLA

dum vel serendum sit, praecipiamus: et primum de his generibus loquendum est, quae possunt duobus seri temporibus, id est autumno et vere. Sunt autem semina brassicae et lactucae, cinarae, erucae, nasturcii, coriandri, chaerephylli, anethi, pastinacae, siseris, papaveris: haec enim vel circa calend. Septembres, vel melius ante calendas Martias
15 Februario seruntur. Locis vero siccis, aut tepidis, qualia sunt Calabriae et Apuliae maritima, possunt circa idus Ianuarias terrae committi. Rursus quae tantum autumno conseri debent (si tamen vel maritimum vel apricum agrum incolimus) haec fere sunt, alium, cepae capitula, ulpicum, sinape. Sed iam potius quo quidque tempore terrae mandari plerumque conveniat, per menses digeramus.[1]

16 Ergo post calendas Ianuarias confestim recte ponetur lepidium.[2] Mense autem Februario vel planta vel semine ruta, atque asparagus, et iterum cepae semen et porri: nec minus si vernum et aestivum fructum voles habere, Syriacae radicis et rapae napique semina obrues. Nam alii, et[3] ulpici
17 ultima est huius temporis positio. At circa calendas Martias locis apricis licet porrum si iam ingranduit transferre. Item panacem ultima parte Martii mensis. Deinde circa calendas Apriles aeque porrum atque inulam, et serotinam plantam rutae. Item ut maturius nascatur, cucumis, cucurbita, capparis serenda est. Nam semen betae, cum Punicum malum florebit, tum demum optime seritur.

---

[1] diceramus *S*.
[2] lepidium *vett. edd.*: lepidum *SAR*.
[3] et *om. SA*.

---

[a] See below § 20.

tivated and sown and at what season, and first of all we must speak of the kinds of seed which can be sown at the two seasons, namely autumn and spring. These are the seeds of cabbage, lettuce, cardoon, rocket, cress, coriander, chervil, dill, parsnip, skirwort and poppy. For these are sown either about September 1st or better in February before March 1st. In dry or warm districts, however, such as the sea-coasts of Calabria and Apulia, they can be put into the ground about January 13th. On the other hand, the seeds which ought to be sown only in the autumn—provided, however, that we inhabit a district either on the coast or exposed to the sun— are commonly these: garlic, small heads of onion, African garlic[a] and mustard. But let us rather now arrange month by month at what season it is generally fitting that each kind should be put into the ground.

Immediately after January 1st it will be proper to plant dittander. In the month of February rue, either as a plant or as a seed, and asparagus, and again the seed of the onion and the leek; likewise, if you want to have the yield in the spring and summer, you will bury in the ground the seeds of radish, turnip and navew. For ordinary garlic and African garlic are the last seeds which can be sown at this season. About March 1st you can transplant the leek in sunny positions, if it has already grown to a good size; and at the end of the month of March you can treat all-heal in the same way, and then about April 1st leeks likewise and elecampane and late plants of rue. Also cucumbers, gourds and capers must be sown that they may come up in good time. The best time to sow beet seed is when the pomegranate blossoms, but heads of leek can still be

LUCIUS JUNIUS MODERATUS COLUMELLA

18 Porri autem caput circa idus Maias tolerabiliter
adhuc transfertur. Post hoc, nihil ingruente aestate
obrui debet, nisi semen apii, si tamen rigaturus es.
Sic enim optime per aestatem provenit. Ceterum
Augusto circa Vulcanalia tertia satio est: eaque
optima radicis et rapae, itemque napi et [1] siseris, nec
minus oleris atri.

19 Atque haec sunt sationum tempora. Nunc de
his, quae aliquam curam desiderant, singulis loquar,
quaeque praeteriero intellegi oportebit nullam postu-
lare operam nisi runcatoris: de qua semel hoc
dicendum est, omni tempore consulendum esse, ut
20 herbae exterminentur. Ulpicum quod quidam alium
Punicum vocant, Graeci autem ἀφροσκόρδον appellant, longe maioris est incrementi quam alium: idque
circa calend. Octobres, antequam deponatur, ex
uno capite in plura dividetur. Habet enim velut
alium plures cohaerentes [2] spicas, eaeque cum sint [3]
divisae, liratim seri debent, ut in pulvinis positae
21 minus infestentur hiemis aquis. Est autem lira [4]
similis ei porcae, quam in sationibus campestribus
rustici faciunt, ut uliginem vitent: sed haec in hortis
minor est facienda, et per summam partem eius,
id est in dorso inter palmaria spatia spicae ulpici
vel allii, nam id quoque similiter conseritur, dispo-
nendae [5] sunt. Sulci lirarum inter se distent semi-
pedali spatio. Deinde cum ternas fibras emiserunt

[1] et om. *SA*.
[2] cohaerentes om. *S*.
[3] sint *AR* : sunt *S*.
[4] libra *S*.
[5] disponendi *SA*.

---

[a] August 23rd.
[b] *Smyrnium olusatrum*.
[c] This African garlic is probably a form of ordinary garlic *Allium sativum* but it may be *Allium nigrum*.

passably well transplanted about May 15th. After this, when the summer is coming on, nothing ought to be put in the ground except parsley-seed, and this only if you intend to water it; for then it comes on very well during the summer. In August about the time of the feast of Vulcan [a] comes the third sowing time; it is the best time for sowing radish and turnips, also navew, skirwort and alexanders.[b]

So much for the times of sowing. I will now speak 19 individually of those kinds which need some special care, and it will have to be understood those which I pass over require no attention except from the weeder; and on this subject I must declare once and for all that at every season steps must be taken to exterminate weeds. The African garlic, which some 20 people call Carthaginian garlic and which the Greeks name ἀφροσκόρδον,[c] is of a much greater growth than the ordinary garlic, and about October 1st, before it is planted, will be divided from one head into several; for, like ordinary garlic, it has a number of cloves sticking together, and these, when they have been separated, ought to be planted on ridges, in order that, being placed in raised beds, they may be less disturbed by winter rains. This 21 ridge is like the "balk" which farmers make when they are sowing their fields, that they may avoid dampness. This must be constructed on a smaller scale in gardens, and along the top of it, that is, on the back of the ridge, cloves of African or ordinary garlic (for the latter is planted in the same manner as the former) must be set at the distance of a hand's breadth from one another. The furrows forming the ridges should be half a foot away from one another. Then when the cloves have sent up three

spicae, sarriantur. Nam quo saepius id factum est, maius semina capiunt incrementum. Deinde ante quam caulem faciant,[1] omnem viridem superficiem intorquere et in terram prosternere conveniet, quo
22 vastiora capita fiant. Regionibus autem pruinosis neutrum horum per autumnum seri debet: nam brumali tempore corrumpuntur quod fere mense Ianuario mitescit: et idcirco frigidis locis tempus optimum est alium vel ulpicum ponendi circa idus praedicti mensis. Sed quandoque vel conseremus, vel iam matura in tabulatum reponemus, servabimus ut in iis horis[2] quibus aut obruentur aut eruentur, luna infra terram sit. Nam sic sata, et rursus sic recondita,[3] existimantur neque acerrimi saporis existere, neque mandentium halitus inodorare.[4]
23 Multi tamen haec ante calend. Ianuarias mediis diebus serunt mense Decembri, si caeli tepor et situs terrae permittit.

Brassica, cum VI foliorum erit, transferri debet, ita ut radix eius liquido fimo prius illita, et involuta[5] tribus algae taeniolis pangatur. Haec enim res efficit, ut in coctura celerius madescat et viridem colo-
24 rem sine nitro conservet. Est autem frigidis et pluviis regionibus positio eius optima post idus Apriles; cuius depressae plantae cum tenuerint,[6] quantum holitoris ratio patitur, saepius sarta et stercorata

[1] faciant *R*: faciunt *SAa*.
[2] horis *S*: locis *AR*.
[3] sic fere condita *a*: fere condita *SA*: sic recondita *R*.
[4] saporis—inodorare *om. A*: inodorare *R*: odorari *S*: moderare *a*.
[5] et involuta *R*: etiam voluta *SA*.   [6] tenuerant *A*.

blades, they should be hoed; for the more often this is done, the greater is the growth of the seedlings. Then, before they form any stalk, it will be a good plan to twist all the green portion above ground and lay it flat on the soil, so that the heads may grow larger. But in districts which are liable to frosts neither of 22 these plants should be sown during the autumn, for they rot in the season of mid-winter, which generally turns milder in the month of January; and for that reason in cold places the best time for planting ordinary or African garlic is about the 13th of the said month. But whenever we are going to sow them or, when they are already ripe, are going to store them in a loft, we shall be careful that the moon is below the earth at the time when they are either put into the ground or taken out of it; for if they are planted or, on the other hand, laid away under these conditions, they are held not to be of a very pungent flavour nor to give an odour to the breath of those who chew them. Nevertheless many people sow them before 23 January 1st in the month of December in the middle of the day if the warmth of the weather and the situation of the ground allows.

A cabbage ought to be transplanted when it consists of six leaves, provided that its root is smeared with liquid manure and wrapped up in three bands of seaweed before it is fixed in the ground; this precaution ensures that in the boiling it becomes soft more quickly and keeps its green colour without the use of soda. But in cold and rainy districts the best 24 time for setting it is after April 13th; and when plants of it have been sunk into the ground and have taken root, the more often they are hoed and manured, as far as the gardener's plan of work allows,

melius convalescit, pleniorisque incrementi et coliculum facit et cymam. Nonnulli hanc eandem locis apricioribus a calend. Martiis deponunt: sed maior pars eius in cymam prosilit, nec postea hibernum caulem amplum facit, cum est semel desecta. Possis autem vel maximos caules bis transferre. Idque si facias, plus seminis, et maioris incrementi praebere dicuntur.

25 Lactuca totidem foliorum quot brassica transferri debet. Locis quidem apricis et maritimis optime autumno ponitur,[1] mediterraneis et frigidis contra: hieme[2] non aeque commode dispangitur.[3] Sed huius quoque radix fimo liniri debet, maioremque copiam desiderat aquae, sicque fit[4] tenerioris folii.

26 Sunt autem complura lactucae genera, quae suo quidque tempore seri[5] oportet: earum quae fusci est vel[6] purpurei, aut etiam viridis coloris et crispi folii, uti Caeciliana, mense Ianuario recte disseritur. At Cappadocia, quae pallido et pexo densoque folio[7] viret, mense Februario: quae deinde candida est et crispissimi folii, ut in provincia Baetica et finibus Gaditani municipii, mense Mart. recte pangitur.

27 Est et Cyprii[8] generis ex albo rubicunda, levi et tenerrimo folio, quae usque in idus Apriles commode disponitur. Fere tamen aprico caeli statu, quibus locis aquarum copia est, paene toto anno lactuca seri potest: quae quo tardius caulem faciat

---

[1] ponitur *R* : ponuntur *SA*.
[2] hieme *R* : hiemē *S* : hiemem *A*.
[3] dispangitur *S* : dispargitur *AR*.
[4] sicque fit *Schneider* : si quo *SAR*.
[5] seri *R* : fieri *SA*.
[6] est vel *Lundström* : et ut *SA* : et velut *R*.
[7] foli *S*¹ : folio *S*² : folii *A*.
[8] Cypri *SAR*.

BOOK XI. III. 24–27

the better the cabbage thrives and the fuller the growth it makes in its stalks and sprouts. Some people set this vegetable in exceptionally sunny spots from March 1st onwards, but the greater part of it runs to sprouts and, when once it has been cut, it does not form thereafter a large winter stalk. But you can transplant even the biggest cabbage-stalks twice, and, if you do so, they are said to produce more seed and greater growth.

Lettuce ought to be transplanted when it has as many leaves as the cabbage. Indeed in sunny districts near the sea it is best set in the autumn, but it is otherwise in inland and cold places, for in winter it cannot be planted out with such good effect. The root of the lettuce ought also to be smeared with dung and it requires rather a large quantity of water, and so comes to have tenderer leaves. There are several kinds of lettuce, which must each be sown at its proper season. That which is of a dark or purple or even of a green colour with curled leaves, like the Caecilian[a] variety, is rightly planted out in the month of January; but the Cappadocian lettuce which grows with a pale, woolly, thick leaf is planted in the month of February; the winter variety, which is white with a very curly leaf, such as grows in the province of Baetica and in the territory of the municipal city of Gades, should be put into the ground in the month of March. There is also that of the Cyprian kind, which is of a reddish white with a smooth and very tender leaf, which is advantageously planted out up to April 15th. Generally, however, in a sunny climate where there is plenty of water, lettuce can be sown almost throughout the year; but, so that it may form its stalk more

[a] See note on Book X. line 182.

cum ¹ aliquod incrementum habuerit, exiguam testam media parte accipiat, eo quasi onere coercita in latitudinem se diffundit. Eadem est ratio etiam intubi, nisi quod hiemem magis sustinet: ideoque vel frigidis regionibus primo autumno seri potest.

28   Cinarae sobolem melius per autumni aequinoctium disponemus;² semen commodius circa calendas Martias seremus; eiusque plantam ante³ calend. Novemb. deprimemus, et multo cinere stercorabimus.⁴ Id enim genus stercoris huic holeri videtur aptissimum.

29   Sinape atque coriandrum, nec minus eruca et ocimum, ita uti sata sunt, sua sede immota permanent: neque est eorum cultus alius, quam ut stercorata runcentur. Possunt autem non solum autumno, sed et vere conseri. Plantae quoque sinapis prima hieme translatae plus cymae vere afferunt. Panax utroque tempore levi et subacta terra rarissime disseritur,⁵ quo maius incrementum capiat: melior⁶ tamen eius

30 verna satio est. Porrum si sectivum facere velis, densius satum praeceperunt priores relinqui: et ita cum increverit, secari. Sed nos docuit usus longe melius fieri, si differas, et eodem more, quo capitatum modicis spatiis, id est, inter quaternos⁷ digitos

31 depangas, et cum convaluerit, deseces. In eo autem quod magni capitis efficere voles, servandum⁸ est, ut ante quam translatum deponas, omnes radiculas amputes et fibrarum summas⁹ partes intondeas. Tum testulae, vel conchae, quasi sedes singulis

¹ cum *R* : tum *SA*.
² disponemus *R* : -imus *SA*.
³ ante *SA* : circa *R*.   ⁴ stercoravimus *SA*.
⁵ disseritur *R* : disserentur *S* : disseruntur *A*.
⁶ melior *S* : melius *AR*.
⁷ quaternos *R* : quattuor *SA*.
⁸ *post* servandum *add.* serendum *S*.   ⁹ sumas *S*.

slowly, it should have a small piece of earthenware placed in its centre, when it has grown to some extent; for being checked by having to support, as it were, the weight of this, it spreads out in breadth. The same method is used also for the endive, except that it stands the winter better, and therefore can be sown even in cold districts in the early autumn.

We shall do better to plant out slips of cardoon during the autumn equinox and sow the seed more advantageously about March 1st and set the plants before November 1st and manure them with plenty of ashes; for this kind of manure seems the most suitable for this vegetable. Mustard and coriander and also rocket and basil are left undisturbed as they were sown and require no other cultivation than manuring and weeding. They can be sown not only in autumn but also in spring. Mustard plants transplanted in the early winter produce more sprouts in the spring. All-heal is planted very thinly at both seasons in light and well-worked soil so as to make greater growth; but the spring is the better season for it. If you wish to grow leeks, the ancients directed that it should be sown rather thickly and left to grow and that then, when it increased, it should be cut; but experience has taught us that it does much better if you space it out, and set it in the same way as the headed leek, with moderate spaces between the plants, that is to say, intervals of four inches, and cut it when it has grown up. As for the leek which you wish to form a large head, you must take care that, before you transplant it and re-set it, you cut off all the small roots and shear off the tops of the fibres; then small pieces of earthenware or shells are buried beneath each of the seedlings

LUCIUS JUNIUS MODERATUS COLUMELLA

subiectae seminibus adobruuntur, ut fiant capita
32 latioris incrementi. Cultus autem porri capitati
assidua sartio et stercoratio est. Nec aliud tamen
sectivi, nisi quod [1] toties rigari, et stercorari, sarriri-
que debet, quoties demetitur. Semen eius locis
calidis mense Ianuario, frigidis Februario seritur:
cuius incrementum quo maius fiat, raris linteolis
complura grana illigantur, atque ita obruuntur.
Enatum autem differri debet in iis locis, quibus aqua
subministrari non potest, circa aequinoctium autumni: at quibus possis humorem praebere, mense
Maio recte transfertur.

33 Apium quoque possis plantis serere, nec minus
semine. Sed praecipue aqua laetatur, et ideo
secundum fontem commodissime ponitur. Quod si
quis id velit lati folii facere, quantum seminis possint
tres digiti comprehendere, raro linteolo illiget,[2] et
ita in areolas dispositum releget. Vel si crispae
frondis id fieri maluerit, semen eius inditum pilae, et
saligneo pilo pinsitum, expolitumque, similiter
34 linteolis ligatum obruat. Potest etiam citra hanc
operam fieri crispum qualitercunque satum, si, cum
est natum, incrementum eius supervoluto cylindro
coërceas. Satio eius est optima post idus Maias
usque in solstitium: nam teporem [3] desiderat. Fere
etiam his diebus ocima seruntur: quorum cum semen
obrutum est, diligenter inculcatur pavicula vel

[1] quod S : quo AR.
[2] illiget R : liget SA.
[3] tempore A : tēpore S¹ : teporē S².

to serve as a sort of base, so that heads of a larger growth may be formed. But the cultivation of headed leeks consists in hoeing and manuring them continually; nor does treatment of cut leek differ from this, except that every time it is cut it ought to be watered, manured and hoed. Its seed is sown in January in warm districts, in February in cold places; in order that it may have greater growth several seeds are tied up in a piece of linen of loose texture and then planted. When it has come up in places where it cannot be supplied with water, it ought to be spaced out about the autumn equinox, but when you can provide moisture, it is correct to move it in the month of May.

You can raise parsley equally well from plants or from seed; it takes a special delight in water and for that reason is most advantageously placed near a spring. But if anyone wishes to grow parsley with a broad leaf, he should tie up as much seed as three of his fingers can hold in a piece of linen of loose texture and plant it out in a regular arrangement in small beds. If he prefers that it should grow with curly leaves, he should put the seed into a mortar and pound it with a pestle of willow-wood and smooth it out and tie it up in the same manner as before in a piece of linen and cover it with ground. It can also be made to grow with curly leaves without all this trouble, in whatever manner it is sown, if, when it has come up, you check its growth by passing a roller over it. The best time for sowing it is from May 15th to the solstice, for it needs warmth. It is generally at this season that basil also is sown; the seed after being put into the ground is carefully pressed down

LUCIUS JUNIUS MODERATUS COLUMELLA

cylindro. Nam si terram suspensam relinquas, plerumque corrumpitur.

35 Pastinaca et siser atque inula convalescunt alte pastinato et stercorato loco: sed quam rarissime ponenda sunt, ut maiora capiant incrementa. Inulam vero intervallo trium pedum seri convenit, quoniam vastos, facit frutices,[1] et radicibus, ut oculus harundinis,[2] serpit. Nec est alius cultus horum omnium nisi ut sartionibus herbae tollantur. Commodissime autem deponentur prima parte Septembris, vel
36 ultima Augusti parte. Atrum olus, quod Graecorum quidam vocant πετροσέλινον, nonnulli σμυρναῖον, pastinato loco semine debet conseri, maxime iuxta maceriam: quoniam et umbra gaudet, et qualicunque convalescit loco: idque cum semel severis, si non totum radicitus tollas, sed alternos frutices in semen submittas, aevo[3] manet, parvamque sartionis exigit culturam. Seritur a Vulcanalibus usque in calendas
37 Septembris, sed etiam mense Ianuario. Menta dulcem desiderat uliginem; quam ob causam iuxta fontem mense Martio recte ponitur. Cuius si forte semina defecerunt, licet de novalibus silvestre mentastrum colligere, atque ita inversis cacuminibus disponere: quae res feritatem detrahit, atque
38 edomitam reddit. Rutam autumno semine satam mense Martio differre oportet in apricum, et cinerem aggerare,[4] runcareque donec convalescat, ne herbis enecetur. Sed velata manu debebit runcari: quam nisi contexeris, perniciosa nascuntur ulcera. Si

---

[1] frutices S: fructices AR.   [2] harundis SA.
[3] saevom S: revom A: sevo a.   [4] adgerare SA.

---

^a So-called because it has an odour like myrrh.
^b August 23rd.
^c The usual cultivated mint of the Romans was water mint,

BOOK XI. III. 34-38

with a rammer or roller, for if you leave the soil loose, the seed usually rots.

The parsnip, skirwort and elecampane thrive in 35 deeply trenched and well-manured soil, but they must be planted far apart so that they may attain to a greater growth. Elecampane should be sown at intervals of three feet, for it grows huge bushes and has creeping roots like the " eyes " of a reed. All these vegetables require no cultivation except that the weeds should be removed by hoeing. It will be most suitable to plant them in the first part of September 36 or the last part of August. Alexanders, which some Greeks call πετροσέλινον (rock-parsley) and others σμυρναῖον,[a] should be sown in well-trenched soil, preferably near a stone wall, since it delights in shade and thrives in any kind of soil. When once you have sown it, if you do not pull it all up by the roots but leave alternate bushes to go to seed, it goes on for ever and only needs a little cultivation by hoeing. It is sown from the feast of Vulcan [b] to September 1st, but also in the month of January. Mint [c] requires a 37 sweet, marshy soil; for which reason it is right to plant it near a spring in the month of March. If the seeds happen to have proved a failure, you can collect wild mint from fallow ground and then plant it with the top inverted, for this draws the wildness out of it and makes it tame. You should sow rue in the 38 autumn and thin it out in March to a sunny place and heap ashes round it and cultivate it until it is strong, that it may not be killed by weeds. It will have to be weeded with the hand covered; for unless you wrap it up, pernicious ulcers form.

Mentha aquatica, while *mentastrum* was usually horse mint, *Mentha silvestris*.

LUCIUS JUNIUS MODERATUS COLUMELLA

tamen per ignorantiam [1] nuda manu runcaveris et prurigo atque tumor incesserit, oleo subinde perungito. Eiusdem frutex pluribus annis permanet innoxius, nisi si mulier, quae in menstruis est, contigerit eum, et ob hoc exaruerit.

39 Thymum, et transmarina cunila, et serpyllum, sicut priore libro iam retuli, magis alvearia [2] curantibus, quam holitoribus studiose conseruntur. Sed nos ea condimentorum causa nam sunt quibusdam esculentis aptissima non alienum putamus etiam in hortis habere. Locum neque pinguem neque stercoratum, sed apricum desiderant, ut quae macerrimo solo per 40 se maritimis plerumque regionibus nascantur.[3] Eae [4] res et semine et plantis circa aequinoctium vernum seruntur.[5] Melius tamen est thymi novellas plantas disponere; quae cum subacto solo depressae fuerint, ne tarde comprehendant,[6] aridi thymi fruticem contundi oportet, atque ita pinsito illo [7] pridie quam volueris uti, aqua medicare; quae cum succum eius perceperit, depositis fruticibus infunditur, donec 41 eos recte confirmet. Ceterum cunila vivacior est, quam ut impensius [8] curanda sit. Lepidium cum ante cal. Martias habueris dispositum, velut porrum sectivum demetere poteris: rarius tamen. Nam post cal. Novemb. secandum non erit, quoniam frigoribus violatum emoritur: biennio tamen sufficiet,

---

[1] ignorantia *A*.
[2] saluaria *AS*: aluearia *R*.
[3] nascantur *S*: nascuntur *AR*.
[4] eae *R*: ea *SAa*.
[5] servatur *SA*.
[6] conprehendat *SAR*.
[7] pinsito illo *Lundström*: in illo *SAa*: insito *R*.
[8] impensius *R*: inpenses *SA*.

---

[a] *Thymus vulgaris*.
[b] Latin *cunila* usually denotes savory, *Satureia hortensis*, but the Greek κονίλη = marjoram.

If, however, through ignorance you have weeded it with the hand bare and itching and swelling have come on, anoint the hand thoroughly from time to time with olive-oil. A shrub of rue lasts for many years without deteriorating, unless a woman, who is in her menstrual period, touches it; in which case it dries up. Thyme,[a] foreign marjoram[b] and wild thyme,[c] as I have remarked already in a previous book,[d] are sown carefully rather by those who look after bees than by gardeners; but we do not consider it out of place to have them in gardens for use as seasonings, for they are most suitable to go with some edibles. They need a spot which is neither rich nor manured but sunny, since they grow of their own accord in very thin soil, generally in sea-coast districts. They are put in the ground in the form both of seed and plants about the time of the spring equinox. It is better, however, to set young plants of thyme, and when they have been pressed down into well-worked soil, in order that they may not be slow in striking root, a stalk of dry thyme should be crushed and after it has been pounded on the day before that on which you wish to use it, you should steep it in water and, when the water has absorbed the juice, it is poured on the stalks of thyme which you have planted, until it makes them quite strong. Marjoram, however, is so full of life that it need not be looked after with any special care. Dittander, when you have planted it out before March 1st, you will be able to cut like leeks, but not so often. After November 1st it will not be possible to cut it, because, if it is damaged in cold weather, it dies; still it will

[c] *Thymus sezpyllum.*
[d] Book IX. 4. 6.

si diligenter sartum et stercoratum fuerit. Multis etiam locis vivacitatem suam usque in annos decem prorogat.

42 Beta florente Punico malo semine obruitur, et simul atque quinque foliorum est, ut brassica, differtur aestate, si riguus est locus : ac si siccaneus, autumno, cum iam pluviae incesserint, disponi debebit.[1] Chaerephyllum, itemque olus atriplicis, quod Graeci vocant ἀνδράφαξιν,[2] circa cal. Octob. obrui oportet non frigidissimo loco. Nam si regio saevas hiemes habet, post idus Februarias semina[3] disserenda sunt, suaque sede patienda. Papaver et anethum eandem habent conditionem sationis, quam chaerephyllum et

43 ἀνδράφαξις. Sativi asparagi, et quam corrudam[4] rustici vocant, semina fere biennio praeparantur. Ea cum pingui et stercoroso loco post idus Februarias sic obrueris, ut quantum tres digiti seminis comprehendere queunt, singulis fossulis deponas, fere post quadragesimum diem inter se implicantur, et quasi unitatem faciunt ; quas radiculas sic illigatas[5] atque connexas holitores sphongias appellant. Eas post quattuor et viginti menses in locum apricum et bene

44 madidum,[6] stercorosumque transferri convenit. Sulci autem inter se pedali mensura distantes fiunt non amplius dodrantalis altitudinis, in quam ita sphongiolae deprimuntur, ut facile superposita terra[7] germinent. Sed in locis siccis partibus sulcorum imis disponenda sunt semina, ut tanquam in alveolis

---

[1] debebunt *SAR*.
[2] ΑΝΔΡΑΦΑΞΙΝ *S* : ΑΝΑΡΑΦΑΖΙΝ *A*.
[3] semine *SAR*.
[4] corundam *SAa*.
[5] illigatas *R* : inligas *SA*.
[6] madidum *R* : validum *SA*.
[7] terrae *S* : terre *Aa*.

last for two years if it is carefully hoed and manured; and in many places it prolongs its natural vigour for as many as ten years.

Beet is put into the ground in the form of seed when the pomegranate flowers, and as soon as it consists of five leaves, it is, like cabbage, transplanted in the summer if the place is well watered; and, if it is dry, it should be planted out in the autumn when the rains have already come on. Chervil and also orach, which the Greeks call ἀνδράφαξις,[a] ought to be put in the ground about October 1st and not in a very cold place; for if the locality has hard winters, the seedlings should be planted out after February 13th and left where they are. For the poppy and the dill the conditions of sowing are the same as for the chervil and orach. The seeds of garden asparagus and what the country folk call *corruda* (wild asparagus) take almost two years to prepare. When you have buried them after February 13th in a rich, well-manured place in such a way that you place in each little trench as much seed as your three fingers can hold, usually after the fortieth day the plants become intertwined and form a kind of single mass, and the little roots thus attached and collected together the gardeners call " sponges." After twenty-four months they should be transplanted to a sunny, well-watered and manured spot. Furrows are then made a foot distant from one another and not more than three-quarters of a foot deep into which the little " sponges " are pressed down so that they may easily spring up when the earth is put over them; but in dry places the seedlings must be placed in the bottom of the furrows, so that they remain as it were in little

[a] *Atriplex hortensis.*

LUCIUS JUNIUS MODERATUS COLUMELLA

maneant. At uliginosis e contrario in summo porcae
45 dorso collocanda, ne humore nimio laedantur. Primo
deinde anno, cum ita consita sunt, asparagum quem
emiserunt,[1] infringi oportet. Nam si ab imo vellere
volueris, adhuc teneris invalidisque[2] radiculis, tota
sphongiola sequetur. Reliquis annis non erit decer-
pendus, sed radicitus vellendus. Nam nisi ita fiat,
stirpes praefractae angunt[3] oculos sphongiarum, et
quasi excaecant, nec patiuntur asparagum emittere.
Ceterum stilus, qui novissime autumnali tempore
nascitur, non omnis est tollendus, sed aliqua pars eius
46 in semen submittenda. Deinde cum spinam fecerit,
electis seminibus ipsis, scopiones[4] ita uti sunt, in suo
loco perurendi sunt, et deinde sulci omnes consarri-
endi, herbaeque eximendae; mox vel stercus, vel cinis
iniciendus, ut tota hieme succus eius cum pluviis
manans ad radicem perveniat. Vere deinde prius
quam coeperit germinare, capreolis, quod genus bi-
cornis ferramenti est, terra commoveatur,[5] ut et
facilius stilus emicet, et relaxata humo plenioris
crassitudinis fiat radix.

47 Raphani radix bis anno recte[6] seritur, Februario
mense, cum vernum fructum expectamus, et Augusto
mense circa Vulcanalia, cum maturum. Sed haec
satio sine dubio melior habetur. Cura est eius, ut

[1] emiserunt *R* : meminit tunc *S* : meminit tc̄ *A*.
[2] validis *SAR*.   [3] angunt *S* : agunt *AR*.
[4] scopiones *Gesner* : scorpiones *SAR*.
[5] commoveantur *SA*.
[6] raphani *et recte om. A*.

---

[a] August 23rd.

troughs. On the contrary, in swampy ground they must be placed on the topmost back of the ridge, so that they may not suffer from too much moisture. Then in the first year after they have been thus planted, the asparagus which they have put forth must be broken off; for if you should pull it up from the bottom while the little roots are still tender and weak, the whole " sponge " will come away as well. In the following years it will have to be pulled up by the roots and not picked; for unless this is done, the broken stalks cause pain to the " eyes " of the " sponges " and as it were blind them and do not allow them to produce any asparagus. However, the stalk, which comes up last of all in the autumn season, should not all be removed but some part should be allowed to grow to provide seed. Then when it has formed a spike, after the seeds themselves have been gathered, the haulms, just as they are, should be burnt in their place, and then all the furrows must be hoed and the weeds removed. Directly thereafter either dung or ashes must be thrown on them, so that throughout the winter the moisture percolating through with the rain may reach the roots. Then, in the spring, before the asparagus begins to sprout, the ground should be stirred up with weeding-hoes, which are a two-pronged iron kind of implement, so that the stem may break through more easily and attain to a greater thickness in the loosened soil.

It is correct to sow the radish twice a year, in the month of February, when we expect the spring crop, and in the month of August about the time of the festival of Vulcan[a] when we expect the late crop. The latter sowing is undoubtedly considered the better of the two. The method of cultivation is to

LUCIUS JUNIUS MODERATUS COLUMELLA

terra stercorata et subacta obruatur : post ubi ceperit aliquod incrementum, subinde aggeretur. Nam si super terram emerserit, dura et fungosa fiet.

48 Cucumis et cucurbita, cum copia est aquae, minorem curam desiderant : nam plurimum iuvantur humore.[1] Sin autem loco sicco seri debuerint, quo rigationem[2] ministrare non expediat, mense Februario sesquipedali altitudine fossa facienda est. Post idus deinde Martias quasi tertia pars altitudinis sulci stramentis inditis tegenda, mox stercorata terra usque in dimidium sulcum aggeranda,[3] positisque seminibus tam diu est aqua praebenda, donec enascantur :[4] atque ubi convalescere coeperint, adiecta humo incrementa eorum prosequenda sunt, donec sulcus coaequetur. Sic exculta semina sine rigatione tota aestate satis valebunt, fructumque iucundioris
49 saporis quam rigua praebebunt. Aquosis autem locis primo quoque tempore, non tamen ante calend. Mart. semen ponendum est, ut differri[5] possit aequinoctio confecto. Idque de media parte cucurbitae semen inverso cacumine ponito, ut fiat incrementi vastioris. Nam sunt ad usum vasorum satis idoneae, sicut Alex-
50 andrinae cucurbitae, cum exaruerunt. At si esculentae merci praeparabis, recto cacumine de collo cucurbitae sumptum semen serendum erit, quo prolixior et tenuior fructus eius nascatur, qui scilicet[6] maius

[1] umores *A*.
[2] rigatione *SR*.
[3] aggeranda *R* : adgerenda *S* : aggerenda *A*.
[4] enascatur *SA*.
[5] differri *a* : differre *SA*.
[6] scilicet *R* : si licet *SA*.

sow it in manured and well-worked soil; then when it has reached some growth, it should be earthed up from time to time; for if it emerges above ground, it will become hard and full of holes.

The cucumber and gourd require less care when there is plenty of water; for they take a very great delight in moisture. But if they have had to be sown in a dry place, where it is not easy to supply running water, a ditch one foot and a half deep must be made in the month of February; then after March 15th about one-third of the depth of the trench must be covered by placing straw in it; next, well-manured earth must be heaped up so as to fill half the trench and, after the seeds have been put in, water must be supplied until the seeds come up. And when the plants begin to thrive, earth must be added to keep pace with their growth, until the trench is filled to the level of the ground. Thus cultivated the seeds will thrive quite well during the whole summer without irrigation and provide fruit of a pleasanter flavour than if they were watered by irrigation. But in watery places the seeds must be planted as early as possible, though not before March 1st, so that they can be transplanted when the equinox is passed. You should take the seed from the middle of a gourd and place it in the ground with its top inverted, so that it may attain a greater growth; for the fruits are quite suitable for use as vessels, like the Alexandrian gourds, when they have been thoroughly dried. But if you are growing them to sell for eating, the seed will have to be taken from the neck of the gourd and sown with its top upright, so that the fruit which grows from it may be longer and narrower; this certainly commands a better price than any

LUCIUS JUNIUS MODERATUS COLUMELLA

ceteris invenit pretium. Sed custodiendum est, ut quam minime ad eum locum,[1] in quo vel cucumeres aut cucurbitae consitae sunt, mulier admittatur. Nam fere contactu eius languescunt incrementa virentium. Si vero etiam in menstruis fuerit, visu[2] quoque suo novellos fetus necabit.[3] Cucumis tener et iucundissimus[4] fit, si ante quam seras, semen eius lacte maceres. Nonnulli etiam quo dulcior existat, aqua mulsa idem faciunt. Sed qui praematurum fructum cucumeris habere volet, confecta bruma stercoratam terram inditam cophinis offerat, modicumque praebeat humorem. Deinde cum enata semina fuerint, tepidis diebus et insolatis iuxta aedificium sub divo ponat, ita ut ab omni afflatu protegantur. Ceterum frigoribus ac tempestatibus sub tectum referat: idque tamdiu faciat, dum aequinoctium vernum confiat. Postea totos cophinos demittat in terram. Sic enim praecoquem fructum habebit. Possunt[5] etiam, si sit operae pretium, vasis maioribus rotulae subici, quo minore labore producantur, et rursus intra tecta recipiantur. Sed nihilo minus specularibus integi debebunt,[6] ut etiam frigoribus serenis diebus tuto producantur ad solem. Hac ratione fere toto anno Tiberio Caesari cucumis praebebatur. Nos autem leviore opera istud fieri apud Aegyptiae gentis Bolum Mendesium[7] legimus, qui praecipit aprico et stercoroso loco alternis

[1] locum *om. SA.*
[2] visu *R* : viso *AS.*
[3] necabit *R* : nacavit *S* : negavit *A.*
[4] incundissimus *R* : -is *SA.*
[5] possint *S.*
[6] integi debebunt ut *R* : inintegi debebuntur *S* : inintegi debebī ut *A.*
[7] volum mendisiũ *A* : volumendesiũ *S.*

BOOK XI. III. 50-53

other. Care, however, must be taken that a woman is admitted as little as possible to the place where the cucumbers and gourds are planted; for usually the growth of green-stuff is checked by contact with a woman; indeed if she is also in the period of menstruation, she will kill the young produce merely by looking at it. A cucumber becomes tender and most pleasant to the taste if you were to soak the seed in milk before you sow it; some people also employ mead in the same way to make the cucumber sweeter. Anyone who wishes to have the fruit of the cucumber ripe earlier than usual should, when mid-winter is past, produce well-manured soil enclosed in baskets and give it a moderate amount of water; then, when the seeds have come up, he should place the baskets in the open air on warm and sunny days near a building, so that they may be protected from any blasts of wind; but, if it is cold and stormy, he should bring them back under cover and continue to do so until the spring equinox is over. He should then sink the whole baskets into the ground. He will thus have early fruits. It is also possible, if it be worth the trouble, for wheels to be put under larger vessels, so that they may be brought out and taken indoors again with less labour. In any case the vessels ought to be covered with panes of glass, so that even in cold weather, when the days are clear, they may safely be brought out into the sun. By this method Tiberius Caesar was supplied with cucumbers during almost the whole year. But we read in Bolus of Mendes,[a] the Egyptian writer, that this end can be achieved with much less trouble; for he advises us to have in our garden, on a site open to the sun and

---

[a] A town on the Egyptian coast.

LUCIUS JUNIUS MODERATUS COLUMELLA

ordinibus ferulas, alternis rubos in hortis consitas habere: deinde eas[1] confecto aequinoctio paululo[2] infra terram secare, et ligneo stilo laxatis vel rubi vel ferulae medullis stercus immittere, atque ita semina cucumeris inserere, quae scilicet incremento suo coeant rubis et ferulis. Nam ita non sua, sed quasi materna[3] radice aluntur: sic insitam stirpem frigoribus quoque cucumeris praebere fructum. Satio secunda eius seminis fere Quinquatribus observatur.

54 Capparis plurimis provinciis sua sponte novalibus nascitur. Sed quibus locis eius inopia est, si serenda fuerit, siccum locum desiderabit. Itaque debebit ante circumdari fossula, quae repleatur lapidibus et calce, vel Punico luto, ut sit quasi quaedam lorica, ne possint eam perrumpere praedicti seminis frutices, qui fere per totum agrum vagantur, nisi 55 munimento aliquo prohibiti sunt. Quod tamen non tantum incommodum est subinde enim possunt extirpari quantum, quod noxium virus[4] habent, succoque suo sterile solum reddunt. Cultu aut nullo aut levissimo[5] contenta est, quippe quae res etiam in desertis agris citra rustici operam convalescat.[6] Seritur utroque aequinoctio.

56 Cepina magis frequenter subactam postulat terram quam altius conversam. Itaque ex calendis Novembribus proscindi solum debet, ut hiemis

[1] eas R: ex SA.
[2] paululo S: -um AR.
[3] macerata SAR.   [4] vires A.
[5] levissimo R: novissimo SA.
[6] convalescat S: -it AR.

---

[a] The greater Quinquatria in honour of Minerva were celebrated on March 19th to 23rd.

well-manured, fennels and brambles planted in alternate rows, and then, when the equinox is past, to cut them a little below the surface of the ground and, after loosening with a wooden prong the pith of the bramble or fennel, to put dung into them and thus insert the cucumber-seeds, so that, as they grow, they may unite with the brambles and fennels, for by this method, they are nourished not from their own roots but from what may be called the mother-root; and the stock thus engrafted yields the fruit of the cucumber even in cold weather. The season observed for the second sowing is about the time of the festival of the Quinquatria.[a]

The caper-plant in most provinces grows of its own accord on fallow land; but in places where it is scarce, if it has to be sown, it will require dry ground. And so it will have to be surrounded beforehand by a small ditch, which should be filled with stones and lime or Carthaginian clay, so as to form a kind of breast-work, so that the bushes which grow from the aforesaid seed may not break their way out; for they usually wander over the whole ground, unless there is some barrier to prevent them. This is not so much an inconvenience in itself, for they can be rooted out from time to time, as that they contain a harmful poison and with their juice make the soil barren. The caper is content with very slight cultivation or none at all; for it thrives even in deserted fields without any effort on the part of the husbandman. It is sown at the time of the two equinoxes.

An onion-bed requires soil that has been frequently broken up rather than turned over to any depth. Therefore from November 1st onwards the ground ought to be cut up, so that it may crumble with the cold

LUCIUS JUNIUS MODERATUS COLUMELLA

frigoribus et gelicidiis putrescat, intermissisque quadraginta diebus tum demum iterari, et interpositis uno ac viginti diebus tertiari, et protinus stercorari: mox bidentibus aequaliter perfossum in areas disponi, deletis [1] radicibus omnibus. Deinde ad calendas Februarias sereno die conveniat semina spargi: quibus aliquid satureiae semen intermiscendum erit, ut eam quoque habeamus. Nam et viridis esui est iucunda, nec arida inutilis ad pulmentaria condienda. Sed cepina vel saepius, certe non minus debet quam quater sarriri. Cuius si semen excipere voles, capita maxima generis ascalonii, quod est optimum, mense Februario disponito,[2] quaternorum vel etiam quinum digitorum spatiis distantia: et cum coeperint [3] virere,[4] ne minus ter consarrito:[5] deinde [6] cum fecerint caulem, humilioribus quasi canteriolis interpositis rigorem [7] stilorum conservato. Nam nisi harundines transversas in modum iugatae vineae [8] crebras disposueris, thalli ceparum ventis prosternentur, totumque semen excutietur: quod scilicet non ante legendum est, quam cum maturescere coeperit, coloremque nigrum habere. Sed nec patiendum est, ut perarescat, ut totum decidat, verum integri thalli vellendi sunt, et sole siccandi.

Napus et rapa duas sationes, habent, et eandem culturam, quam raphanus. Melior est tamen satio

---

[1] disponi dilectus *S* : dispondilectus *Aa*.
[2] dispositio *SA*.
[3] coeperit *SA*.
[4] virere *R* : vivere *S* : videre *A*.
[5] consarito *R* : inconserito *SA*.
[6] deinde *R* : die in die *SA*.
[7] rigoribus *SAR*.
[8] vinea *SA*.

and frosts of winter; then after an interval of forty days, and not before, the process should be repeated, and again carried out for a third time twenty-one days later, and the ground manured immediately afterwards; then, after having been forked up equally all over, it should be arranged in beds, all the roots having been destroyed. Next towards February 1st on a calm day it would be a suitable time for the seed to be scattered, with which some savory-seed will have to be mixed, so that we may have some of this also; for it is both pleasant to eat when it is green and, when dried, is not without its uses for seasoning relishes. The onion-bed ought to be hoed not less than four times or even oftener. If you wish to procure seed, plant in the month of February the largest heads of the "un-dug"[a] variety, which is the best, four or even five inches apart, and when they have begun to show green, hoe them not less than three times; then when they have formed a stalk, preserve the stiffness of their shoots by placing rather low trellis-work among them. For unless you place reeds crosswise near to one another, like frames in a vineyard, the stalks of the onions will be blown down by the wind and all the seed will be shaken out, which certainly ought not to be gathered until it has begun to ripen and to have a black colour. But you must not allow it to dry up entirely, so that it all falls on to the ground, but the stalks must be pulled off intact and dried in the sun.

There are two sowings of the navew and turnip, and they have the same cultivation as the radish; the better sowing-time, however, is the month of

[a] See page 207, note *b*.

LUCIUS JUNIUS MODERATUS COLUMELLA

mensis Augusti. Iugerum agri quattuor sextarios seminis eorum poscit, sed ita ut radicis Syriacae [1] super hanc mensuram paulo plus quam heminam 60 seminis recipiat. Qui aestate ista seret, caveat, ne propter siccitates pulix adhuc tenera folia prorepentia consumat. Idque ut vitetur, pulvis, qui supra cameram [2] invenitur, vel etiam fuligo, quae supra focos tectis inhaeret, colligi debet: deinde pridie quam satio fiat, commisceri cum seminibus, et aqua conspergi, ut tota nocte succum trahant. Nam sic 61 macerata postero die recte obserentur.[3] Veteres quidam auctores, ut Democritus, praecipiunt, semina omnia succo herbae quae sedum appellatur medicare, eodemque remedio adversus bestiolas uti: quod verum esse nos experientia docuit. Sed frequentius tamen, quoniam huius herbae larga non [4] est facultas, fuligine et praedicto pulvere [5] utimur, satisque commode tuemur his incolumitatem plan- 62 tarum. Rapae semina Hyginus putat post trituram iacentibus adhuc in area paleis [6] inspergi debere, quoniam fiant [7] laetiora capita, cum [8] subiacens soli duritia non patiatur [9] in altum descendere. Nos istud saepe frustra tentavimus: itaque sicut rapha-

[1] surice *R* : urice *SA*.
[2] caram *SA*¹ : cameram *A*².
[3] obserentur *Sc* : -untur *AR* : observētur *a*.
[4] larga non *SA* : minus larga *R*.
[5] pluere *S* : plure *A*.
[6] paleas *A*.
[7] fiant *S* : fiunt *AR*.
[8] cum *om. SA*.
[9] patiatur *SAa* : patitur *R*.

BOOK XI. III. 59–62

August. A *iugerum* of land requires four *sextarii* of their seed, provided that it receives in addition to this amount a little more than a *hemina* of Syrian radish-seed. He who sows this seed in the summer must beware lest, as a result of drought, the ground-flea [a] does not devour the still tender leaves as they creep forth. To prevent this, the powder which is found above an arched roof or even the soot which adheres to the ceiling above the hearth should be collected; then, the day before the sowing takes place, it should be mixed with the seeds and sprinkled with water, so that they may have a whole night to absorb the juice; for if they are soaked in this way they will be in good condition for sowing the next day. Some ancient authorities, Democritus [b] for example, advise that all seeds should be doctored with the juice of the herb which is called house-leek [c] and that the same remedy should be used against small vermin, and experience has taught us that this is true. For all that, since there is not a large quantity of this plant available, we more frequently use soot and the above-mentioned powder and thus satisfactorily ensure that the plants are not damaged. Hyginus [d] is of opinion that, after threshing, turnip-seeds ought to be scattered on the chaff that is still lying on the threshing-floor, since they make the heads grow more luxuriantly, because the hardness of the soil underneath does not allow the roots to penetrate deep into the earth. We have often tried this experiment without success, and so hold to the opinion that

[a] See Pliny, *N.H.* XIX. § 117.
[b] See note on Ch. 3. 2.
[c] *Sempervivum tectorum* and *S. arboreum*.
[d] See note on Book I. 1. 13.

LUCIUS JUNIUS MODERATUS COLUMELLA

num et napum melius existimamus subacta terra obrui,[1] servantque[2] adhuc antiquorum consuetudinem religiosiores agricolae, qui cum ea serunt, 63 precantur, ut et sibi et vicinis nascantur.[3] Locis frigidis, ubi timor est, ne autumnalis satio[4] hiemis gelicidiis peruratur,[5] harundinibus humiles canterii fiunt virgaeque, et virgis stramenta supra iaciuntur, et sic a pruinis semina defenduntur. Ubi vero apricis regionibus post pluvias noxia incesserunt animalia, quae a nobis appellantur erucae, Graece autem κάμπαι nominantur, vel manu colligi debent, vel matutinis temporibus frutices holerum concuti. Sic enim dum adhuc torpent nocturno frigore, si deciderint, non amplius in superiorem partem prore-64 punt. Id tamen supervacuum est facere, si ante sationem semina, uti iam praedixi, succo herbae sedi macerata sunt. Nihil enim sic medicatis nocent erucae. Sed Democritus in eo libro, qui Graece inscribitur Περὶ ἀντιπαθῶν,[6] affirmat has[7] ipsas bestiolas enecari, si mulier, quae in menstruis est, solutis crinibus et nudo pede unamquamque aream ter circumeat: post hic enim decidere omnes vermiculos,[8] et ita emori.

65 Hactenus praecipiendum existimavi de cultu hortorum et officiis villici; quem quamvis instructum atque[9] eruditum omni opere rustico esse oportere prima parte huius exordii censuerim; quoniam tamen plerumque evenit, ut eorum quae didice-

---

[1] obruit A.
[2] servantq; R : servaritque SA.
[3] nascatur SAR.
[4] autumnali spatio SAR.   [5] perurantur SA.
[6] ΠΕΡΙΑΝΤΙΠΑΦΟΝ S : ΠΕΡΑΝΤΙΠΑΤωΝ A.
[7] has R : hos SA.   [8] vermiculas Aa.
[9] instructum atque om. S.

radishes and navew are better sown in well-worked soil; and the more pious husbandmen still observe the custom of the men of old time and pray, when they sow them, "that they may grow both for themselves and for their neighbours." In cold regions, where men are afraid that the autumn sowing may be nipped by the frosts of winter, low trellises are constructed with reeds, and rods are thrown in and straw on the top of the rods, and thus the plants are protected from frosts. But in sunny districts where after rain the noxious animals which we call caterpillars and the Greeks name κάμπαι, attack the plants, they ought either to be picked by hand or the bushes of vegetable ought to be well shaken in the early morning; for, if the caterpillars thus fall to the ground when they are still torpid from the night's cold, they no longer creep into the upper part of the plants. However, it is unnecessary to do this, if, as I have already said, the seeds, before they are sown, have been soaked in the juice of the herb known as the house-leek; for when the seeds have been thus treated the caterpillars do not harm them. Democritus in the book entitled "On Antipathies" declares that these selfsame little vermin are killed if a woman, who is in the condition of menstruation, walks three times round each bed with her hair loose and her feet bare; for after this all the little worms fall to earth and so die.

Thus far I have deemed it proper to give advice about the cultivation of gardens and the duties of a bailiff, and though I have expressed the opinion in the first part of this treatise that he ought to be instructed and learned in every operation of agriculture, nevertheless, since it generally happens that the recollec-

rimus, memoria nos deficiat, eaque saepius ex commentariis renovanda sint, omnium librorum meorum argumenta subieci, ut cum res exegisset, facile reperiri possit, quid in quoque quaerendum, et qualiter quidque faciendum sit.

---

[a] The final paragraph of this book seems to show that the author intended to have concluded his work at this point and therefore added a kind of index or summary of the contents of the first eleven books, like those given by Pliny the Elder

tion of the things which we have learned fails us and must be renewed rather often from written notes, I have added below a list of the contents of all my books in order that, whenever necessity arises, it may be easily possible to discover what is to be found in each of them and how each task should be carried out.[a]

and Aulus Gellius. If this was so, it appears that, when he added Book XII, he did not transfer the summary to the end of Book XII, nor did he alter the final paragraph of Book XI. The index has not survived.

BOOK XII

LIBER XII

PRAEFATIO

Xenophon Atheniensis eo libro, P. Silvine, qui Oeconomicus inscribitur, prodidit maritale coniugium sic comparatum esse natura, ut non solum iucundissima, verum, etiam utilissima vitae societas iniretur: nam primum,[1] quod etiam Cicero ait, ne genus humanum temporis longinquitate occideret, propter hoc marem cum femina esse coniunctum: deinde ut ex hac eadem societate mortalibus adiutoria senectutis, nec minus propugnacula, praeparentur. 2 Tum etiam, cum victus et cultus humanus non uti feris in propatulo ac silvestribus locis, sed domi sub tecto accurandus erat, necessarium fuit alterutrum foris et sub dio esse, qui labore et industria compararet, quae tectis reconderentur. Siquidem vel rusticari, vel navigare, vel etiam genere alio negotiari necesse erat, ut aliquas facultates acquireremus. Cum vero paratae res sub tectum essent congestae, alium esse oportuit, qui et illatas custodiret, et ea conficeret opera, quae domi deberent administrari. 3 Nam et fruges ceteraque alimenta terrestria in-

[1] iam pridem *SAac*.

---

[a] Ch. VII. 19 ff.
[b] In his translation (which has not survived) of Xenophon's *Economicus*.

BOOK XII

PREFACE

Xenophon, the Athenian, in the book, Publius Silvinus, which is entitled *Economicus*,[a] declared that the married state was instituted by nature so that man might enter what was not only the most pleasant but also the most profitable partnership in life. For in the first place, as Cicero[b] also says, man and woman were associated to prevent the human race from perishing in the passage of time; and, secondly, in order that, as a result of the same association, mortals might be provided with help and likewise with defence in their old age. Furthermore, since 2 man's food and clothing had to be prepared for him, not in the open air and in woods and forests, as for the wild animals, but at home and beneath a roof, it became necessary that one of the two sexes should lead an outdoor life in the open air, in order that by his toil and industry he might procure provisions which might be stored indoors, since indeed it was necessary to till the soil or to sail the sea or to carry on some other form of business in order that we acquire some worldly substance. When, however, the goods thus secured had been stored under cover, there had to be someone else to guard them after they had been brought in and to carry on the operations which ought to be performed at home. For 3 corn and the other forms of food provided by the

LUCIUS JUNIUS MODERATUS COLUMELLA

digebant tecto,[1] et ovium ceterarumque pecudum fetus, atque fructus clauso custodiendi erant, nec minus reliqua utensilia, quibus aut alitur hominum 4 genus, aut etiam excolitur. Quare cum et operam et diligentiam ea quae proposuimus, desiderarent, nec exigua cura foris acquirerentur, quae domi custodiri oporteret: iure, ut dixi, natura comparata est[2] mulieris ad domesticam diligentiam, viri autem ad exercitationem forensem et extraneam. Itaque viro calores et frigora perpetienda, tum etiam itinera et labores pacis ac belli, id est rusticationis et mili- 5 tarium stipendiorum deus tribuit: mulieri deinceps, quod omnibus his rebus eam fecerat inhabilem, domestica negotia curanda tradidit. Et quoniam hunc sexum custodiae[3] et diligentiae assignaverat, idcirco timidiorem reddidit quam virilem. Nam metus plurimum confert ad diligentiam custodiendi. 6 Quod autem necesse erat foris et in aperto victum quaerentibus nonnumquam iniuriam propulsare, idcirco virum quam mulierem fecit audaciorem. Quia vero partis opibus aeque fuit opus memoria[4] et diligentia, non minorem feminae quam viro earum rerum tribuit possessionem. Tum etiam quod simplex natura non omnes res commodas amplecti volebat,[5] idcirco alterum alterius indigere voluit: quoniam quod alteri deest,[6] praesto plerumque est alteri.

7 Haec in Oeconomico Xenophon, et deinde Cicero, qui eum Latinae consuetudini tradidit, non[7] inutiliter

[1] tecto *ac*: tecti *SA*.
[2] opera *post* comparata est *add. SAac*.
[3] custodie *ac*: custodia *SA*.
[4] memoriam *SA*: memoria *ac*.
[5] volebat *SAac*: valebat *Schneider*.
[6] quoniam quod alteri deest *ac*: quod quia alteri id est *SA*.
[7] non *ac*: quod *SA*.

BOOK XII, PREFACE 3-7

earth needed a roof over them, and the young of sheep and of the other kinds of cattle, and fruits, and also all else that is useful for the sustenance and tending of mankind had to be safely kept in security. Wherefore, since the duties which we have described call for both labour and diligence, and since the acquisition of those things which have to be safeguarded at home calls for no small amount of attention out of doors, it is only right, as I have said, that the female sex has been provided for the care of the home, the male for out-of-doors and open-air activities. God, therefore, has assigned to man the endurance of heat and cold and the journeys and toils of peace and war, that is, of agriculture and military service, while he has handed over to woman, since he had made her unsuited to all these functions, the care of domestic affairs. And since he had assigned to the female sex the duties of guardianship and care, he made woman on this account more timid than man, since fear conduces very greatly to careful guardianship. On the other hand, since it was necessary for those who sought for food out of doors in the open air sometimes to repel attacks, God made man bolder than woman. But seeing that, after they had acquired substance, memory and attention were equally necessary to both sexes, God granted no smaller a share of these qualities to women than to men. Then, too, because nature in her simplicity did not wish either sex to enjoy the possession of every advantage, she desired that each should have need of the other, since what one lacks is generally present in the other.

These were the views not unprofitably expressed by Xenophon in the *Economicus* and by Cicero, who translated his work into the Latin language. For

disseruerunt. Nam et apud Graecos, et mox apud Romanos usque in patrum nostrorum memoriam fere domesticus labor matronalis fuit, tamquam ad requiem forensium exercitationum omni cura deposita patribusfamilias [1] intra domesticos penates se recipientibus. Erat enim summa reverentia cum concordia et diligentia mixta, flagrabatque mulier pulcherrima aemulatione,[2] studens negotia viri cura
8 sua maiora atque meliora reddere. Nihil conspiciebatur in domo dividuum, nihil quod aut maritus, aut femina proprium esse iuris sui diceret: sed in commune conspirabatur [3] ab utroque,[4] ut cum forensibus negotiis, matronalis industria rationem parem faceret. Itaque nec villici quidem aut villicae magna erat opera, cum ipsi domini quotidie
9 negotia sua reviserent atque administrarent. Nunc vero cum pleraeque [5] sic luxu et inertia diffluant, ut ne lanificii [6] quidem curam suscipere dignentur, sed domi confectae vestes fastidio sint, perversaque cupidine maxime placeant, quae grandi pecunia et totis paene censibus redimuntur: nihil mirum est, easdem ruris et instrumentorum agrestium cura gravari, sordidissimumque negotium ducere pau-
10 corum dierum in villa moram. Quam ob causam cum in totum non solum exoleverit, sed etiam occiderit vetus ille matrumfamiliarum mos Sabinarum atque Romanarum, necessaria irrepsit villicae cura, quae tueretur officia matronae: quoniam et villici

---

[1] patribus familias *ac* : patribus familiis *SA*.
[2] aemulatione *ac* : -es *SA*.
[3] conspirabatur *c* : conspiciebatur *SAa*.
[4] utroque *ac* : utraque *SA*.
[5] plereque *a* : plerique *c* : pleraque *SA*.
[6] lanificii *Ac* : lanifici *a* : lanificiis *S*.

both amongst the Greeks and afterwards amongst the Romans down to the time which our fathers can remember domestic labour was practically the sphere of the married woman, the fathers of families betaking themselves to the family fireside, all care laid aside, only to rest from their public activities. For the utmost reverence for them ruled in the home in an atmosphere of harmony and diligence, and the most beauteous of women was fired with emulation, being zealous by her care to increase and improve her husband's business. No separate ownership was to 8 be seen in the house, nothing which either the husband or the wife claimed by right as one's own, but both conspired for the common advantage, so that the wife's diligence at home vied with the husband's public activities. Thus there was not much for the bailiff or the bailiff's wife to do, since the master and mistress themselves daily watched over and directed their own affairs. Nowadays, however, when most 9 women so abandon themselves to luxury and idleness that they do not deign to undertake even the superintendence of wool-making and there is a distaste for home-made garments and their perverse desire can only be satisfied by clothing purchased for large sums and almost the whole of their husband's income, one cannot be surprised that these same ladies are bored by a country estate and the implements of husbandry, and regard a few days' stay at a country house as a most sordid business. Therefore, since 10 the ancient practice of the Sabine and Roman mistresses of households has not only become entirely out of fashion but has absolutely died out, management by a bailiff's wife has of necessity crept in to carry out the duties of the proprietor's wife, just as

quoque successerunt in locum dominorum, qui quondam prisca consuetudine non solum coluerant, sed habitaverant rura. Verum, ne videar intempestive censorium opus[1] obiurgandis[2] moribus nostrorum temporum suscepisse, iam nunc officia villicae persequar.

I. Ea porro (ut institutum ordinem teneamus, quem[3] priore volumine inchoavimus) iuvenis esse debet, id est non nimium puella, propter easdem causas, quas de aetate villici retulimus: integrae quoque valitudinis, nec foedi habitus, nec rursus pulcherrima. Nam illibatum robur et vigiliis et aliis sufficiet laboribus: foeditas fastidiosum, nimia species desidiosum faciet eius contubernalem. Itaque curandum est, ut nec vagum villicum et aversum a contubernio suo habeamus, nec rursus intra tecta desidem, et complexibus adiacentem feminae. Sed nec haec tantum, quae diximus, in villica custodienda sunt. Nam in primis considerandum erit, an a vino, ab escis, a superstitionibus, a somno, a viris remotissima sit, et ut cura[4] eam subeat, quid meminisse, quid in posterum prospicere debeat, ut fere eum morem servet, quem villico praecepimus: quoniam pleraque similia esse debent in viro atque femina, et tam malum vitare, quam praemium recte factorum sperare. Tum elaborare, ut quam minimam operam

[1] censorium opus *ac* : censuri *SA*.
[2] obiurgandibus *A*.
[3] quem *ac* : quae *SA*.
[4] cura *ac* : cure *SA*.

bailiffs too have succeeded to the position of the owners of property who formerly had followed the ancient custom of not only cultivating their estates but living on them. But that I may not appear to have unseasonably taken upon myself the task of censor in reproving the manners of our own times, I will now describe the duties of a bailiff's wife.

I. Moreover, (that we may keep to the order laid down, which we first employed in the preceding book) a bailiff's wife ought to be young, that is not too much of a girl, for the same reasons as we mentioned when speaking of the age of a bailiff. She ought also to have sound health and neither have an ugly appearance nor on the other hand be very beautiful; for unimpaired strength will suffice for long vigils and other toils, and ugliness will disgust her mate, while excessive beauty will make him slothful. So care must be taken that our bailiff is not of a wandering nature and does not avoid his wife's company, and that, on the other hand, he does not waste his time indoors and never far from her embraces. But these points which I have indicated are not the only ones to be looked for in a bailiff's wife; for it is of the first importance to observe whether she is far from being addicted to wine, greediness, superstition, sleepiness, and the society of men, and whether she readily grasps what she ought to remember and what she ought to provide for the future, in order that she may in general maintain the manner of life which we have laid down for the bailiff; for the husband and wife ought to resemble one another in most respects and, they should as much avoid evil, as they hope for the reward of work well done. Then she ought so to labour that the bailiff

The choice and duties of a bailiff's wife.

LUCIUS JUNIUS MODERATUS COLUMELLA

villicus intra tectum impendat, cui et primo mane cum familia prodeundum est, et crepusculo peractis operibus fatigato redeundum.

4 Nec tamen instituendo villicam domesticarum rerum villico remittimus curam, sed laborem eius adiutrice data levamus. Ceterum munia, quae domi capessuntur, non in totum muliebri officio relinquenda sunt, sed ita deleganda ei, ut identidem oculis villici custodiantur. Sic enim diligentior erit villica, si meminerit ibi[1] esse, cui ratio frequenter reddenda 5 sit. Ea porro persuasissimum habere debebit aut in totum aut certe plurimum domi se morari oportere: tum quibus aliquid in agro faciendum erit servis, eos foras emittere; quibus autem in villa quid agendum videbitur, eos[2] intra parietes continere, atque animadvertere, ne diurna cessando[3] frustrentur opera: quae domum autem inferuntur, diligenter inspicere, ne delibata sint, et ita explorata atque inviolata recipere: tum separare, quae consumenda sunt, et quae superfieri possunt, custodire, ne 6 sumptus annuus[4] menstruus fiat. Tum siquis ex familia coeperit adversa valetudine affici, videndum erit ut is quam commodissime ministretur. Nam ex huiusmodi cura nascitur benevolentia, nec minus obsequium. Quinetiam fidelius quam prius servire student, qui convaluerint, cum est aegris adhibita diligentia.

[1] villica si meminerit ibi *om. SA.*
[2] eos : *SAac.*
[3] cessendo *SA.*
[4] annuus *ac* : annus *SA.*

BOOK XII. i. 3-6

may have as little as possible to do indoors, since he ought to go forth early in the morning with the slaves and return weary at twilight, when his work is done.

However, while we fix the duties of the bailiff's 4 wife, we do not relieve the bailiff of responsibility for household matters but merely lighten his task by giving him an assistant. In fact the duties which are undertaken at home must not be left entirely to be carried out by the woman but must be so delegated to her that they are at the same time watched over by the bailiff's eye. For she will be more diligent if she remembers that there is someone on the spot to whom a frequent account must be rendered. Also 5 she will have to be absolutely convinced that she must remain entirely, or at any rate for the most part, at home; and that she must send out of doors those slaves who have some work to do in the fields, and keep within the walls those for whom it seems that there is some duty to perform in the villa; and she must see that the daily tasks are not spoilt by inaction. She must carefully inspect everything that is brought into the house to see that it is not damaged, and receive it after it has been examined and found intact; then she must set apart what has to be consumed and guard what can be placed in reserve, so that the provision for a year may not be spent in a month. Again, if any member of the household is beginning 6 to be affected by bad health, she will have to see that he is given the most suitable treatment; for attention of this kind is a source of kindly feeling and also of obedience. Moreover, those who have recovered their health, after careful attention has been given them when they were ill, are eager to give more faithful service than before.

LUCIUS JUNIUS MODERATUS COLUMELLA

II. Post haec meminisse debebit, quae inferantur, ut idoneis et salubribus locis recondita sine noxa permaneant. Nihil enim magis curandum est, quam praeparare, ubi quidque reponatur, ut, cum opus sit, promatur. Ea loca qualia esse debeant, et in primo volumine, cum villam constitueremus et in undecimo, cum de officio villici [1] disputaremus, iam dicta sunt.
2 Sed ne nunc quidem demonstrare breviter pigebit. Nam quod excelsissimum est conclave, pretiosissima vasa et vestem desiderat: quod denique horreum siccum atque aridum, frumentis habetur idoneum: quod frigidum, commodissime vinum custodit: quod bene illustre, fragilem supellectilem atque ea
3 postulat opera, quae multi luminis indigent. Praeparatis igitur receptaculis, oportebit suo quidque loco generatim, atque etiam specialiter nonnulla disponere: quo facilius, cum quid expostulabit usus, reperiri [2] possit. Nam vetus est proverbium, paupertatem certissimam esse, cum alicuius indigeas,[3] uti eo non posse, quia ignoretur, ubi proiectum iaceat quod desideratur. Itaque in re familiari laboriosior
4 est negligentia, quam diligentia. Quis enim dubitet nihil esse pulchrius in omni ratione vitae dispositione atque ordine? quod etiam ludicris spectaculis licet saepe cognoscere. Nam ubi chorus canentium non ad certos modos neque numeris praeeuntis magistri consensit, dissonum quiddam ac tumultuosum audi-

---

[1] de officio villici *ac*: de villico *SA*.
[2] recipere *SAac*.
[3] indigeas *ac*: -at *SA*.

BOOK XII. II. 1-4

II. Next she will have to remember what is brought into the house that these things be stored in suitable and healthy places and so remain without damage; for nothing is more important than to be sure to provide a place where each article may be laid by, so that it may be produced when required. We have already described in the first book, when we dealt with the structure of the farm, and in the eleventh book, when we discussed the duties of the bailiff, what should be the nature of these storage-places; but it will do no harm briefly to describe them again here. The highest room in the house claims the most precious vessels and clothing; a granary which is dry and free from moisture is considered suitable for cereals; a cold place is best adapted for keeping wine; a well-lighted room calls for the use of delicate tools and the performance of tasks which require plenty of light. When, therefore, the receptacles have been prepared, it will be necessary to arrange each article in its place according to its kind, and some things, too, in separate sections each of the same kind, so that they may be more easily found when they are required for use. For there is an ancient proverb that the surest kind of poverty occurs when you need something but cannot have the use of it, because you do not know where the object which you require lies carelessly thrown down. Thus in family life carelessness gives more trouble than care. For who can doubt that there is nothing more beautiful in the whole conduct of life than arrangement and order? We can often remark this at public performances. For when a choir of singers does not agree in certain measures nor keep time with the master who directs them, they seem to their audience to be singing something dis-

The management of provisions and store-houses.

entibus canere videtur: at ubi certis numeris ac pedibus velut facta conspiratione consensit atque concinuit, ex eiusmodi vocum concordia non solum ipsis canentibus amicum quiddam et dulce resonat, verum etiam spectantes audientesque laetissima voluptate permulcentur. Iam vero in exercitu neque miles neque imperator sine ordine ac dispositione quidquam valet explicare, cum armatus inermem, eques peditem, plaustrum equitem, si sint permixti, confundant. Haec eadem ratio praeparationis atque ordinis etiam in navigiis plurimum valet. Nam ubi tempestas incessit, et est rite disposita navis, suo quidque ordine locatum armamentum sine trepidatione minister promit, cum est a gubernatore postulatum. Quod si tantum haec possunt vel in theatris vel in exercitibus vel etiam in navigiis, nihil dubium est, quin cura villicae ordinem dispositionemque rerum, quas reponit, desideret. Nam et unumquidque[1] facilius consideratur, cum est assignatum suo loco, et siquid forte abest, ipse vacuus locus admonet, ut quod deest requiratur. Siquid vero curari aut concinnari oportet, facilius intellegitur, cum ordine suo recensetur.[2] De quibus omnibus M. Cicero auctoritatem Xenophontis secutus in Oeconomico[3] sic inducit Ischomachum sciscitanti Socrati narrantem.

III. Praeparatis idoneis locis instrumentum et supellectilem distribuere coepimus: ac primum ea

---

[1] unumquitque *SA* : unumquodque *ac*.
[2] recensetur *Aac* : recensentur *S*.
[3] Oeconomico *ac* : -os *SA*.

BOOK XII. II. 4–III. 1

cordant and confused; but when the singers agree with definite numbers and beats, as though they had conspired together to do so, and sing together, from such a concord of voices not only is a sound produced which is pleasant and delightful to the singers themselves, but the spectators and audience are also charmed by a feeling of delightful pleasure. In an army, too, neither the soldier nor the general can carry out any movement if order and arrangement are lacking, since, if they are mixed together, the armed man will throw the unarmed into disorder, the horseman the foot soldier and the transport the cavalry. This same system of preparation and order is also of great value on ship-board; for, when a storm comes on and a ship is properly organized, the boatswain produces without any confusion each article of tackling which is arranged in its proper place, on the demand of the helmsman. Now if such conditions are so effective in theatres, armies and also in ships, there can be no doubt that the duties of a bailiff's wife require order and arrangement in the things which she stores away. For, when each individual object is assigned to its own place, it attracts the attention more easily, and, if anything happens to be absent, the empty space itself warns us to look for what is lacking. Indeed if a thing requires attention and adjustment, it is more easily perceived when it is surveyed in its own proper order. All these are points on which Marcus Cicero, following the authority of Xenophon in his *Economicus*, introduces Ischomachus as holding forth in reply to the questions of Socrates.

III. Having prepared suitable storerooms we proceeded to distribute the utensils and furniture; and, first of all, we set aside the objects which we are in

On the storage of household furniture and utensils.

LUCIUS JUNIUS MODERATUS COLUMELLA

secrevimus, quibus ad res divinas uti solemus, postea mundum muliebrem, qui ad dies festos comparatur, deinde ad bella [1] virilem, item dierum solemnium ornatum, nec minus calciamenta utrique sexui convenientia: tum iam seorsum arma ac tela separabantur, et in altera parte, quibus ad [2] lanificia utun-
2 tur. Post quibus ad cibum comparandum vasis uti assolent [3] constituebantur: inde quae ad lavationem, quae ad exornationem, quae ad mensam quotidianam, atque epulationem pertinent, exponebantur. Postea ex iis quibus quotidie utimur, quod menstruum esset seposuimus, annuum quoque in duas partes divisimus: nam sic minus fallit, qui
3 exitus futurus sit. Haec postquam omnia secrevimus, tum suo quaeque loco disposuimus: deinde quibus quotidie servuli utuntur, quae ad lanificia, quae ad cibaria coquenda et conficienda pertinent, haec ipsis, qui his uti solent, tradidimus, et ubi exponerent, demonstravimus, et ut salva essent,
4 praecepimus. Quibus autem ad dies festos et ad hospitum adventum utimur et ad quaedam rara negotia, haec promo [4] tradidimus, et loca omnium demonstravimus, et omnia annumeravimus, atque annumerata ipsi exscripsimus, eumque admonuimus,[5] ut quodcumque opus esset, sciret unde daret, et meminisset atque annotaret, quid et quando et cui dedisset, et cum recepisset, ut quidque suo loco reponeret.

[1] bella *edd.* : *om. SAac.*
[2] ad *om. SAac.*
[3] vasa adsolent A^1S : vasa ut adsolent A^2 : vasa ut assolent *ac.*
[4] primo *SAac.*

the habit of using for the worship of the gods, after that the women's apparel, which is provided for festal days, and the men's apparel for war and also their dress for solemn occasions, and likewise foot-ware for both sexes. Next arms and weapons were stored apart, and in another place implements used for manufacturing wool. After this a place was found 2 for the vessels which are generally used for keeping food, and then those connected with washing and the toilet and with ordinary daily meals and with banquets were set out. Next of the things which we use daily we set apart what would suffice for a month, and what would suffice for a year we divided into two portions; for then there is less likely to be a mistake as to what the outcome might be. After we had separated all these things, we arranged 3 them each in its proper place. Next we handed over to the actual people who are in the habit of using them the things which are used daily by the slaves, namely, those connected with the making of wool and the cooking and preparation of food, and pointed out where they should put them and charged them to keep them safe. As for the things which we 4 use on days of festival and on the arrival of guests and on certain rare occasions, these we handed over to the steward and pointed out the places where they all were and numbered them all and we ourselves wrote out a list of what we had numbered and warned him that he must know whence to produce whatever was needed and that he must remember, and note down what he has given out and when and to whom, in order that he might put back each article in its proper place.

---

⁵ admonuimus *ac* : monuimus *SA*.

LUCIUS JUNIUS MODERATUS COLUMELLA

5 Igitur haec nobis antiqui per Ischomachi personam praecepta industriae ac diligentiae tradiderunt, quae nunc nos villicae demonstramus. Nec tamen una eius cura esse debebit, ut clausa custodiat, quae tectis illata receperit, sed subinde recognoscat atque consideret, ne aut supellex vestisve condita situ dilabatur, aut fruges, aliave utensilia neglegentia desidiaque sua corrumpantur.

6 Pluviis vero diebus, vel cum frigoribus aut pruinis mulier sub dio rusticum opus obire non potuerit, ut ad lanificium reducatur, praeparatae[1] sint et pectitae lanae, quo facilius iusta lanificio persequi atque exigere possit. Nihil enim nocebit, si sibi atque actoribus et aliis in honore servulis vestis domi confecta fuerit, quo minus patrisfamilias[2] rationes
7 onerentur. Illud vero etiam in perpetuum custodiendum habebit, ut eos, qui foris rusticari debebunt, cum iam e villa familia processerit, requirat, ac siquis, ut evenit, curam contubernalis eius intra tectum tergiversans fefellerit, causam desidiae sciscitetur, exploretque utrum adversa valetudine inhibitus restiterit, an pigritia delituerit. Et si compererit, vel simulantem languorem sine cunctatione in valetudinarium deducat: praestat enim opere fatigatum

[1] praeparatae quae *SA* : praeparataeque *ac*.
[2] patrisfamiliae *SA* : patris familie *a* : patris fa. *c*.

190

BOOK XII. III. 5-7

These precepts of industry and care, therefore, the 5 ancients have handed down to us through the medium of Ischomachus and we now use them to instruct the bailiff's wife. She ought not, however, to limit her care to the locking up and guarding the goods which have been brought into the house and which she has received, but she ought to inspect them from time to time and take care that the furniture and clothing which have been stored away do not fall to pieces with decay and that the fruits of the earth and other things in general use are not ruined through her neglect and laziness.

But in order that she may have recourse to wool- 6 work on rainy days or when, owing to cold or frost, a woman cannot be busy with field-work under the open sky, there should be wool prepared and combed out ready, so that she may be able more easily to carry out the task of spinning and demand this work also from others. For it will not be a bad plan if clothing is made at home for herself and the overseers and other slaves of good position, so that the account of the master of the house may be less heavily charged. She will also have to be perpetually on 7 the watch, when the slaves have left the villa, and seek out those who ought to be doing agricultural work outside, and if anyone, as sometimes happens, has managed to skulk indoors and escape the vigilance of her mate, she must inquire the reason for his laziness and find out whether he has stayed behind because bad health has prevented him from working or whether he has hidden himself through idleness. If she finds him even pretending to be ill, she must without delay conduct him to the infirmary; for, if he is worn out by his work, it is better that he should

LUCIUS JUNIUS MODERATUS COLUMELLA

sub custodia requiescere unum aut alterum diem, quam pressum nimio labore veram noxam concipere.

8 Denique uno loco quam minime oportebit eam consistere, neque enim sedentaria eius opera est, sed modo ad telam debebit accedere, ac siquid melius sciat, docere: si minus, addiscere ab eo qui plus intelligat, modo eos qui cibum familiae conficiunt, invisere: tum etiam culinam et bubilia, nec minus praesepia mundanda curare: valetudinaria quoque vel si vacent[1] ab imbecillis, identidem aperire, et immunditiis liberare, ut cum res exegerit, bene ordinata et salubria languentibus praebeantur:

9 promis quoque et cellariis aliquid appendentibus[2] intervenire: nec minus interesse pastoribus in stabulis fructum cogentibus, aut fetus ovium, aliarumve pecudum subrumantibus: tonsuris vero earum utique interesse, et lanas diligenter percipere, et vellera ad numerum pecoris recensere: tum insistere atriensibus, ut supellectilem exponant, et aeramenta[3] detersa nitidentur,[4] atque rubigine liberentur, ceteraque quae refectionem desiderant, fabris concinnanda tradantur.

10 Postremo his rebus omnibus constitutis, nihil hanc arbitror distributionem profuturam, nisi, ut iam dixi, villicus saepius, et aliquando tamen dominus aut matrona consideraverit, animadverterit-

---

[1] vagent *S*.
[2] appendentibus *edd*.: attendentibus *SAac*.
[3] aeramenta *SAc*: aramenta *a*: ferramenta *edd*.
[4] nitidentur *ac*: nidentur *SA*.

rest for a day or two under observation than that he come to some real harm by being forced to overwork himself.

In a word, it will be her duty to remain as little as possible in one place, for hers is not a sedentary task; but at one moment she will have to visit the loom and impart any superior knowledge which she possesses, or, failing this, learn from one who understands the matter better than she does; at another moment she will have to look after those who are preparing the food for the family. Then too she will have to see that the kitchen and the cowsheds and also the mangers are cleaned, and she will have to open the sick-wards from time to time, even if they contain no patients, and keep them free from dirt, so that, when necessary, the sick may find them in an orderly and healthy condition. She ought also to come unawares upon stewards and cellar-men when they are weighing out anything, and likewise to be present when the shepherds are milking the ewes at the sheep-folds or putting the lambs or the young of other cattle to suck their dams. But she must also be there when the sheep are being sheared and keep a watchful eye on the wool and count the fleeces, comparing them with the number of sheep. Again she must insist that the servants in charge of the hall put out the furniture to air and that bronze utensils are scoured and polished and freed from rust, and see that anything else which needs mending is handed over to the craftsmen for repair.

Lastly, when all these arrangements have been made, this distribution of goods will, I think, be of no benefit, unless, as I have already said, the bailiff rather frequently, and the master or mistress of the house

que, ut ordinatio instituta conservetur. Quod etiam in bene moratis civitatibus [1] semper est observatum; quarum primoribus atque optimatibus non satis visum est bonas leges habere, nisi custodes earum diligentissimos cives creassent,[2] quos Graeci νομο-
11 φύλακας appellant. Horum erat officium, eos qui legibus parerent, laudibus prosequi, nec minus honoribus: eos autem qui non parerent, poena multare. Quod nunc scilicet faciunt magistratus, assidua iurisdictione vim legum custodientes. Sed haec in universum administranda tradidisse abunde sit.

IV. Nunc de ceteris rebus, quae omissae erant prioribus libris, quoniam villicae reservabantur officiis, praecipiemus, et ut aliquis ordo custodiatur, incipiemus a verno tempore, quoniam fere maturis atque trimestribus consummatis sationibus, vacua tempora iam contingunt ad ea exequenda, quae
2 deinceps docebimus. Parvarum rerum curam non defuisse Poenis Graecisque auctoribus atque etiam Romanis, memoria tradidit. Nam et Mago Carthaginiensis, et Hamilcar,[3] quos secuti videntur Graecae gentis non obscuri scriptores Mnaseas atque Paxamus,[4] tum demum [5] nostri generis, postquam a bellis vacuum fuit, quasi quoddam tributum victui humano

[1] civilibus S.   [2] creassent *om.* SAac.
[3] Halcar A.   [4] phaxamus Aac : -es S.
[5] deinde S.

---

a See Book I. 1. 10 and note.

b Perhaps the well-known Carthaginian, father of Hannibal, though he is not mentioned by other authors as interested in agriculture.

c Mnaseas is mentioned by Varro, Pliny and Athenaeus, and was a native of Patara in Lycia.

from time to time, carry out an inspection and take note that arrangement once determined on is maintained. This practice was also always observed in well-regulated cities, whose chief and noblest men did not consider it enough to have good laws, unless they had elected the most careful citizens as guardians of those laws, whom the Greeks call νομοφύλακες (guardians of the laws). The duty of these men was to bestow upon those who obeyed the laws praise and likewise distinctions, and to inflict punishment on those who did not obey them. This is now, of course, the function of magistrates, who safeguard the power of the laws by the constant administration of justice. Let the statement of these precepts suffice as far as general management is concerned.

IV. We will now give instructions about the other subjects which had been omitted in the earlier books because they were reserved for treatment among the duties of the bailiff's wife, and that some order may be preserved, we will begin with the season of spring, because then, the early and the three-monthly sowings being practically completed, an empty period occurs when we can carry out the operations which we next explain. Tradition has told us that the Carthaginian and Greek authors, and the Roman also, did not neglect to pay attention to small details. For Mago,[a] the Carthaginian, and Hamilcar,[b] whom Mnaseas[c] and Paxamus,[d] by no means obscure authors of Greek nationality, seem to have followed, and finally authors of our own race, when there was rest from war, did not disdain to contribute something to the subject of feeding mankind, such as

On the vessels to be used for preserves.

[d] Paxamus is otherwise unknown.

LUCIUS JUNIUS MODERATUS COLUMELLA

conferre dedignati non sunt, ut M. Ambivius, et Maenas[1] Licinius, tum etiam C. Matius; quibus studium fuit pistoris et coci nec minus cellarii diligentiam suis praeceptis instituere.

3 His autem omnibus placuit, eum, qui rerum harum officium susceperit, castum esse continentemque oportere, quoniam totum in eo sit, ne contractentur pocula vel cibi, nisi aut ab impubi, aut certe abstinentissimo rebus venereis. Quibus si fuerit operatus vel vir vel femina, debere eos flumine aut perenni aqua, priusquam penora contingant, ablui. Propter quod his necessarium esse pueri vel virginis ministerium, per quos promantur, quae usus postulaverit.

4 Post hoc praeceptum locum et vasa idonea salgamis[2] praeparari iubent: locum esse debere aversum a sole, quam frigidissimum et siccissimum, ne situ penora mucorem contrahant. Vasa autem fictilia vel vitrea plura potius quam ampla, et eorum alia recte picata, nonnulla tamen pura, prout conditio 5 condiuturae exegerit. Haec vasa dedita opera fieri oportet patenti ore, et usque ad imum aequalia, nec in modum doliorum formata, ut exemptis ad usum salgamis quidquid superest aequali pondere usque ad fundum deprimatur, cum ea res innoxia penora conservet, ubi non innatent, sed semper sint iure submersa. Quod in utero dolii fieri vix posset propter

---

[1] Mecenas *S*: Bascenas *Aac*.
[2] salgamis *SAac*.

---

[a] M. Ambivius and Maenas are otherwise unknown.
[b] A special kind of apple was called after Matius (Book V. 10. 19: XII. 47. 5). It is possible that he was the favourite of Augustus mentioned by Tacitus (*Ann.* XII. 60).

BOOK XII. iv. 2-5

Marcus Ambivius [a] and Maenas Licinius [a] and also Gaius Matius,[b] whose object it was by their instructions to prescribe the activities of the baker, the cook and the cellar-man.

All these writers held that he who undertakes the performance of these duties ought to be chaste and continent, because it was of prime importance that neither drinking vessels nor food should be handled except by one who had not reached puberty or, at any rate, only by one who was most abstemious in sexual intercourse. Any man or woman who indulged in it ought, they thought, to wash in a river or running water before touching food; consequently in their view one must employ the services of a boy or maiden to give out the food required for daily use.

Their next precept is that they order that a special place and suitable vessels should be prepared for storing preserves, and this place ought not to be exposed to the sun but should be as cold and dry as possible, so that the food may not become mouldy and decay. Vessels should be either of earthenware or glass and should be numerous rather than large, and some of them should be properly treated with pitch but some in their natural state as the condition of the material preserved demands. Great care ought to be taken in the making of these vessels that they have a wide mouth and that they are of the same width right down to the bottom and not shaped like wine-jars, so that, when the preserved food is removed for use, what remains may be pressed with equal weight to the bottom, since the food is kept fresh when it does not float on the surface but is always covered by liquid. This can scarcely happen in the globular shape of a wine-jar, because of the

LUCIUS JUNIUS MODERATUS COLUMELLA

inaequalitatem figurae. Maxime autem ad hoc[1] necessarium esse aceti et durae muriae usum,[2] quae utraque sic confieri.

V. Quemadmodum ex vino vapido acetum fiat. In sextarios quadraginta octo[3] fermenti libram, fici aridae pondo quadrantem, salis sextarium haec subterito, et subtrita cum quartario mellis aceto diluito, atque ita in praedictam mensuram adicito. Quidam hordei tosti sextarios quattuor, et nuces ardentes iuglandes quadraginta, et mentae viridis pondo selibram in eandem mensuram adiciunt.
2 Quidam ferri massas exurunt, ita ut ignis speciem habeant, easque in eandem mensuram demittunt.[4] Tum etiam exemptis nucleis ipsas nuces pineas vacuas[5] numero quinque vel sex incendunt, et ardentes eodem demittunt. Alii nucibus sapineis ardentibus idem faciunt.

VI. Muriam duram sic facito: dolium quam patentissimi oris locato in ea parte villae, quae plurimum solis accipit. Id dolium aqua caelesti repleto; ea est enim huic rei aptissima; vel si non fuerit pluvalis, certe fontana dulcissimi saporis. Tum indito sportam iunceam, vel sparteam, quae replenda est sale candido, quo candidior muria fiat. Cum salem per aliquot dies videbis liquescere,[6] ex eo
2 intelliges nondum muriam esse maturam. Itaque subinde alium salem tamdiu ingeres, donec in sporta permaneat integer, nec minuatur. Quod cum animadverteris, scias habere muriam maturitatem

---

[1] haec *SA* : hoc *a* : *om. c.*   [2] usuum *S.*
[3] quadraginta octa *SA* : duodequinquaginta *ac.*
[4] dimittunt *SAac.*
[5] quas *post* vacuas *add. SA* : *om. ac.*
[6] et *post* liquescere *add. SAac.*

BOOK XII. iv. 5–vi. 2

irregularity of its form. The use of vinegar and hard brine is very necessary, they say, for the making of preserves; that the following are the ways in which each of these is manufactured.

V. How to make vinegar from wine that has gone flat. Take forty-eight *sextarii* of wine, and crush together a pound of yeast and a *quadrans* of dried fig and a *sextarius* of salt, and after they have been crushed add a quarter pound of honey and dilute with vinegar and add the mixture to the aforesaid quantity of wine. Some people add four *sextarii* of toasted barley and forty burning walnuts and a half pound of green mint to the same quantity of wine; some 2 heat lumps of iron, so that they look like fire, and plunge them into the same measure, and then after taking out the kernels they set fire to five or six empty pine-nut shells and plunge them still burning into the wine. Others do the same thing with burning fir-cones.

VI. The following is the way to make hard brine. Place a wine-jar with as wide a mouth as possible in the part of the farm-house which gets the most sun. Fill the jar with rain-water, for this is most suitable for the purpose, or, if rain-water is not available, let it be at any rate spring-water of very sweet flavour. Then place in the water a basket made of rushes or of broom which must be filled with white salt, so that the brine may be whiter. When in the course of several days you see that the salt continues to melt, you will know from this fact that the brine is not yet ready. You will, therefore, put in more salt from 2 time to time until it remains unchanged in the basket and does not grow less. When you notice that it does so, you can be sure that the brine has come to

LUCIUS JUNIUS MODERATUS COLUMELLA

suam. Et si facere aliam volueris, hanc in vasa bene picata diffundes, et opertam in sole habebis. Omnem enim mucorem vis solis aufert, et odorem bonum praebet. Est et [1] aliud muriae maturae experimentum. Nam ubi dulcem caseum demiseris in eam, si pessum ibit, scies esse adhuc crudam: si innatabit, maturam.

VII. His praeparatis circa vernum aequinoctium herbas in usum colligi et reponi oportebit, cymam, caulem, capparim, apii coliculos, rutam, holeris atri cum suo cole florem antequam de folliculo exeat: item ferulae cum coliculo silentem [2] florem: pastinacae agrestis vel sativae cum coliculo [3] silentem florem: vitis albae et asparagi et rusci et tamni et digitelli et puleii et nepetae et lapsanae et batis et eius coliculum, qui milvinus pes appellatur; quin 2 etiam tenerum coliculum faeniculi. Haec omnia una conditura commode servantur, id est aceti duas partes, et tertiam durae muriae si miscueris. Sed vitis alba, ruscus, et tamnum et asparagus et pastinaca et nepeta et batis generatim in alveos componuntur, et sale conspersa biduo sub umbra, dum consudent, reponuntur: deinde si tantum remiserint 3 humoris, ut suo sibi iure ablui possint: si minus, superfusa dura muria lavantur, et pondere imposito

---

[1] et est *SA*: est *ac*.
[2] quam tenerrimum *post* silentem (silentes) *add. ac*: *om. SA*.
[3] coliculo *SA*: coliculis *a*: colliculis *c*.

---

[a] *Vitis alba*, common bryony; *tamnus* = black bryony.
[b] *Satureia calamintha*, esp. var. *nepeta*.
[c] *Hirschfeldia incana*.
[d] Perhaps a type of plantain.

maturity. And if you wish to make more brine, you will pour the brine already made into vessels well daubed with pitch and keep it covered up in the sun; for the action of the sun takes away all mustiness and causes a pleasant odour. There is also another method of proving that the brine is ripe; for, when you plunge a piece of fresh cheese into it, if it sinks to the bottom, you will know that it is still crude, but, if it floats on the surface, you will know that it has reached maturity.

VII. When the vinegar and birne have been got ready about the time of the spring equinox it will be necessary to collect herbs and store them up for use, namely, sprouts and stalks of cabbage, capers, stalks of parsley, rue, the flower of alexanders with its stalk before it comes out of its sheath, also the unopened flower of the giant fennel with its stalk, and the unopened flower of the wild or the cultivated parsnip with its stalk, the flower of the bryony,[a] asparagus, butcher's-broom, black bryony,[a] house leek, pennyroyal, calamint,[b] hoary mustard [c] and samphire, and the little stalk of what is called " kite's foot ",[d] also the tender little stalk of the fennel. All these are conveniently preserved by one method of pickling, that is to say, a mixture of two-thirds of vinegar and one-third of hard brine. But bryony, butcher's broom, black bryony, asparagus, parsnip, calamint, and samphire are also placed according to their kinds on trays, and having been sprinkled with salt are placed for two days in the shade until they yield up their moisture; then if they have produced enough liquid to enable them to be washed in their own juice, well and good; but if not, they are washed by having hard brine poured over them and have the liquid

How to pickle

LUCIUS JUNIUS MODERATUS COLUMELLA

exprimuntur: tum suo quidque vase conditur, et ius, ut supra dixi, quod est mixtum duabus partibus aceti et una muriae, infunditur, faeniculique aridi, quod est per vindemiam proximo anno lectum, spissamentum imponitur, ita ut herbas deprimat, et ius usque[1] in summum[2] labrum fideliae perveniat.

4 Olusatrum et ferulam et faeniculum cum legeris, sub tecto exponito,[3] dum flaccescat: deinde folia et corticem omnem coliculorum detrahito. Caules si fuerint pollice crassiores, harundine secato, et in duas partes dividito. Ipsos quoque flores, ne sint immodici, diduci[4] et partiri oportebit, atque ita in vasa condi. Deinde ius, quod supra scriptum est, infundi, et paucas radiculas laseris, quod Graeci σίλφιον vocant, adici, tum spissamento faeniculi aridi 5 contegi,[5] ut ius superveniat. Cymam, caulem, capparim, pedem milvi, puleium, digitellum, compluribus diebus sub tecto siccari, dum flaccescat, et tum eodem modo condiri convenit, quo ferulam, rutam, satureiam, cunilam. Sunt qui rutam muria tantum dura sine aceto condiant,[6] deinde, cum usus exigit, aqua vel etiam vino abluant, et superfuso oleo utantur. Haec conditura possit commode satureia viridis, et aeque viridis cunila servari.

VIII. Oxygalam sic facito. Ollam novam sumito, eamque iuxta fundum terebrato: deinde cavum, quem feceris, surculo obturato, et lacte ovillo quam

[1] et ius usque *ac*: eusque *S*: deprimate usque *A*.
[2] summo *S*.   [3] reponito *SAac*.   [4] deduci *SAac*.
[5] contegi *a*: conteri *SA*: contingi *c*.
[6] condiant *c*: condant *a*: condiunt *SA*.

---

[a] *Cf.* p. 200 note *d*.

squeezed out of them by having a weight placed over them. Then each is stored in its own vessel, as I have said above, and a fluid which is a mixture of two parts of vinegar and one of brine is poured on it, and a stopper of dry fennel, which was pickled during the vintage in the previous year, is inserted so as to press down the herbs and make the liquid come up to the brim of the jar.

When you have picked alexanders and giant-fennel and ordinary fennel, lay them out indoors until they wither; then strip off all the leaves and the outer coverings of the stalks. If the stalks are thicker than one's thumb cut them with a reed and divide them into two portions; also the flowers themselves will have to be pulled apart and divided, that they may not be too large, and then stored in vessels. Next the liquid, as described above, must be poured in and a few small roots of laser, which the Greeks call σίλφιον, added, then the whole must be covered with a stuffing of dry fennel so that the liquid comes above it. The sprouts and stalks of cabbage, capers, "kite's-foot",[a] pennyroyal and house-leek must be put indoors to dry for several days until they wither, and then be preserved in the same manner as fennel, rue, savory and marjoram. Some people preserve rue with hard brine only, without vinegar; then, when it is required for use, they wash it with water or even wine and pour oil over it before they use it. Green savory and likewise green marjoram can be conveniently preserved by this method.

VIII. The following is the way to make sour milk. Take a new vessel and bore a hole in it near the base; then fill up the hole which you have made with a small stick and fill the vessel with the freshest

How to make sour milk.

LUCIUS JUNIUS MODERATUS COLUMELLA

recentissimo vas repleto, eoque adicito viridium [1]
condimentorum fasciculos, origani, mentae, cepae,
coriandri. Has herbas ita in lacte demittito, ut
2 ligamina earum extent. Post diem quintum surcu-
lum, quo cavum obturaveras, eximito, et serum
emittito. Cum deinde lac coeperit manare, eodem
surculo cavum obturato, intermissoque triduo, ita ut
supra dictum est, serum emittito, et fasciculos condimentorum exemptos abicito: deinde exiguum
aridi thymi et cunilae aridae super lac destringito,
concisique sectivi porri quantum videbitur adicito,
et permisceto: mox intermisso biduo rursus emittito
serum, cavumque obturato, et salis triti quantum
satis erit adicito, et misceto, deinde operculo imposito et oblinito, non ante aperueris ollam, quam
usus exegerit.

3 Sunt qui sativi vel etiam silvestris lepidii herbam
cum collegerunt in umbra siccent, deinde folia eius
abiecto caule die et nocte muria macerata expressa-
que, lacti misceant sine condimentis, et salis quantum satis arbitrantur adiciant: tum cetera, quae
supra praecepimus faciant. Nonnulli recentia folia
lepidii cum dulci lacte in olla miscent, et post diem
tertium, quemadmodum praecepimus, serum emit-
tunt: deinde compertam satureiam viridem, tum
etiam arida semina coriandri atque anethi et thymi
et apii in unum bene trita adiciunt, salemque bene
coctum cribratum permiscent. Cetera eadem quae
supra faciunt.

[1] viridium *ac*: viridum *SA*.

possible sheep's milk and add to this small bunches of green seasonings, marjoram, mint, onion and coriander. Plunge these herbs in the milk so that the strings with which they are tied are not submerged. After 2 the fifth day take out the small stick with which you had blocked up the hole and drain off the whey; then when the milk begins to flow, block up the hole with the same stick and, after an interval of three days, let out the whey in the manner described above and take out and throw away the bunches of seasonings; then shred a little dry thyme and dry marjoram over the milk and add as much leek cut small as you think fit. Next, after an interval of two days, again let out the whey and block up the hole and add as much pounded salt as shall suffice, and mix the whole together. Then put the cover on the vessel and seal it up, and do not open the vessel until the contents are required for use.

Some people, when they have picked garden or 3 even wild pepperwort, dry it in the shade, then, after throwing away the stalk, they steep the leaves for a day and a night in brine and after squeezing the liquid out of them, mix them with milk without any seasoning and add as much salt as they think fit; then they go through the other operations which we have described above. Some people mix the young leaves of dittander in a vessel with sweet milk and, after the third day, draw off the whey as we have directed; they then find some green savory and also some dry seeds of coriander, dill, thyme and parsley well pounded into a mass and add them to it and mix in some salt which has been well roasted and sifted. The rest of the operations are the same as in the above recipe.

LUCIUS JUNIUS MODERATUS COLUMELLA

IX. Conditura lactucae. Caules lactucae ab imo depurgatos eatenus, qua tenera folia videbuntur, in alveo salire [1] oportet, diemque unum et noctem sinere, dum muriam remittant: deinde muriam eluere,[2] et expressos in cratibus pandere, dum assiccescant: tum substernere anethum aridum [3] et faeniculum rutaeque aliquid et porri concidere, atque ita miscere: tum siccatos coliculos ita componere, ut faseoli [4] virides integri interponantur, quos ipsos ante dura muria die et nocte macerari 2 oportebit, similiterque assiccatos cum fasciculis lactucarum condi, et superfundi ius quod sit aceti duarum partium atque unius muriae: deinde arido spissamento faeniculi sic comprimi, ut ius supernatet. Quod ut fiat,[5] is [6] qui huic officio praeerit, saepe suffundere ius debebit, nec pati sitire salgama,[7] sed extrinsecus munda spongia vasa pertergere, et aqua fontana quam recentissima refrigerare. Simili ratione intubum et cacumina ruti,[8] qua lactucam condire oportet, nec minus thymi et satureiae et origani tum etiam armoraciorum cymam. Haec autem, quae supra scripta sunt, verno tempore componuntur.

X. Nunc quae per aestatem circa messem vel etiam exactis iam messibus colligi et reponi debeant, praecipiemus. Pompeianam, vel ascaloniam cepam,

[1] sallire *SAac*.
[2] muriam eluere *scripsi*: in muria eluere *ac*: in muriae luere *SA*.
[3] aridi dum *S*.
[4] passioli *SA* : fasioli *a* : faseoli *c*.
[5] ut fiat *SAac*.   [6] his *S*.
[7] nec pati sitire *ac* : neque patis nec tanere *SA*.
[8] cacumina ruti *scripsi*: cacumina rub *A* : cacumina rubis *ac* : cacuminarum *S*.

IX. Lettuce is preserved in the following manner. The stalks of the lettuce should be stripped off from the bottom upwards up to a point where the leaves appear to be tender, and salted in a basin and let stand a day and a night till they yield up the brine; then wash out the brine and after squeezing them spread them out on hurdles till they become dry; then strew underneath them dry dill and fennel and cut up a little rue and leek and mix it in; then, when the stalks are dry, you should so arrange them that entire green calavance may be placed between them, which themselves will have to be steeped for a day and a night in hard brine and, after being similarly dried, must be pickled with the bunches of lettuce, and a liquid, consisting of two parts of vinegar and one of brine, must be poured over them; then they must be pressed down with a dried plug of fennel in such a manner that the liquid floats above them. To do this effectively, the person charged with the task will have frequently to pour the liquid on to them and not allow the pickle to become dry, but he must wipe the outside of the vessels with a clean sponge and cool them with the freshest possible spring-water. Chickory and tops of rue should be pickled in the same way as lettuce, also the tops of thyme, savory and marjoram and likewise the sprouts of the wild radish. All the above preparations are made during the spring.

The preservation of lettuce.

X. We will now give instructions about things which ought to be collected and stored during the summer about the time of the harvest or even when it is over. Choose a Pompeian [a] or "undug" [b] or

The preservation of onions, apples, pears and other fruit

---

[a] Called after Pompeii, the city near Naples.
[b] Related to the Greek word ἄσκαλος = "not dug up".

LUCIUS JUNIUS MODERATUS COLUMELLA

vel etiam Marsicam simplicem, quam vocant unionem rustici, eligito: ea est autem, quae non fruti-
2 cavit, nec habuit soboles adhaerentes. Hanc prius in sole siccato, deinde sub umbra refrigeratam substrato thymo vel cunila componito in fidelia, et infuso iure, quod sit aceti trium partium et unius muriae, fasciculum cunilae superponito, ita ut cepa deprimatur: quae cum ius combiberit, simili mixtura vas suppleatur. Eodem tempore corna, et pruna onychina, et pruna silvestria, nec minus genera pirorum et malorum condiuntur.[1]
3 Corna, quibus pro olivis utamur, item pruna silvestria, et pruna onychina adhuc solida nec maturissima legenda sunt, nec tamen nimium cruda. Deinde uno die umbra siccanda: tum aequis partibus acetum et sapa[2] vel defrutum misceatur et infundatur. Oportebit autem aliquid salis adicere, ne vermiculus aliudve[3] animal innasci possit. Verum commodius servantur, si duae partes sapae cum aceti una parte misceantur.
4 Pira Dolabelliana, Crustumina, regia, veneria, volema, Naeviana, Lateritiana,[4] Decimiana, laurea, myrapia,[5] purpurea, cum immatura, non tamen percruda legeris, diligenter inspicito, ut sint integra

[1] conduntur *SAac*.
[2] sapam *SAac*.
[3] aliusve *S*.
[4] laterisiana *SAac*.
[5] myrapira pruna *S* : myraprima *A* : myrapia prima *a* : mirapia pruna *c*.

---

[a] The Marsi were a people of Latium.
[b] Dolabellian, Crustuminian, royal, Venus, warden, Naevian and purple pears are mentioned in Book V. 10. 18, where see notes.

BOOK XII. x. 1–4

even the plain Marsian [a] onion which the country folk call *unio*. This is the kind which has not sprouted or had offshoots attached to it. First dry it in the sun, then, after it has been cooled in the shade, arrange it in a pot with thyme or marjoram strewn underneath it and, after pouring in a liquid consisting of three parts of vinegar and one of brine, put a bunch of marjoram on the top, so that the onion may be pressed down. When it has absorbed the liquid, let the vessel be filled up with similar liquid. At the same season cornel-berries and onyx-coloured plums and wild plums and likewise various kinds of pears and apples are preserved.

Cornel-berries, which we use instead of olives, also wild plums and onyx-coloured plums should be picked while they are still hard and not very ripe; they must not, however, be too unripe. They should then be dried for a day in the shade; then vinegar and must boiled down to half or one-third of its original volume should be mixed and poured in, but it will be necessary to add some salt, so that no worms or other form of animal life can be engendered in them, but the better method of preservation is when two parts of must boiled down to half its original volume are mixed with one part of vinegar.

When you have gathered Dolabellian,[b] Crustumi- nian, royal and Venus pears, warden pears, Naevian, Lateritian,[c] Decimian,[d] bay-leaf pears and myrrh-pears and purple pears, before they are ripe but when they are no longer quite raw, examine them carefully

[c] Perhaps called after the Villa of Quintus Cicero at Arpinum.
[d] An adjective formed from the gentile name Decimius.

LUCIUS JUNIUS MODERATUS COLUMELLA

sine vitio aut vermiculo: tum in fictili picata fidelia componito, et aut passo aut defruto completo, ita ut omne pomum submersum sit, operculum deinde impositum gypsato. Illud in totum praecipiendum existimavi, nullum esse genus pomi, quod non possit melle servari. Itaque cum sit haec res interdum aegrotantibus salutaris, censeo vel pauca poma in melle custodiri,[1] sed separata generatim. Nam si commisceas, alterum ab altero genere corrumpitur. 6 Et quoniam opportune mellis fecimus mentionem, hoc eodem tempore alvi castrandae, ac mel conficiendum, cera facienda est: de quibus nono libro iam diximus: nec nunc aliam curam exigimus a villica, quam ut administrantibus intersit, fructumque custodiat.

XI. Ceterum cum eodem tempore mella nec minus aqua mulsa in vetustatem reponi debeat, meminisse oportebit, ut cum secundarium mel de favis fuerit exemptum, cerae statim minute resolvantur, et aqua fontana vel caelesti macerentur. Expressa deinde aqua coletur,[2] et in vas plumbeum defusa decoquatur, omnisque spurcitia[3] cum spumis eximatur: quae decocta, cum tam crassa fuerit quam defrutum, refrigeretur, et bene picatis lagoenis

---

[1] custodire *SAac*.
[2] coloretur *SA* : coleretur *a* : coletur *c*.
[3] spurcitiae *S*.

---

a Mella is a feminine substantive; it was a liquor made with fresh bees' wax which was steeped in water and then boiled; *mulsa* was a mixture of water and honey.

to see that they are sound and free from blemish or worms, and then arrange them in an earthenware vessel that has been treated with pitch and fill it with raisin-wine or must boiled down to one-third of its original volume, so that all the fruit is submerged; then put a cover on the top and plaster it up. I have thought it necessary to lay down the general principle that there is no kind of fruit which cannot be preserved in honey; therefore, since such food is sometimes salutary for those who are sick, I think that a few fruits at least should be preserved in honey but with the different kinds separately from one another; for if you mix them up together, one kind is spoilt by another. And since we have had an opportune occasion of mentioning honey, this same season is the time when the hives should be cleared of honey-combs and the honey prepared and the wax made—subjects with which we have dealt already in our ninth book; and we do not now demand any other service from the bailiff's wife than that she should be present with those who are carrying out these duties and take the fruit into her custody.

XI. But since bee's-wax-water [a] and likewise honey-water ought to be stored at the same time so as to become old, it will be necessary to remember that when the second yield of honey has been removed from the combs the wax should be immediately broken up into small pieces and steeped in spring-water or rain-water. Then the water should be squeezed out and strained and poured into a leaden vessel and boiled, and all impurities removed with the scum. When it has boiled down and is of the same consistency as must reduced to a third part of its volume, it should be allowed to cool and be

How to make bee's-wax-water for preserving fruit.

2 condatur. Hac quidam mella pro aqua mulsa utuntur, nonnulli etiam pro defruto in condituras olivarum; quibus quidem magis idoneam censeo, quia cibarium saporem habet, nec potest languentibus pro aqua mulsa remedio esse, cum si bibatur, inflationem stomachi et praecordiorum facit.[1] Itaque seposita ea et ad condituras destinata, per se facienda erit optimo[2] melle aqua mulsa.

XII. Haec autem non uno modo componitur. Nam quidam multos ante annos caelestem aquam vasis includunt, et sub dio in sole habent: deinde cum saepius eam in alia vasa[3] transfuderunt et eliquaverunt nam quoties aqua post longum tempus diffunditur, aliquod crassamentum in imo simile faeci reperitur veteris aquae sextarium cum libra mellis 2 miscent. Nonnulli tamen qui austeriorem volunt efficere gustum, sextarium aquae cum dodrante pondo mellis diluunt, et ea portione repletam lagoenam gypsatamque patiuntur per Caniculae ortum in sole quadraginta diebus esse; tum demum in tabulatum, quod fumum accipit, reponunt. 3 Nonnulli, quibus non fuit curae caelestem inveterare aquam, recentem sumunt, eamque usque in quartam partem decoquunt: deinde cum refrixerit, sive dulciorem mulsam[4] facere volunt, duobus aquae sextariis sextarium mellis permiscent: sive austeriorem,[5] sextario aquae dodrantem mellis adiciunt, et his portionibus factam in lagoenam diffundunt:

[1] faciat *S*. [2] optimo *ac* : -a *SA*.
[3] aliam vasam *S*. [4] mulseam *SAac*.
[5] austeriorem *ac* : -e *SA*.

BOOK XII. xi. 2–xii. 3

stored in flagons well daubed with pitch. Some people use this bee's-wax-water instead of honey-water, others use it instead of boiled-down must for preserving olives, for which purpose I regard it as more suitable, because it has a flavour which goes well with food; it cannot be used as a remedy instead of honey-water for those who are sick, because, if it be drunk, it causes inflation of the stomach and intestines. It is, therefore, set apart and kept for preserving, while honey-water will have to be made separately with the best honey.

XII. There is more than one way of making honey-water. Some people put up rain-water many years before in vessels and keep it out of doors in the sun; then, when they have poured it into several vessels one after another and clarified it, for however often water is poured from one vessel into another over a long period, thick matter resembling dregs is found at the bottom, they mix a *sextarius* of the stale water with a pound of honey. Some people, however, who wish to produce a harsher flavour, dilute a *sextarius* of water with a *dodrans* of honey and allow a flagon filled with a mixture in these proportions and covered with plaster to remain in the sun for forty days during the rising of the Dog-star; finally they store it in a loft where smoke can reach it. Others, who have not taken the trouble to let rain water become stale, take fresh water and boil it down to a quarter of its volume; then when it has cooled, if they wish to make rather sweet honey-water, they mix a *sextarius* of honey with two *sextarii* of water, or, if they wish to make it rather harsh, they add a *dodrans* of honey to a *sextarius* of water, and, having made it in these proportions, pour it into

How to make honey-water.

eamque, sicut supra dixi, quadraginta diebus insolatam postea in tabulatum, quod suffumigatur, reponunt.

XIII. Caseo usibus domesticis[1] praeparando hoc maxime idoneum tempus est, quod et caseus seri minimum remittit: et ultimo tempore, cum iam exiguum lactis est, non tam expedit operas morari ad forum[2] fructibus deferendis: et sane saepe deportati per[3] aestum acore vitiantur. Itaque praestat eos hoc ipso tempore in usum conficere. Id autem ut quam optime fiat opilionis officium est, cui septimo libro praecepta dedimus, quae sequi debeat.

2 Sunt etiam quaedam herbae, quas appropinquante vindemia condire possis, ut portulaca,[4] et holus cordum,[5] quod quidam sativam batim vocant. Hae herbae diligenter purgantur, et sub umbra expanduntur: deinde quarto die sal in fundis fideliarum substernitur, et separatim unaquaeque earum componitur, acetoque infuso iterum sal superponitur: nam his herbis muria non convenit.

XIV. Hoc eodem tempore, vel etiam primo mense Augusto, mala et pira dulcissimi saporis mediocriter matura eliguntur, et in duas aut tres partes harundine vel osseo cultello divisa in sole ponuntur, donec arescant. Eorum si est multitudo, non minimam partem cibariorum per hiemem rusticis vindicant.

---

[1] domesticis *ac* : om. *SA*.
[2] eorum *SAac*.
[3] propter *SAac*.
[4] portulacam *SAac*.
[5] olus cordeum *c* : holos cordum *SA* : holus cordium *a*.

---

[a] Book VII. Ch. 8.

a flagon. When the flagon has been placed in the sun for forty days, as I said above, they store it in a loft which receives smoke from below.

XIII. This is also the most suitable time for preparing cheese for domestic uses, because then the cheese gives out very little whey; and at the end of the season, when the yield of milk is small, it is not so expedient to take up the labourers' time in carrying fruit to the market, and, indeed, fruit which has been transported during the heat is often spoilt by going sour. It is better therefore that during this season they should only prepare fruit which is required for use. It is the duty, however, of the shepherd to make cheese in the best possible manner and we have given him directions in the seventh book [a] which he should follow.

There are also certain herbs which you could preserve when the vintage is approaching, such as purslane and the late pot-herb which some call garden *batis* (samphire). These herbs are carefully cleaned and spread out in the shade; then on the fourth day a layer of salt is spread underneath them at the bottom of the jars and each of the herbs is arranged separately; then another layer of salt is placed on the top after vinegar has been poured in, for brine does not suit these herbs.

XIV. At this same season, or even at the beginning of August, apples and pears of the sweetest flavour are picked when they are moderately ripe, and, after they have been cut into two or three parts with a reed or small bone knife, are placed in the sun until they become dry. If there is a large quantity of them they provide the country-folk with not the least part of their food during the winter; for they serve in-

LUCIUS JUNIUS MODERATUS COLUMELLA

Nam pro pulmentario cedit, sicuti ficus, quae cum arida seposita est hiemis temporibus rusticorum cibaria adiuvat.

XV. Ea porro neque nimium vieta neque immatura legi debet, et in eo loco expandi, qui toto die solem accipit. Pali autem quattuor pedibus inter se distantes figuntur, et perticis iugantur. Factae deinde in hunc usum cannae iugis superponuntur, ita ut duobus pedibus absint a terra, ne humorem, quem fere noctibus remittit humus, trahere possint: tunc ficus inicitur, et crates pastorales culmo vel carice vel filice textae ex utroque latere super terram planae disponuntur, ut cum sol in occasu fuerit, erigantur, et inter se acclines testudineato tecto more tuguriorum, viescentem[1] ficum a rore,[2] et interdum a pluvia defendant.[3] Nam utraque res praedictum fructum corrumpit. Cum deinde aruerit, in orcas bene picatas meridiano tepore calentem ficum condere et calcare diligenter oportebit, subiecto tamen arido faeniculo, et iterum repletis vasis superposito: quae vasa confestim operculare, et oblinire convenit, et in horreum siccissimum reponi, quo melius ficus perennet.

Quidam lectis ficis pediculos adimunt, et in sole eas expandunt: cum deinde paulum siccatae sunt, antequam indurescant, in labra fictilia vel lapidea congerunt eas: tum pedibus lautis in modum farinae proculcant, et admiscent torrefacta sesama cum

---
[1] viescentem *SA* : virescentem *ac*.
[2] arbore *A*.
[3] defendant *ac* : -at *SA*.

stead of a relish, as does the fig, which is dried and stored away and helps to feed the country-folk in time of winter.

XV. Figs ought not to be picked when they are either too much shrivelled up or unripe, and they should be spread out in a place which is in the sunshine all day long. Stakes are fixed in the ground four feet apart and joined with poles; then on the top of these frames are placed reed canes made for the purpose, so as to be two feet from the ground, that they cannot attract the moisture which the earth generally gives out at night. On these the figs are thrown, and shepherd's hurdles woven of straw, rushes or ferns are placed flat on the ground on both sides, so that, when the sun is setting, they may be erected and, leaning against one another, may form an arched roof, like that of a hut, and protect the shrivelling figs from the dew and at times from the rain, both of which spoil these fruits. When the figs are dry, they will have to be stored while still warm from the midday heat in wide vessels, which have been well daubed with pitch, and carefully trodden down, dried fennel, however, being strewed underneath and laid on the top of the vessels when they are full. It is well immediately to put the covers on these vessels and seal them up and place them in a very dry store-house, so that the figs may keep better.

Some people remove the pedicles from the figs after they have been picked and spread the figs in the sun; then, when they have become slightly dry and before they have begun to harden, they heap them in basins of earthenware or stone; then, after washing their feet, they tread them as they do meal and mix

The drying of figs.

LUCIUS JUNIUS MODERATUS COLUMELLA

aniso[1] Aegyptio et semine faeniculi et cymini.
4 Haec cum bene proculcaverint, et totam massam comminutae fici permiscuerint, modicas offas foliis ficulneis involvunt, ac religatas iunco vel qualibet herba offas reponunt in crates,[2] et patiuntur siccari: deinde cum peraruerint, picatis vasis eas condunt. Nonnulli hanc ipsam farinam fici orcis sine pice includunt, et oblita vasa clibano vel furno torrefaciunt, quo celerius omnis humor excoquatur: siccatam in tabulatum reponunt, et cum exegerit usus, testam comminuunt: nam duratam massam fici aliter eximere non possunt.

5 Alii pinguissimam quamque viridem ficorum eligunt, et arundine vel digitis divisam dilatant, atque ita in sole viescere patiuntur, quas deinde bene siccatas meridianis teporibus, cum calore solis emollitae sunt, colligunt, et, ut est mos Afris atque Hispanis, inter se compositas comprimunt in figuram stellarum flosculorumque, vel in formam panis redigentes: tum rursus in sole assiccant, et ita in vasis recondunt.

XVI. Similem curam uvae desiderant, quas dulcissimi saporis albas, maximis acinis, nec spissis, luna decrescente, sereno et sicco caelo post horam quintam legi oportet, et in tabulis paulisper porrigi, ne inter se pondere suo pressae collidantur: deinde aeno vel in olla nova fictili ampla praeparatam lixi-

---

[1] anesol aegyptio *S* : anesolegyptio *A* : aneso *ac*.
[2] crates *ac* : grates *SA*.

with them toasted sesame and Egyptian anise and the seed of fennel and cumin. When they have 4 trodden these thoroughly and mixed in the whole mass of crushed figs, they wrap up balls of the mixture of moderate size in fig-leaves and, after tying them up with rushes or some other vegetable matter, place them on hurdles and let them dry; then when they are thoroughly dried, they place them in vessels treated with pitch. Some people enclose this same fig-paste in vessels which are not covered with pitch and after smearing them heat them on an oven or furnace, so that the moisture may be more speedily cooked out of them. When the fig-paste is dry they store it in a loft and, when it is required for use, they break the earthenware vessel, since they cannot otherwise extract the hardened mass of fig. Other people pick out all the 5 richest figs while they are green and after splitting them with a reed or with the fingers spread them out and let them shrivel up in the sun; when they have been well dried by the midday warmth and softened by the heat of the sun, they collect them and, as is the custom among the Africans and Spaniards, arrange them together and press them and reduce them to the shapes of stars and small flowers or the form of a loaf of bread; they then dry them again in the sun and store them in vessels.

XVI. Grapes require similar care. White grapes of the sweetest flavour, with large berries which do not grow close together, should be picked after the fifth hour, when the moon is waning and the weather is clear and dry, and be spread for a short time on boards, so that they may not be pressed together by their own weight and become bruised. After this it is

On the making of raisins and service-apples.

LUCIUS JUNIUS MODERATUS COLUMELLA

viam cineris sarmenti calefieri convenit; quae cum fervebit, exiguum olei quam optimi adici, et ita permisceri: deinde uvas pro magnitudine binas, vel ternas inter se colligatas in aenum fervens demitti, et exiguum pati, dum decolorentur: nec rursus committere ut excoquantur: nam quadam modera-
2 tione temperamentoque opus est. Cum deinde exemeris, in crate disponito rarius quam ut altera alteram contingat. Post tres deinde horas unamquamque uvam convertito, nec in eodem vestigio reponito, ne in humore, qui defluxerit, corrumpatur: noctibus autem contegi debent quemadmodum fici, ut a rore vel pluvia tutae sint. Cum deinde modice aruerint, in vasa nova sine pice operculata et gypsata sicco loco reponito.
3 Quidam uvam passam foliis ficulneis involvunt et assiccant: alii foliis vitigineis nonnulli plataninis semivietas [1] uvas contegunt, et ita in amphoras recondunt. Sunt qui culmos fabae exurant, et ex eo, quod cremaverint, cineream lixiviam faciant, deinde in lixiviae sextarios decem salis tres cyathos [2] et olei cyathum adiciant, tum [3] adhibito igne calefaciant, et cetera eodem modo administrent. Quod si videbitur in aeno parum inesse olei, subinde quantum satis erit adiciatur, quo sit pinguior et nitidior uva passa.
4 Eodem tempore sorba manu lecta curiose in urceolos picatos adicito, et opercula picata imponito, et gypso

[1] semivietas *ac*: semivitas *SA*.
[2] quiathos . . . quiathum *S*.
[3] cum *S*.

BOOK XII. xvi. 1-4

a good plan to heat up lye prepared beforehand, made with cinders of brushwood in a brazen cauldron or a large new earthenware pot, and when it boils to add a little of the best possible oil and mix it in. Then two or three clusters of grapes according to their size should be tied together and plunged into the boiling cauldron and left there for a short time until they are discoloured, but you must not, on the other hand, give them time to be thoroughly boiled; for you must use a certain restraint and moderation. Then after you have taken them out, arrange them on a hurdle so far apart as not to touch one another. Next, after three hours turn each bunch over and do not replace them on the same spot lest they spoil by standing in the juice which has flowed from them. But at night they should be covered in the same way as figs, so that they may be safe from dew and rain. Next, when they are moderately dry, store them in a dry place in new vessels with covers and plastered, not treated with pitch.

Some people roll up raisins in fig-leaves and dry them; others cover the bunches of grapes, when they are half-shrivelled up, with vine- or plane-leaves and then store them in pitchers. Some people burn the stalk of beans and make cinder-lye with the burnt matter, then they add three *cyathi* of salt and a *cyathus* of oil to ten *sextarii* of lye and then apply fire and heat to the mixture, carrying out the rest of the process as before. But if it appears that there is not enough oil in the cauldron, let the necessary quantity be added in small amounts, so that the raisins may be richer and brighter in colour.

At the same season place service-apples carefully picked by hand in little pots treated with pitch and place covers also treated with pitch on the top and

linito, tum scrobibus bipedaneis sicco loco intra tectum factis, urceolos ita collocato, ut oblita ora eorum deorsum spectent: deinde terram congerito, et modice desuper[1] calcato. Melius est autem pluribus scrobibus pauciora vasa distantia inter se disponere. Nam in exemptione eorum dum unum tollis, si reliqua commoveris, celeriter sorba vitiantur. Quidam hoc idem pomum in passo, quidam etiam in defruto commode servant, adiecto spissamento aridi[2] faeniculi, quo deprimantur ita sorba, ut semper ius supernatet, ac nihilominus picata opercula diligenter gypso linunt, ne possit spiritus introire.

XVII. Sunt quaedam regiones, in quibus vini ideoque etiam aceti penuria est. Itaque hoc eodem tempore ficus viridis quam maturissima legenda est, utique si iam pluviae incesserunt, et propter imbres in terram decidit: quae cum sublecta est, in dolium vel in amphoras conditur, et ibi sinitur fermentari: deinde cum exacuit, et remisit liquorem, quicquid est aceti diligenter colatur, et in vasa picata bene olida diffunditur. Hoc primae notae acerrimi aceti usum praebet, nec unquam situm aut mucorem contrahit, si non humido loco positum est. Sunt qui multitudini studentes aquam[3] ficis permisceant; et subinde maturissimas ficus recentes adiciant, et patiantur in eo iure tabescere, donec satis acris aceti

---

[1] modice super *S*: mode super *A*: modico desuper *a*: mo ... desuper *c*.
[2] aridi *edd.*: spissi *SAac*.
[3] aquam *a*: quam *SAc*.

daub them with plaster; then make trenches two feet deep in a dry place under cover and place the pots in them so that their mouths, which have been plastered up, face downwards; then pile earth on the top and tread it down slightly from above. It is better to place fewer vessels in several trenches at a distance from one another; for if, in removing one of the vessels when you are taking them out, you disturb the rest, the service-apples quickly go bad. Some people preserve this same fruit conveniently 5 in raisin-wine and others in boiled-down must, putting on the top a plug of dry fennel, by which the service-apples may be pressed down, so that the liquid always covers them; notwithstanding, they carefully smear with plaster the covers which have been treated with pitch to prevent the air from entering.

XVII. There are some regions in which wine, and therefore vinegar also, is lacking. So in the same season of the year green figs in as ripe a state as possible must be picked, especially if the rains have already come on and the figs have fallen to the ground owing to the showers. When they have been gathered, they are stored in jars or pitchers and allowed to ferment there. When they have become sour and have yielded up their moisture, whatever vinegar there is, is carefully strained and poured into sweet-smelling vessels treated with pitch. This liquid takes the place of very sharp vinegar of the first quality and is never affected by decay or mustiness as long as it is not kept in a damp place. Some 2 people, who prefer quantity to quality, mix water with the figs and from time to time add very ripe fresh figs and allow them to dissolve in the liquid until the flavour of sufficiently sharp vinegar results; after

On the making of vinegar from figs.

sapor fiat: postea in iunceis fiscellis vel sparteis saccis percolant, liquatumque acetum infervefaciunt, dum spumam[1] et omnem spurcitiam eximant: tum torridi salis aliquid adiciunt, quae res prohibet vermiculos aliave innasci animalia.

XVIII. Quamvis priore libro, qui inscribitur Villicus, iam diximus quae ad vindemiam praeparanda sunt,[2] non tamen alienum est etiam villicae de iisdem rebus praecipere, ut intellegat suae curae esse debere, quaecunque sub tecto administrantur 2 circa vindemiam. Si ager amplus, aut vineta aut arbusta grandia sunt, perenne fabricandae decemmodiae[3] et trimodiae et fiscellae texendae et picandae: nec minus falculae, et ungues ferrei quam plurimi parandi et exacuendi sunt, ne vindemitor manu destringat uvas, et non minima fructus portio 3 dispersis acinis in terram dilabatur. Funiculi quoque fiscellis aptandi sunt, et lora trimodiis: tum lacus vinarii et torcularii et fora omniaque vasa, si vicinum est mare, aqua marina, si minus, dulci eluenda sunt, et communanda, et diligenter assiccanda, ne humorem habeant.[4] Cella quoque vinaria omni stercore liberanda, et bonis odoribus suffienda, ne 4 quem redoleat foetorem acoremve. Tum sacrificia Libero Liberaeque et vasis pressoriis[5] quam sanctis-

[1] spumam c: spumen SAa.
[2] sunt c: sint SAa.
[3] decēmodia S: decemodia A: decemo die a: decimo die c.
[4] habeant c: habeat SAa.
[5] pressoriis c: pressoris SAa.

---

[a] Apparently the part of the wine-press in which the grapes were laid (Varro, R.R. I. 54. 2).
[b] Old Italian deities afterwards identified with Bacchus and Proserpina.

this they let it percolate through small rush baskets or sacking made from broom and boil the clarified vinegar until they get rid of the scum and every kind of impurity. They then add a little grilled salt which prevents the production of worms or other animals in it.

XVIII. Although in the preceding book, entitled the Bailiff, we have already spoken of the preparations which must be made for the vintage, it is, nevertheless, not irrelevant to give instructions to the bailiff's wife on the same subject, so that she may understand that whatever operations in connexion with the vintage are carried out in the house ought to come within her province. If the farm is large and the vineyards and plantations extensive, it is necessary continually to make vessels containing ten and three *modii* and to weave small baskets and treat them with pitch; also as many small sickles and iron hooks as possible must be procured and sharpened, so that the vintager may not strip off the bunches of grapes with his hand, which causes no small part of the fruit to fall to the ground and the grapes to be scattered. Cords too must be attached to the small baskets and thongs to those containing three *modii*. Next vats to hold the wine and others connected with the wine-presses and grape-tanks [a] and all the vessels must be washed out with sea-water, if the sea is nearby, or, failing that, with fresh water, and must be thoroughly cleaned and carefully dried, that no moisture remains in them. The wine-cellar must be cleansed of all filth and fumigated with pleasant odours, that it may not smell at all mouldy or sour. Next sacrifices must be offered in the greatest piety and purity to Liber and Libera [b] and to the vessels of the wine-

Of preparations to be made for the vintage.

LUCIUS JUNIUS MODERATUS COLUMELLA

sime castissimeque facienda: nec per vindemiam ab torculari aut vinaria cella recedendum est, ut omnia, qui mustum conficiunt, pure mundeque faciant: nec furi locus detur partem fructuum intercipiendi.

5 Dolia quoque et seriae ceteraque vasa ante quadragesimum vindemiae diem picanda sunt, atque aliter ea quae demersa sunt humi, aliter quae stant supra terram. Nam ea quae demersa sunt, ferreis lampadibus ardentibus calefiunt, et cum pix [1] in fundum destillavit,[2] sublata lampade, rutabulo ligneo et ferrea curvata radula [3] ducitur, quod destillavit,[4] aut quod in lateribus haesit: deinde penicillo detergitur, et ferventissima pice infusa novo alio rutabulo 6 et scopula picatur. At quae supra terram consistunt, complures dies antequam [5] curentur in solem producuntur. Deinde cum satis insolata sunt, in labra convertuntur, et subiectis parvis tribus lapidibus suspenduntur, atque ita ignis subicitur, et tamdiu incenditur, donec ad fundum calor tam vehemens perveniat, ut apposita manus patiens eius non sit: tum dolio in terram demisso, et in latus deposito, pix ferventissima infunditur, volutaturque, ut omnes 7 dolii partes linantur. Sed haec die quieto a ventis fieri debent, ne admoto igne cum afflaverit ventus vasa rumpantur. Sunt autem satis [6] sesquiculle-

---

[1] picem S : pixim Ac : paxim a.
[2] destinavit ac : destinabit SA.
[3] radula ac : radule SA.
[4] destillavit ac : -abit SA.
[5] antequam $edd.$: qua S : quam A : q̃ c : om. a.
[6] satis autem SAa : sunt autem c.

press, and during the vintage neither the wine-press nor the wine-cellar must ever be deserted so as to ensure that those who are making the new wine may do everything in a pure and clean manner and that no opportunity may be offered to a thief to carry off part of the fruit.

The barrels too and the jars and the other vessels should be treated with pitch forty days before the vintage, a different process being applied to those which are sunk into the earth and those which stand above ground. For those which are sunk into the earth are heated with burning iron torches, and when the pitch has dripped down into the bottom of the vessel, the torch is removed, and what has dripped to the bottom or stuck to the sides is spread with a wooden ladle and a curved iron scraper. The vessel is then wiped with a brush, and very hot pitch having been poured in, is covered with pitch by means of another new ladle and with a small broom. But vessels which stand above ground are put out in the sun for several days before they are treated; then, when they have been sufficiently exposed to the sun, they are turned with their openings downwards and raised from the ground by the placing of three small stones underneath them; then a fire is placed underneath and allowed to burn until so strong a heat reaches the bottom that a hand placed there cannot endure it. Then the vessel is let down on to the ground and laid on its side, and very hot pitch is poured into it and it is rolled round and round that every part of it is coated with pitch. This operation, however, ought to be carried out on a day which is quiet and windless, so that the vessels may not burst because the fire has been applied to them in a blowing wind.

LUCIUS JUNIUS MODERATUS COLUMELLA

aribus doliis picis durae pondo vicenaquina. Nec dubium, quin si[1] quinta pars picis Brutiae in universam cocturam adiciatur, utilissimum sit omni vindemiae.

XIX. Cura quoque adhibenda est, ut expressum mustum perenne sit, aut certe usque ad venditionem durabile. Quod quemadmodum fieri debeat, et quibus condituris adiuvari, deinceps subiciemus. Quidam partem quartam eius musti, quod in vasa plumbea coniecerunt, nonnulli tertiam decoquunt. Nec dubium, quin ad dimidium si quis excoxerit, meliorem sapam facturus sit, eoque usibus utiliorem, adeo quidem, ut etiam vice defruti sapae, mustum, quod est ex veteribus vineis, condire possit.

2 Quaecunque vini nota sine condimento valet perennari,[2] optimam esse eam censemus, nec omnino quidquam permiscendum, quo naturalis sapor eius infuscetur.[3] Id enim praestantissimum est, quod suapte natura placere potuerit. Ceterum cum aut regionis vitio, aut novellarum vinearum mustum laborabit, eligenda erit pars vineae, si est facultas, Amineae, si minus, quam bellissimi vini, quaeque 3 erit et vetustissima et minime uliginosa. Tum observabimus decrescentem lunam, cum est sub terra, et sereno siccoque die uvas quam maturissimas legemus,[4] quibus proculcatis mustum quod defluxerit, ante quam prelo pes eximatur, satis de

---

[1] sic *SA* : sit *ac*. [2] perennare *S*.
[3] infucetur *S*. [4] legimus *SAac*.

---

[a] The Bruttians lived in the extreme south of Italy.
[b] Aminaea was in the territory of the Piceni and produced the best wine; see note on Book III. 2. 27.

Twenty-five pounds of hard pitch is enough for vessels containing a *culeus* and a half. No doubt if a fifth part be added of Bruttian [a] pitch to the whole quantity boiled, it would be of great advantage to every vintage.

XIX. Care should also be taken so that the must, when it has been pressed out, may last well or at any rate keep until it is sold. We will then next set forth how this ought to be brought about and by what preservatives the process should be aided. Some people put the must in leaden vessels and by boiling reduce it by a quarter, others by a third. There is no doubt that anyone who boiled it down to one-half would be likely to make a better thick form of must and therefore more profitable for use, so much so that it can actually be used, instead of must boiled down to one-third, to preserve the must produced from old vineyards.

How to preserve and strengthen wine.

We regard as the best wine any kind which can keep without any preservative, nor should anything at all be mixed with it by which its natural savour would be obscured; for that wine is most excellent which has given pleasure by its own natural quality. But when the must labours under a defect due to the district which produced it or to the newness of the vineyard, we must choose part of a vineyard of Aminaean [b] grapes, if we have the opportunity of doing so, or, if not, of one which produces the finest wine and which is very old and the least marshy. We shall then watch for the waning of the moon and the time when it is under the earth, and on a calm, dry day, we shall pick the ripest possible grapes, and, when they have been trodden, we shall carry from the vat to the boiling-vessels as much as we require of the

LUCIUS JUNIUS MODERATUS COLUMELLA

lacu in vasa defrutaria deferemus,[1] lenique primum igne et tenuibus[2] admodum lignis, quae cremia[3] rustici appellant, fornacem incendemus, ut ex com-
4 modo mustum ferveat. Isque qui praeerit huic decoquendo, cola iuncea vel spartea sed crudo, id est non malleato sparto praeparata habeat: itemque fasciculos faeniculi fustibus illigatos, quos possit usque ad fundum vasorum demittere, ut quidquid faecis subsederit, exagitet et in summum reducat: tum colis omnem spurcitiam, quae redundarit, expurget. Nec absistat id facere, donec videbitur eliquatum omni faece mustum carere. Tum sive mala cydonia, quae percocta sublaturus sit, seu quoscunque voluerit convenientes odores adiciat, et nihilo minus subinde faeniculo peragitet, nequid
5 subsederit, quod possit plumbeum perforare. Cum deinde iam acriorem potuerit ignem vas sustinere, id est, cum aliqua iam parte mustum excoctum in se fervebit, tum codices et vastiora ligna subiciantur, sed ita ne fundum contingant. Quod nisi vitatum fuerit, saepe vas ipsum pertundetur; vel si id factum non erit, utique aduretur mustum, et amaritudine
6 concepta condituris fiet inutile. Oportebit autem antequam mustum in vasa defrutaria coniciatur, oleo bono plumbea intrinsecus imbui, et bene fricari,

---

[1] conferemus *S*: conferamus *Aa*: deferamus *c*.
[2] tenuimus *S*.
[3] gremia *SAac*.

must which has flowed from them before the pedicles of the grapes are removed from the wine-press, and we shall heat the furnace at first with a gentle fire and with only very small pieces of wood, which the country people call *cremia* (brushwood), so that the must may boil in a leisurely manner. The man in charge of this boiling should have ready prepared strainers made of rushes or broom, but the latter should be in a raw state, that is to say, not beaten with a hammer. He should also have bundles of fennel attached to the ends of sticks which he can let down right to the bottom of the vessels, so that he can stir up any dregs which have settled at the bottom and bring them up to the top; he should then clear away with the strainers any scum which remains on the surface, and he should go on doing this until the must seems cleared of all lees. Then he should add either some quinces, which he will remove when they are thoroughly boiled, or any other suitable scents which he likes, continuing nonetheless to stir the liquid from time to time with the fennel to prevent anything from sinking to the bottom which might perforate the leaden vessel. Next, when the vessel can stand a fiercer fire, that is, when the must, being partly boiled away, is in a state of internal seething, stems of trees and larger pieces of wood should be put underneath, without, however, actually touching the bottom; for unless this contact is avoided, the vessel itself will not infrequently be pierced, or, if this does not happen, the must will certainly be burnt, and having acquired a bitter taste will be rendered useless as a preservative. But, before the must is poured into the boiling-vessels, it will be well that those which are made of lead should be coated inside with

LUCIUS JUNIUS MODERATUS COLUMELLA

atque ita mustum adici. Ea res non patitur[1] defrutum aduri.

XX. Quinetiam diligenter factum defrutum, sicut vinum, solet acescere: quod cum ita sit, meminerimus anniculo[2] defruto, cuius iam bonitas explorata est, vinum condire. Nam vitioso[3] medicamine[4] fructus, qui perceptus est, vitiatur. Ipsa autem vasa, quibus sapa aut defrutum coquitur, plumbea potius quam aenea esse debent. Nam in coctura aeruginem remittunt aenea, et medicaminis saporem vitiant. Odores autem vino fere apti sunt, qui cum defruto coquuntur, iris, faenum Graecum, schoenum: harum rerum singulae librae in defrutarium, quod ceperit musti amphoras nonaginta, cum iam deferbuerit, et expurgatum erit, tum adici debent. Deinde si natura tenue mustum erit, cum ad tertiam partem fuerit decoctum, ignis subtrahendus est, et fornax protinus aqua refriger anda. Quod etiam si fecerimus, nihilo minus defrutum infra tertiam partem vasis[5] considit sed id quamvis aliquid detrimenti habeat, prodest tamen: nam quanto plus decoquitur (si modo non est adustum) melius et spissius fit. Ex hoc autem defruto, quod sic erit coctum, satis est singulos sextarios singulis amphoris immisceri. Cum amphoras musti nonaginta in defrutario decoxeris, ita ut iam exiguum supersit de toto[6] (quod[7] significat decoctum ad tertias): tum demum

[1] patiatur *S*.
[2] anniculos *S*.
[3] vitiosum *S*.
[4] medicamento tunc *S*: medicamentum tunc *Aa*: medicamentũ tũ *c*.
[5] vasi *SAac*.
[6] toto *scripsi*: tota *SAa*: decocta *c*.
[7] quae *SA*: que *ac*.

BOOK XII. xix. 6–xx. 3

good oil and be well rubbed, and that then the must should be put in. This prevents the boiled-down must from being burnt.

XX. Furthermore, boiled-down must, though carefully made, is, like wine, apt to go sour. This being so, let us be mindful to preserve our wine with boiled-down must of a year old, the soundness of which has been already tested; for the fruit which has been gathered in is corrupted by bad methods of preservation. The vessels themselves in which the thickened and boiled-down must is boiled should be of lead rather than of brass; for, in the boiling, brazen vessels throw off copper-rust and spoil the flavour of the preservative. The odours boiled with the must which are generally speaking suitable for wine are iris, fenugreek and sweet-rush; a pound of each of them ought to be put in the boiling-cauldron, which has received ninety *amphorae* of must, when it has just gone off the boil and has been cleared of scum. Then if the must is naturally thin, when it has been boiled down to a third of its original quantity, the fire must be removed from below it and the furnace immediately cooled with water; even if we have done this, the boiled-down must nevertheless sinks to a level lower than the third of the vessel. But, although this is some disadvantage, it is nevertheless beneficial; for the more the must is boiled down, —provided it be not burnt—the better and the thicker it becomes. Of this boiled-down must, when it has been thus treated, it is enough if one *sextarius* is mixed with one *amphora* of wine. When you have boiled ninety *amphorae* of must in the boiling-cauldron to such an extent that only a little of the whole remains (which means that it has been boiled down to a third),

LUCIUS JUNIUS MODERATUS COLUMELLA

medicamina adicito, quae sunt aut liquida, aut resinosa, id est, picis liquidae Nemeturicae, cum eam diligenter ante aqua marina decocta perlueris, decem sextarios: item resinae terebinthinae sesquilibram. Haec cum [1] adicies, plumbeum peragitabis, ne adurantur. Cum deinde ad tertias subsederit coctura, subtrahe ignem, et plumbeum subinde agitabis, ut defrutum et medicamenta coeant: deinde cum videbitur mediocriter calere defrutum, reliqua aromata contusa et cribrata paulatim insperges, et [2] iubebis rutabulo ligneo agitari quod decoxeris, eousque dum [3] defrigescat. Quod si non ita ut praecepimus, permiscueris, subsident aromata, et adurentur. Ad praedictum autem modum musti adici debent ii odores, nardi folium, iris Illyrica, nardum Gallicum, costum, palma, cyperum, schoenum, quorum singulorum selibrae satisfacient: item myrrhae [4] quincunx, calami pondo libram, casiae selibram, amomi pondo quadrans, croci quincunx, cripae pampinaceae [5] libram. Haec, ut dixi, arida contusa et cribrata debent adici, et his commisceri rasis, quod est genus crudae picis: eaque quanto est vetustior, tanto melior habetur. Nam longo tempore durior facta, cum est contusa, in pulverem redigitur, et his medicaminibus admiscetur. Satis est autem praedictis ponderibus sex libras eius

[1] cum *om. SA.*   [2] insparge sed *S.*
[3] cum *SAac.*   [4] murram *SAac.*
[5] eris praepanpinaceae *SA*: eripe *ac*: sertae Campanicae *Barbarus.*

---

[a] The Nemeturii were a people living in the Ligurian Alps.
[b] *Valeriana celtica.*
[c] An Indian aromatic plant, *Saussurea lappa* (Pliny, *N.H.* XII. § 41).
[d] This plant has not been identified. The text is uncertain.

BOOK XII. xx. 3-6

then, and not before, add the preservatives which are either liquid or resinous, namely, ten *sextarii* of liquid Nemeturican [a] pitch, after you have carefully washed it with boiled sea-water, and also a pound and a half of turpentine resin. When you will add these things, you will stir the leaden vessel thoroughly, so that they may not be burnt. Then when the boiling liquid has sunk to a third of its original quantity, withdraw the fire and stir the leaden vessel from time to time, so that the boiled-down must and the preservatives may mingle together. Then when the boiled-down must seems to be moderately hot, you will gradually sprinkle into it the rest of the spices after they have been bruised and sifted, and you will give orders that what you have boiled down is to be stirred with a wooden ladle until it begins to cool. If you do not mix them as we have directed, the spices will sink to the bottom and be burnt. To the aforesaid quantity of must the following spices ought to be added: the leaf of spikenard, the Illyrian iris, the French spikenard,[b] the costus,[c] the date, the angular rush and the sweet-rush, of every one of which half-a-pound will suffice; also a *quincunx* of myrrh, a pound of sweet reed, half-a-pound of cassia, a *quadrans* of cardamom, a *quincunx* of saffron, and a pound of vine-leafed *cripa*.[d] These, as I have said, ought to be added after having been pounded when dry and sifted, and *rasis*, a kind of crude pitch, should be mixed with them, which is considered to be better the older it is; for in the course of time it grows harder and, when it is pounded, it is reduced to powder and mixes with these preservatives. But it is enough for six pounds of it to be mixed with the quantities already mentioned. It is uncertain how

misceri. Ex hac compositione, quantum in sextarios musti quadragenos octonos adiciendum sit, incertum est, quoniam pro natura vini[1] aestimari oportet, quod satis sit: cavendumque est, ne conditus sapor 7 intelligatur. Nam ea res emptorem fugat. Ego tamen, si humida fuerit vindemia, trientem; si sicca, quadrantem medicaminis in binas amphoras miscere solitus sum ita, ut quattuor urnarum esset musti modus, urna autem quattuor et viginti sextariorum. Nonnullos agricolas singulis amphoris quadrantem medicaminis indidisse scio, sed hoc coactos fecisse propter nimiam infirmitatem vini eiusmodi, quod 8 vix triginta diebus integrum permanebat. Hoc tamen mustum, si sit lignorum copia, satius est infervefacere, et omnem spumam cum faecibus expurgare: quo facto decima pars decedet, sed reliqua perennis est. At si lignorum penuria est, marmoris vel gypsi, quod flos appellatur, uncias singulas, item ad tertias decocti defruti sextarios binos singulis amphoris miscere oportebit. Ea res etiamsi non in totum perennem, at certe usque in alteram vindemiam plerumque vini saporem servat.

XXI. Mustum quam dulcissimi saporis decoquetur ad tertias, et decoctum, sicut supra dixi, defrutum vocatur; quod cum defrixit, transfertur in vasa, et reponitur, ut post annum sit in usu. Potest tamen etiam post dies novem, quam refrixerit, adici in vinum: sed melius est, si anno requieverit. Eius defruti[2] sextarius in duas urnas musti adicitur, si mustum ex vineis collinis est: sed si ex campestribus,

[1] vinea *S*.   [2] defruti *Aa*: defrutis *S*: fruti *c*.

much of this preparation ought to be added to forty-eight *sextarii* of must, because the calculation of the right amount must be based on the quality of the wine, and care must be taken that the flavour of the preservative is not noticeable, for that drives away the purchaser. I personally, if the vintage is wet, usually mix a *triens* of the preservative in two *amphorae*; if it is dry, a *quadrans*, so that the quantity of must is four *urnae*, an *urna* being twenty-four *sextarii*. I am aware that some husbandmen have put a *quadrans* of the preservative in each *amphora*, but that they were obliged to do so owing to the excessive weakness of the wine, which scarcely kept sound for thirty days. However, if there is plenty of wood, it is better to boil the must and clear off all the scum with the dregs; if this is done a tenth part will be lost, but the rest keeps good forever. But if there is a scarcity of wood, it will be necessary to mix an ounce of what is called the "flower" of marble or plaster and also two *sextarii* of must boiled down to a third of its original quantity with each *amphora*. This, though it does not make the flavour of the wine last forever, yet at any rate generally preserves it until another vintage.

XXI. Must of the sweetest possible flavour will be boiled down to a third of its original volume and when boiled down, as I have said above, is called *defrutum*. When it has cooled down, it is transferred to vessels and put in store that use may be made of it after a year. But it can also be added to wine nine days after it has cooled; but it is better if it has remained undisturbed for a year. A *sextarius* of this boiled-down must is added to two *urnae* of must if the latter comes from vineyards in the hills; but if it comes

The use of boiled-down must as a preservative of wine.

LUCIUS JUNIUS MODERATUS COLUMELLA

2 tres heminae adiciuntur. Patimur autem, cum de lacu mustum [1] sublatum est, biduo defervescere, et purgari. Tertio die defrutum adicimus. Deinde interposito biduo, cum id mustum pariter cum defruto deferbuerit, purgatur, et ita eo adicitur in binas urnas ligula cumulata, vel mensura semunciae bene plenae salis cocti et triti. Sal autem quam candidissimus conicitur in urceo fictili sine pice, qui urceus, cum recipit salem, diligenter totus oblinitur luto paleato, et ita igni admovetur, ac tamdiu torretur, quamdiu strepitum edit. Cum silere coepit, finem habet cocturae.

3 Praeterea faenum Graecum maceratur in vino vetere per triduum: deinde eximitur, et in furno siccatur vel in sole: idque cum est aridum factum, molitur, et ex eo molito post salituram musti cochlear cumulatum, vel simile genus poculi eius [2], quod [3] est quarta pars cyathi, adicitur in binas urnas. Deinde cum iam perfecte mustum deferbuit et constitit, tantundem gypsi floris miscemus, quantum salis adieceramus: atque ita postero die purgamus dolium, et nutritum vinum operimus atque oblinimus.

4 Hac conditura Columella patruus meus, illustris agricola, uti solitus est in iis fundis in quibus palustres vineas habebat. Sed idem, cum collina vina [4] condiebat, aquam salsam decoctam ad tertias pro

---

[1] mustus *S*.
[2] puius *S* : huius *Aa* : eius *c*.
[3] quod *scripsi* : qui *SAac*.
[4] vina *om. S*.

---

[a] Marcus Columella (see also Ch. XLIII), who farmed in the Baetic province of Spain and with whom much of the author's youth seems to have been spent.

from vineyards in the plains, three *heminae* are added. When the must has been removed from the vat, we allow it to cool off for two days and to become clear. On the third day we add the boiled-down must; then, after an interval of two days, when the must together with the boiled-down must has finished fermenting, it is purified, and a heaped spoonful or a half-ounce measure generously filled with roasted and pounded salt should be added to each two *urnae* of must. For this purpose the whitest possible salt is thrown into an earthenware ewer which has not been treated with pitch; this, when it receives the salt, is carefully smeared all over with clay mixed with chaff and then put on the fire, and allowed to roast as long as it continues to crackle; when it begins to be silent, the process of cooking is complete.

Moreover, fenugreek is steeped in old wine for three days and then taken out and dried in an oven or in the sun, and when it is dry it is ground up; and of this, when it is ground up, after the must has been salted, a heaped spoonful or a similar kind of drinking-vessel, which holds a fourth part of a *cyathus*, is added to two *urnae* of it. Then, when the must has quite left off fermenting and has been standing for some time, we mix in the same amount of the flower of gypsum as we had already added of salt; then, on the next day, we cleanse the wine-jar and cover up the wine which has been preserved and seal it up.

My paternal uncle Columella,[a] the distinguished agriculturist, used to employ this method of preservation on the farms where he had vineyards on marshy ground; but, when he was preserving wines from vineyards on the hills, he added salt water boiled

sale adiciebat. Ea porro facit sine dubio maiorem mensuram et odoris melioris: sed periculum habet, ne vitietur vinum, sı male cocta sit aqua. Sumitur autem haec, ut iam dixeram, quam longissime a littore. Nam liquidior et purior[1] est, quantum altiore mari hausta est. Eam si quis ut Columella faciebat reponat, et post triennium in alia vasa eliquatam transfundat: deinde post alterum triennium decoquat usque ad partem tertiam: longe meliorem habebit condituram vini, nec ullum periculum erit, ne vina vitientur. Satis est autem sextarios singulos adicere salsae aquae in binas musti urnas: quamvis multi etiam binos immisceant, nonnulli etiam ternos sextarios: idque ego facere non recusem, si genus vini tantum valeat, ut aquae salsae non intellegatur sapor, Itaque diligens paterfamiliae cum paraverit fundum, statim prima[2] vindemia tres aut quattuor notas condiurae totidem amphoris musti experietur, ut exploratum habeat, quantum plurimum salsae vinum, quod fecerit, sine offensa gustus pati possit.

XXII. Picus liquidae alterum medicamen, quo mustum condias. Picis liquidae Nemeturicae metretam adde in labrum aut in alveum, et in eodem infundito cineris lixiviae[3] congios duos, deinde permisceto spatha lignea. Cum requieverit, eliquato lixiviam: deinde iterum tantundem lixiviae addito, eodem pacto permisceto, et eliquato. Tertio quoque idem facito. Cinis autem odorem picis aufert, et

---

[1] turpior *S.*   [2] primum *S.*
[3] lexivae *SAa*: lixiviae *c.*

down to one-third of its original volume instead of salt. This, moreover, certainly increases the amount of wine and improves its bouquet, but it is attended by the risk that the wine may be spoilt if the water is not properly boiled. This water, as I have already said, is taken as far as possible from the shore; for the greater the distance in the open sea from which it is drawn, the clearer and the purer it is. If one stores it, as Columella did, and after three years strains it and transfers it to other vessels and then after another three years boils it down to one-third of its volume, he will have a much better preservative for his wine and there will be no risk of his wines being spoilt. But it is enough to add one *sextarius* of salt water to two *urnae* of must, although many people indeed mix in two and some even three *sextarii*. I should have no objection to doing this if the type of wine had such strength that the taste of the salt water would not be noticed. Therefore a careful landed proprietor, when he has secured an estate, will at the first vintage immediately make trial of three or four different kinds of preservative in as many different jars, so that he may discover how much salt water the wine which he has made can stand without spoiling the taste.

XXII. There is another method of preserving must by means of liquid pitch. Put a *metreta* of liquid Nemeturican [a] pitch in a tub or trough and pour into the same two *congii* of lye-ash and then mix them with a wooden spatula. When the mixture has settled, strain off the lye; then put in again the same quantity of lye and mix and strain in the same manner; then repeat the process for a third time. The ashes take away the smell of the pitch and wash away impurities.

How to preserve wine with liquid pitch.

[a] See note on Ch. XX. 3.

LUCIUS JUNIUS MODERATUS COLUMELLA

2 eluit spurcitiam. Post eodem addito picis Bruttiae, si minus, alterius notae quam purissimae quinque libras. Haec minute concidito, et admisceto pici Nemeturicae. Tum aquae marinae quam vetustissimae, si erit; si minus, ad tertiam partem recentis aquae marinae decoctae congios duos inicito. Apertum labrum sinito in sole per Caniculae ortum, et spatha lignea permisceto quam saepissime, usque eo, dum ea quae addideris, in pice colliquescant, et unitas fiat. Noctibus autem labrum operire con-
3 veniet, ne irroretur. Deinde cum aqua marina, quam addideris, sole consumpta videbitur, sub tectum vas totum ferre curabis. Huius autem medicaminis quidam pondo quadrantem in sextarios XLVIII miscere soliti sunt, et hac conditura contenti esse. Alii cyathos tres eius medicamenti adiciunt in totidem sextarios, quot [1] supra diximus.

XXIII. Pix corticata appellatur, qua utuntur ad condituras Allobroges. Ea sic conficitur, ut dura sit, et quanto facta est vetustior, eo melior [2] in usu est. Nam omni lentore misso, facilius in pulverem resolvitur atque cribratur. Hanc ergo conteri et cribrari oportet: deinde cum bis mustum deferbuerit, quod plerumque est intra quartum diem, quam de lacu sublatum est, diligenter manibus expurgatur, et tunc demum praedictae picis sextans et semuncia in sextarios quinque et quinquaginta adicitur, et rutabulo ligneo permiscetur, nec postea

[1] quod *Sa* : quot *A* : quot *c*.
[2] melior *a* : mellior *A* : mollior *S*.

---

[a] See note on Ch. XVIII. 7.

BOOK XII. xxii. 2–xxiii. 1

After that add five pounds of Bruttian [a] pitch, or, 2 failing this, some other brand which is as pure as possible. Cut it up into small pieces and mix it with the Nemeturican pitch. Next pour in two *congii* of very old sea-water, if you have it, but, if not, of fresh sea-water boiled down to one-third of its original volume. Leave the tub uncovered in the sun during the rising of the Dog-star and mix it as often as possible with a wooden spatula until the ingredients which you have added dissolve in the pitch and become one with it. You will do well, however, to cover the tub at night, so that the dew may not fall into it. Then 3 when the sea-water which you have put in seems to have been consumed by the sun, you will be sure to put the vessel completely under cover. Some people have been in the habit of mixing a quarter of a pound of this preservative in forty-eight *sextarii* of wine and to content themselves with this method of preservation. Others add three *cyathi* of this preservative to the number of *sextarii* mentioned above.

XXIII. The pitch which the Allobroges [b] use for preserving is called "bark-pitch." It is made so as to be hard, and the older it has become, the better it is for use; for having lost all its toughness, it is more easily reduced to powder and sifted. It must therefore be bruised and sifted; then, when the must has ceased fermenting for the second time, which is generally before the end of the fourth day after it was taken from the vat, it is carefully cleansed by hand and then, and not until then, a *sextarius* and half an ounce of the said pitch are added to fifty-five *sextarii* and thoroughly mixed in with a wooden ladle and not

Of the pitch used by the Allobroges for preserving wine.

[b] A people inhabiting Gallia Narbonensis, the modern Savoy.

LUCIUS JUNIUS MODERATUS COLUMELLA

tangitur, dum confervescat: quod tamen non amplius diebus quattuordecim a conditura patiendum 2 est. Nam oportebit post hunc numerum dierum confestim vinum emundare, et si quid faecis aut labris vasorum aut lateribus inhaesit, eradi, ac suffricari, et protinus operculis impositis oblini.

At si ex eadem pice totam vindemiam condire volueris, ita ne gustus picati vini possit intellegi, sat erit eiusdem picis sex scripula in sextarios quinque et quadraginta tum demum miscere, cum mustum de- 3 ferbuerit, et faeces expurgatae fuerint. Oportebit autem salis decocti contritique semunciam in eundem modum musti adicere. Nec solum huic notae vini sal adhibendus est; verum, si fieri possit, in omnibus regionibus omne genus vindemiae hoc ipso pondere saliendum est: nam ea res mucorem vino inesse non patitur.

XXIV. Pix Nemeturica in Liguria conficitur. Ea deinde ut fiat condituris idonea, aqua marina quam longissime a littore de pelago sumenda est, atque in dimidiam partem decoquenda: quae cum in tantum refrixerit, quantum ne contacta corpus urat, partem aliquam eius, quae satis videbitur, praedictae pici immiscebimus, et diligenter lignea spatha vel etiam manu peragitabimus, ut si quid inest 2 vitii eluatur. Deinde patiemur picem considere, et cum siderit, aquam eliquabimus: postea bis aut ter

touched again until it ferments, though this process should not be allowed to go on for more than fourteen days from the time when the preservative was added; for after this number of days it will be time to purify 2 the wine without delay, and if any dregs have stuck to the brim or the sides of the vessels, they will have to be scraped and rubbed off, and the covers must be immediately put on and the vessels sealed up.

But if you wish to preserve the whole vintage with the same pitch in such a way that it is impossible to tell from the taste that it has been preserved with pitch, it will be enough to mix six *scripula* of the same pitch with forty-five *sextarii* of wine when at length it has ceased to ferment and the dregs have been cleared away. But you will have to add half- 3 an-ounce of roasted and powdered salt to the same quantity of must. Nor should salt be put into this kind of wine only, but, if possible, every sort of vintage in every district ought to be salted with this same quantity; for this prevents there being any mouldy taste in the wine.

XXIV. Nemeturican pitch is made in Liguria. In order that it may be rendered fit for preserving wine, sea-water must be taken from the open sea as far as possible from the shore and boiled down to half its original volume; and when it has cooled to such a temperature that it does not burn the body when it is in contact with it, we shall mix such a quantity of the said pitch as shall seem sufficient and carefully stir it with a wooden spatula or even with the hand, in order that any blemish in it may be washed away. We shall then allow the pitch to settle and, when it 2 has done so, strain the water away; afterwards we shall wash it two or three times with some of the

Of the use of Nemeturican pitch for preserving wine.

ex reliqua parte aquae decoctae tamdiu lavabimus [1] et subigemus [2] eam, donec rutila fiat: tum [3] eliquatam in sole quattuordecim diebus patiemur esse, ut quisquis ex aqua humor remansit, assiccetur. Noctibus autem vas tegendum erit, ne irroretur. Cum hoc modo picem praeparaverimus, et vina, cum iam bis deferbuerint, condire voluerimus, in musti sextarios octo et quadraginta cyathos duos picis praedictae sic adiciemus. Ex ea mensura, quam condituri sumus, sextarios duos musti sumere oportebit, deinde ex his sextariis in picis sextantem paulatim mustum infundere, et manu tamquam mulsum subigere, quo facilius coëat.[4] Sed ubi toti duo sextarii cum pice coierint, et quasi unitatem fecerint, tum eosdem in id vas [5] unde sumpseramus, perfundere, et ut permisceatur medicamen, rutabulo ligneo peragitare conveniet.

XXV. Quoniam quidam, immo etiam fere omnes Graeci, aqua salsa vel marina mustum condiunt, eam quoque partem curae non omittendam putavi. In mediterraneo, quo non est facilis aquae marinae invectio, sic erit ad condituras conficienda muria. Huic rei maxime est idonea caelestis aqua: si minus, ex fonte liquidissimo profluens. Harum ergo alterutram curabis quam plurimam et quam optimis vasis conditam ante quinquennium in sole ponere: deinde cum computruerit, tamdiu pati, donec ad pristinum

---

[1] laudabimus *SAac*.
[2] subicebimus *S*: subigebimus *Aa*: subigemus *c*.
[3] dum *SAac*.
[4] quoeat *S*.   [5] duas *SAac*.

BOOK XII. xxiv. 2–xxv. 2

remaining boiled water and turn it over and over until it becomes bright-red in colour; then, after straining it, we shall allow it to stand in the sun for fourteen days, so that any moisture which has remained from the water may be dried up. At night, however, the vessel must be covered so that the dew may not fall upon it. When we have prepared the pitch in this manner and wish to preserve wines when they have finished fermenting for the second time, we shall add two *cyathi* of the above-mentioned pitch to forty-eight *sextarii* of must in the following manner: we shall have to take two *sextarii* of must from the quantity which we are about to preserve, and then gradually pour must from these two *sextarii* into a *sextarius* of pitch and turn it over and over with the hand, as one does honey-water, to make it amalgamate more easily, and when the whole of the two *sextarii* have amalgamated with the pitch and formed, as it were, a single substance, it will then be proper to pour them into the vessel from which we had taken them and to stir them with a wooden ladle so that the preservative may be thoroughly mixed in.

XXV. Since some people—and indeed almost all the Greeks—preserve must with salt or sea-water, I thought that part of the process ought not to be passed over in silence. In an inland district, to which sea-water is not easily conveyed, brine for preserving purposes will have to be made as follows. Rainwater is most suitable for this process, or, failing that, water flowing from a very clear spring. Therefore you will take care to place as much as possible of one or other of these in the best vessels available five years beforehand in the sun; then, when it has putrefied, you must leave it alone till it has returned to its former

How to preserve must and wine with salt-water.

modum perveniat. Quod cum factum fuerit, alia vasa habeto, et in ea sensim aquam eliquato, donec ad faecem pervenias. Semper enim in requieta aqua crassa-
3 men aliquod in imo reperitur. Sic curata cum fuerit, in modum defruti ad tertias decoquenda est. Adiciuntur autem in aquae dulcis sextarios quinquaginta salis candidi sextarius et mellis optimi unus sextarius. Haec pariter decoqui, et omnem spurcitiam expurgari oportet. Deinde cum refrixerit, tum quantumcunque humoris[1] est, tantam in amphoram musti portionem adici.

4 Quod si ager maritimus est, silentibus ventis de alto quam quietissimo mari sumenda est aqua, et in tertiam partem decoquenda, adiectis, si videbitur, aliquibus aromatis ex iis quae supra retuli, ut sit odoratior vini curatio. Mustum autem antequam de lacu tollas, vasa rore marino vel lauro vel myrto suffumigato, et large repleto, ut in effervescendo
5 vinum se bene purget. Postea vasa nucibus pineis suffricato. Quod vinum volueris dulcius esse, postero die: quod austerius, quinto die quam sustuleris, condire oportebit, et ita supplere, et oblinire vasa. Nonnulli etiam suffumigatis seriis prius condituram addunt, et ita mustum infundunt.

XXVI. In quo agro vinum acescere[2] solet, curandum est, ut cum uvam legeris et calcaveris,

[1] moris *S*. [2] acessere *S*.

BOOK XII. xxv. 2–xxvi. 1

condition. When it has done so, you must have other vessels ready, and gradually strain the water into them, until you reach the dregs; for some sediment is always found in water which has been allowed to stand. When the water has been thus treated, it 3 must be boiled down to a third of its original volume, in the same way as must which is boiled down. Then a *sextarius* of white salt and a *sextarius* of the best honey are added to fifty *sextarii* of fresh water, which must be likewise boiled down and all impurities removed. Then, when the mixture has cooled down, whatever liquid there is; that amount must be added to an *amphora* of must.

If, however, your property is near the sea, water 4 should be taken from the open sea when the winds are silent and the sea is as quiet as possible, and should be boiled down to a third of its original volume, some of the spices detailed above having been added, if it shall be thought fit, so that the wine may have more flavour after it has been treated. But before you draw the must from the vat, you should fumigate the vessels with rosemary or bay or myrtle and fill them to the brim, so that the wine may purify itself in the process of fermentation; afterwards you should rub the vessels with pine-cones. Wine which you wish to 5 have rather sweet you will have to preserve on the day after you take it out of the vat, but, if you want it rather harsh in flavour, on the fifth day. You must then fill up the vessels and seal them. Some people, after fumigating the jars, put the preservative in first and then pour in the must.

XXVI. On any estate where the wine often turns acid care must be taken, when you have gathered and trodden out the grapes and before you squeeze the

How to treat wine which has gone sour.

LUCIUS JUNIUS MODERATUS COLUMELLA

prius quam vinacea torculis exprimantur, mustum in cortinam defundas, et aquae dulcis puteanae [1] ex eodem agro partem decimam adicias, et coquas, donec ea aqua, quam adieceris, decocta sit. Postea cum refrixerit, in vasa defundas, et operias, et oblinas: ita diutius durabit, et detrimenti nihil fiet.

2   Melius est, si veterem servatam compluribus annis aquam addideris; longeque melius si aquae nihil addideris, et decimam musti decoxeris, frigidumque in vasa transtuleris, et si in sextarios VII [2] musti heminam gypsi miscueris, posteaquam decoctum refrixerit. Reliquum mustum, quod e vinaceis fuerit expressum, primo quoque tempore absumito, aut aere commutato.

XXVII. Vinum dulce sic facere oportet. Uvas legito, in sole per triduum expandito,[3] quarto die meridiano tempore calidas uvas proculcato, mustum lixivum, hoc est, antequam prelo pressum sit, quod [4] in lacum musti fluxerit tollito: cum deferbuerit, in sextarios quinquaginta irim bene pinsitam nec plus unciae pondere addito, vinum a faecibus eliquatum diffundito. Hoc vinum erit suave, firmum, corpori salubre.

XXVIII. Irim quam candidissimam pinsito, faenum Graecum vetere vino macerato: deinde in sole exponito aut in furno, ut siccescat: tum commolito minutissime. Item odoramenta trita, id est, irim cribratam, quae sit instar pondo quincuncem et trientem, faeni Graeci pondo quincuncem et trien-

---

[1] puta ea ne *S*.
[2] ·L· *S* : ī *A* : vel *a* : ut *c*.
[3] expandito *A*.
[4] quod *om. S*.

grape-skins in the wine-presses, to pour the must into the cauldron and to add a tenth part of fresh well-water from the same estate and boil it until the water which you have added is boiled away. Afterwards, when it has cooled, you should pour it into vessels, cover it and seal it up; in this way it will keep longer and no harm will befall it.

It is better if you add old water which has been kept for several years, and better still by far if you add no water at all but boil down a tenth part of the must and transfer it cold into the vessels, and if you mix a *hemina* of gypsum with seven *sextarii* of must, after it has been boiled down and gone cold. The rest of the must which has been squeezed out of the grape-skins, you should use up as soon as possible or else turn it into money.

XXVII. The following is the way to make sweet wine. Gather the grapes and spread them out in the sun for three days; on the fourth day at noon tread the grapes while still warm; remove the must from the untrodden grapes, that is the must which has flowed into the must-vat before it has been squeezed in the wine-press and, when it has ceased to ferment, add well-crushed iris, but not more than an ounce of it to fifty *sextarii* and pour off the wine after straining it free from dregs. This wine will be pleasant to the taste and will keep in good state and is wholesome for the body.

XXVIII. Crush an iris which should be as white as possible and soak fenugreek in old wine, and then expose it to the sun or else put it in an oven to dry; then pound it up very small. Also mix pounded up perfumes together, namely, sifted iris, about the quantity of a *quincunx* and a *triens*, and a *quincunx* and a *triens* of

tem, schoeni pondo quincuncem in unum permisceto:
tum in serias singulas quae sint amphorarum septe-
num, addito medicaminis pondo unciam et scripula
2 octo: gypsi, cum ex locis palustribus mustum erit,
in serias singulas ternas heminas: cum de novellis
vineis erit, sextarium; cum de veteribus et locis siccis,
heminas singulas adicito. Tertio die quam calcaveris, condituram infundito, sed antequam condias,
musti aliquantum in seriam de seria transferto, ne in
condiendo cum medicamento effervescat et effluat.[1]
3 Sic autem curatum[2] gypsum et medicamentum in
labello permisceto, quantum seriis singulis fuerit
necessarium, idque medicamentum musto diluito, et
ipsa ad serias addito et permisceto: cum deferbuerit,
statim repleto et oblinito.

Omne vinum cum condieris, nolito statim diffundere,
sed sinito in doliis liquescere: postea cum de doliis
aut de seriis diffundere voles, per ver[3] florente rosa,
defaecatum[4] quam limpidissimum in vasa bene picata
4 et pura transferto. Si in vetustatem servare voles,
in cado duarum urnarum quam optimi vini sextarium,
aut faecis generosae recentis sextarios tres addito:
aut si vasa recentia, ex quibus vinum exemptum sit,
habebis, in ea confundito. Si horum quid feceris,
multo melius et firmius erit vinum. Etiam si bonos
odores addideris, omnem malum odorem et saporem

[1] effluat *Warmington.*   [2] curato *S*: curatio *Aac.*
[3] puer *S*: vere *Aac.*   [4] effecatum *S.*

… fenugreek, and a *quincunx* of sweet rush. Then into each vessel, which should contain seven *amphorae*, put an ounce and eight *scripula* of the preparation, and, if the must comes from marshy ground, three *heminae* of gypsum in each jar, or, if it comes from newly-established vineyards, a *sextarius*, or, if from old-established and dry places, one *hemina*. On the third day after you have trodden the grapes, pour in the preservative, but before you begin the process of preserving, transfer a small quantity of must from one vessel to another, lest in the process of preservation it ferments with the preservative and overflows. But mix thoroughly the gypsum thus prepared and the preservative in a pan, as much as shall be necessary for each vessel, and dilute this preservative with must and put them into the vessels and mix thoroughly. When the must has ceased fermenting, immediately fill up the vessels and seal them.

When you preserve any kind of wine, do not pour it out immediately but allow it to become clear in the wine-casks; afterwards when you wish to pour it from the casks and vessels, during the spring when the rose is flowering, after clearing out the dregs and making the wine as transparent as possible transfer it to clean vessels well treated with pitch. If you wish to preserve the wine until it is old, put a *sextarius* of the best possible wine into a jar containing two *urnae* and add three *sextarii* of lees fresh and from a fine wine; or, if you have fresh vessels, out of which the wine has been emptied, pour it into these. Whichever of these things you do the wine will be much better and will keep longer. Also if you put in good perfumes, you will prevent any bad odour or flavour; for there

prohibueris: nam nulla res alienum odorem celerius ad se ducit, quam vinum.

XXIX. Mustum ut semper dulce tamquam recens permaneat, sic facito.[1] Antequam prelo vinacea subiciantur, de lacu quam recentissimum addito mustum in amphoram novam, eamque oblinito, et impicato diligenter, ne quicquam [2] aquae introire possit: tunc in piscinam frigidae et dulcis aquae totam amphoram mergito, ut nequa pars exstet; deinde post dies XL eximito: sic usque in annum dulce permanebit.

XXX. Ab eo tempore quo primum dolia operculaveris, usque ad aequinoctium vernum semel in diebus XXXVI vinum curare satis est, post aequinoctium vernum [3] bis, aut si vinum florere incipiet, saepius curare oportebit: ne flos eius pessum eat, et saporem vitiet. Quanto maior aestus erit, eo saepius convenit nutriri refrigerarique, et ventilari: nam quamdiu bene frigidum erit, tamdiu recte manebit.
2 Labra vel fauces doliorum semper suffricari nucibus pineis oportebit, quoties vinum curabitur. Siqua vina erunt duriora aut minus bona, quod agri vitio aut tempestate sit factum, sumito faecem vini boni, et panes facito, et in sole arefacito, et coquito in igne: postea terito, et pondo quadrantem amphoris singulis infricato, et oblinito, bonum fiet.

[1] facto *S*.
[2] ne quicquam *Aac* : nequiquam *S*.

BOOK XII. xxviii. 4–xxx. 2

is nothing which attracts to itself the odour of something else more quickly than wine.

XXIX. That must may remain always as sweet as though it were fresh, do as follows. Before the grape-skins are put under the press, take from the vat some of the freshest possible must and put it in a new wine-jar; then daub it over and cover it carefully with pitch, that thus no water may be able to get in. Then sink the whole flagon in a pool of cold, fresh water so that no part of it is above the surface. Then after forty days take it out of the water. The must will then keep sweet for as much as a year.

How to keep must sweet.

XXX. From the time when you have first put the covers on the wine-jars until the spring equinox, it is enough to attend to the wine once every thirty-six days, and after the spring equinox twice in that period, or, if the wine begins to " flower " you will have to attend to it more often, lest the " flower " sink to the bottom and ruin the flavour. The greater the heat, the more often should the wine be attended to and cooled and ventilated; for as long as it is properly cold, so long will it remain in good condition. The brims and necks of the vessels will 2 always have to be rubbed with pine-cones, whenever the wine is being attended to. If any wine is rougher and less good than you could wish, which may be due to the bad quality of the ground or to the weather, take dregs of good wine and make it into cakes and dry them in the sun and bake them on the fire; then grind them up and rub a *quadrans* of them into each wine-jar and seal it up, and good wine will be the result.

---

³ semel—vernum *ac* : *om. SA.*

LUCIUS JUNIUS MODERATUS COLUMELLA

XXXI. Si quod animal in mustum ceciderit, et interierit, uti[1] serpens aut mus sorexve, ne mali odoris vinum faciat, ita ut repertum corpus fuerit, id igne adoleatur, cinisque eius in vas, quo deciderat, frigidus infundatur, atque rutabulo ligneo permisceatur: ea res erit remedio.

XXXII. Vinum marrubii multi utile putant ad omnia intestina vitia, et maxime ad tussim. Cum vindemiam facies, marrubii caules teneros[2] maxime de locis incultis et macris legito, eosque in sole siccato: deinde fasciculos facito, et tomice palmea aut iuncea ligato, et in seriam mittito, ita ut vinculum exstet: in musti dulcis sext. cc, marrubii libras VIII adicito, ut simul cum musto defervescat: postea eximito marrubium, et purgatum vinum diligenter oblinito.

XXXIII. Vinum scilliten ad[3] concoquendum et ad corpus reficiendum, item ad veterem tussim et ad stomachum hoc modo condire oportet. Primum ante dies quadraginta quam vinum voles vindemiare, scillam legito, eamque secato quam tenuissime, sicut raphani radicem, taleolasque sectas suspende[4] in umbra, ut adsiccentur: deinde cum aridae erunt, in musti Aminei sextarios XLVIII scillae aridae adde pondo libram, eamque inesse patere diebus XXX, postea eximito, et defaecatum vinum in amphoras bonas adicito. Alii scribunt in musti sextarios XLVIII, scillae aridae pondo libram et quadrantem adici oportere: quod et ipsum non probo.

[1] uti *om. SAac.*
[2] teneros *ac* : -as *SA.*
[3] ad *om. SAac.*
[4] suspendere *SAac.*

XXXI. If any creature has fallen into the must and died there, such as a snake or a mouse or a shrew-mouse, in order that it may not give the wine an evil odour, let the body in the condition in which it was found be burnt and its ashes when cool be poured into the vessel into which it had fallen and stirred in with a wooden ladle; this will cure the trouble.

How to treat must in which an animal has been drowned.

XXXII. Many people consider horehound-wine beneficial for all internal complaints, and particularly for a cough. When you are carrying out the vintage, collect tender stalks of horehound chiefly from uncultivated and lean places and dry them in the sun; then make small bundles of them and tie them up with palm- or rush-string and let them down into a wine-jar so that the band sticks out. Put eight pounds of horehound into two hundred *sextarii* of sweet must, so that the horehound may ferment with the must; afterwards, take out the horehound and, after cleansing the wine, carefully seal it up.

How to make horehound wine.

XXXIII. Squill-wine for promoting digestion and invigorating the body and also for a cough of long standing and for the stomach should be preserved in the following manner. First of all, forty days before you wish to begin the vintage, pick the squills and cut them up as thinly as possible, as you do radishes, and hang up the little sections in the shade to dry; then, when they are dry, put a pound of dry squill into forty-eight *sextarii* of Aminean must and let it remain there for thirty days; afterwards take it out and, after getting rid of the dregs, put the wine into sound wine-jars. Another prescription is 2 that a pound and a *quadrans* of dry squill should be put into forty-eight *sextarii* of must; in this I see nothing of which to approve.

How to make squill wine.

LUCIUS JUNIUS MODERATUS COLUMELLA

XXXIV. Hoc ipsum scillae pondus, quod supra, in aceti duas urnas adiciunt, et per XL dies inesse patiuntur, qui scilliticum acetum facere volunt. Ad embammata [1] in tres amphoras musti mittis aceti acris congium aut duplum, si non est acre,[2] et in ollam, quae fert amphoras tres, decoquis ad palmum, id est, ad quartas: aut si non est dulce mustum, ad tertias: despumetur. Sed mustum desub massa et limpidum sit.

XXXV. Vinum absinthiten, et hyssopiten, et abrotoniten, et thymiten, et marathriten, et glechoniten sic condire oportet. Pontici absinthii pondo libram cum musti sextariis IV decoque usque ad quartas: reliquum quod erit, id frigidum adde in musti Aminei urnam. Idem ex reliquis rebus quae suprascripta sunt, facito. Possunt etiam puleii aridi tres librae cum congio musti ad tertias decoqui, et cum refrixerit liquor, exempto puleio in urnam musti adici: idque mox tussientibus per hiemem recte datur: vocaturque vini nota glechonites.

XXXVI. Mustum tortivum est, quod post primam pressuram vinaceorum circumciso pede exprimitur. Id mustum conicies in amphoram novam, et implebis ad summum. Tum adicies ramulos roris marini aridi lino colligatos, et patieris una defervescere per dies septem: deinde eximes ramulorum fasciculum, et purgatum diligenter vinum gypsabis. Sat erit

---

[1] ad embammata *Schneider* : decem bambatae *SAac*.
[2] acris *c* : acri *SAa*.

---

[a] The MS. reading *desub massa* is probably corrupt.
[b] *I.e.* from the shores of the Black Sea.

XXXIV. Those who wish to make squill-vinegar put just the same weight of squill as is given above into two *urnae* of vinegar and allow it to remain there for forty days. For sauces you put into three *amphorae* of must a *congius* of sharp vinegar, or double the amount if it is not sharp, and boil it in a pot which holds three *amphorae*, down to a *palmus*, that is, a quarter of its original volume, or, if the must is not sweet, to a third; the scum should then be removed. The must should be under pressure of a heavy weight [a] and transparent.

How to make squill-vinegar.

XXXV. The following is the way to make up wines flavoured with wormwood, hyssop, southern-wood, thyme, fennel and pennyroyal. Take a pound of Pontic [b] wormwood and four *sextarii* of must and boil down to a quarter of the original quantity and put what remains when it is cold into an *urna* of Aminean must. Do the same with the other things mentioned above. Three pounds of dry pennyroyal in a *congius* of must can also be boiled down to a third part, and the liquid, when it has cooled, and the pennyroyal is removed, can be added to an *urna* of must. This is presently given, and properly, to those who have coughs in the winter; it is called the pennyroyal brand of wine.

How to make wine flavoured with wormwood, etc.

XXXVI. "Squeezed" must is that which is pressed out after the first pressing of the grape-skins, when the pedicles have been cut away. This must you will pour into a new wine-jar and fill it to the brim; you will then add little branches of dried rosemary tied together with flax and allow them to ferment together for seven days. Then you will take out the bundle of little branches and carefully plaster up the wine after cleansing it. It will be enough

"Squeezed" must.

autem roris marini sesquilibram in duas urnas musti adicere. Hoc vino post duos menses possis pro remedio uti.

XXXVII. Vinum simile Graeco facere. Uvas praecoquas quam maturissimas legito, easque per triduum in sole siccato. Quarto die calcato, et mustum quod nihil habeat ex tortivo, conicito in seriam, diligenterque curato, ut cum deferbuerit, faeces expurgentur: deinde quinto die cum purgaveris mustum, salis cocti et cribrati duos sextarios, vel, quod est minimum, adicito unum sextarium in sextarios musti XLVIII.[1] Quidam etiam defruti sextarium miscent: nonnulli etiam duos adiciunt, si existimant[2] vini notam parum esse firmam.

XXXVIII. Vinum myrtiten ad tormina, et ad alvi proluviem, et ad imbecillum stomachum sic facito. Duo sunt genera myrti, quorum alterum est nigrum, alterum album. Nigri generis bacae, cum sunt maturae leguntur, et semina earum eximuntur, atque ipsae sine seminibus in sole siccantur, et in fictili fidelia sicco loco reponuntur. Deinde per vindemiam ex vetere arbusto, vel si id non est, ex vetustissimis vineis Amineae[3] bene maturae uvae sole calido leguntur, et ex his mustum adicitur in seriam, et statim primo die antequam id ferveat, bacae myrti, quae fuerant repositae, diligenter conteruntur, et totidem earum librae contusarum appenduntur, quot amphorae condiri debent: tum exiguum musti sumitur ex ea seria, quam medi-

---

[1] XLVIII S: XLIX Aac. [2] existimant a: -at SAc.
[3] aminee ac: aminne SA.

to add half a pound of rosemary to two *urnae* of must. After two months you could use this wine as a medicine.

XXXVII. To make wine like Greek wine, gather early grapes as ripe as possible and dry them for three days in the sun; tread them on the fourth day and pour the must, which should have none of the last squeezing in it, into a jar and be careful that, when it has finished fermenting, the lees are cleared out. Then on the fifth day when you have purified the must, add two *sextarii*—or at the very least one *sextarius*—of toasted and sifted salt to forty-eight *sextarii* of must. Some people also mix in a *sextarius* of must boiled down to a third of its original volume; others even add two *sextarii*, if they think the brand of wine is not likely to keep well.

How to make wine like Greek wine.

XXXVIII. You should make myrtle-wine for cholic, looseness of the bowels and a weak stomach in the following manner. There are two kinds of myrtle-tree, one black and the other white. The berries of the black kind are picked when they are ripe and their seeds removed, and are dried by themselves without the seeds in the sun and stored in an earthenware pot in a dry place; then, during the vintage, well ripened Aminean grapes are gathered while the sun is hot from an old vine supported by a tree or, if that is impossible, from very old vineyards, and the must from them is put into a jar and immediately on the first day, before the must ferments, the myrtle-berries which were stored away are carefully crushed and as many pounds of these bruised berries are weighed out as there are *amphorae* of wine to be prepared. Then a little must is taken from the vessel in which we are going to

How to make myrtle-wine.

LUCIUS JUNIUS MODERATUS COLUMELLA

caturi sumus, et tanquam farina conspergitur, quidquid contusum et appensum est. Post haec complures ex ea massulae fiunt, et ita per latera seriae in mustum demittuntur, ne altera offa super alteram
3 perveniat. Cum deinde bis mustum deferbuerit, et bis curatum est, rursus eodem modo, et tantumdem ponderis bacae sicut supra dixi, contunditur: nec iam ut prius massulae fiunt, sed in labello mustum de eadem seria sumitur, praedicto ponderi permiscetur, sic ut sit instar iuris crassi: quod cum est permixtum, in eandem seriam confunditur, et rutabulo ligneo
4 peragitatur. Deinde post nonum diem quam id factum est, vinum purgatur, et scopulis aridae myrti seria[1] suffricatur, operculumque superimponitur, ne quid eo decidat. Hoc facto post septimum diem rursus vinum purgatur, et in amphoras bene picatas et bene olidas diffunditur: sed curandum est, ut cum diffundis, liquidum et sine faece diffundas.
5 Vinum aliud myrtiten sic temperato.[2] Mel Atticum ter infervere facito, et toties despumato: vel si Atticum non habueris, quam optimum mel eligito, et quater vel quinquies despumato. Nam[3] quanto est deterius, tanto plus habet spurcitiae: cum deinde mel refrixerit, bacas albi generis myrti quam maturissimas legito, et perfricato, ita ne interiora
6 semina conteras. Mox fiscello lineo inclusas ex-

---

[1] seria *ac* : serie *SA*.
[2] vinum—temperato *om. SAac*.
[3] cum *ac* : qūo *SA*.

BOOK XII. xxxviii. 2–6

make our preparation and the berries which have been bruised and weighed out are sprinkled like flour upon it; after this a number of small lumps are made of the mixture and are let down the sides of the jar into the must in such a way that one pellet does not fall on the top of another. Next, after the must has 3 ceased fermenting for the second time and has been twice treated, the same weight of berries as I have mentioned above is crushed in the same manner, but this time no lumps are made, as before, but must is taken with a ladle from the same vessel and thoroughly mixed with the above-mentioned quantity of berries so as to resemble a thick juice, and after being thoroughly mixed it is poured into the same vessel and is stirred well with a wooden ladle. Then, nine 4 days after this has been done, the wine is purified and the jar is rubbed with small brushes of dry myrtle, and the lid is put on to prevent anything from falling into the jar. When this has been done, after the seventh day the wine is again purified and poured into jars which have been thoroughly treated with pitch and smell very sweet, but care must be taken that, when you pour it out, it is clear and free from lees.

Another kind of myrtle-wine should be com- 5 pounded as follows. Take Attic honey and make it boil three times and remove the scum the same number of times, or, if you have no Attic honey, choose the best honey you can get and remove the scum four or five times; for the worse it is, the more impurities it contains. Then, when the honey has cooled down, gather the ripest possible berries of the white kind of myrtle and break them up in such a way as not to bruise the seeds inside; next enclose them 6 in a bag made of flax and press them and mix the

primito, succumque earum qui sit sextariorum sex, cum mellis decocti sextario[1] immisceto,[2] et in lagunculam diffusum oblinito. Sed hoc mense Decembri fieri debebit, quo fere tempore matura sunt myrti semina: custodiendumque erit, ut ante quam bacae legantur, si fieri potest, septem diebus, sin autem, ne minus triduum serenum fuerit, aut certe non pluerit; et ne rorulentae legantur cavendum.

7   Multi nigram vel albam myrti bacam, cum iam maturuit, destringunt,[3] et duabus horis eam cum paululum in umbra expositam siccaverunt, perterunt ita, ut quantum fieri potest, interiora semina integra permaneant. Tum[4] per lineum fiscum, quod pertriverant, exprimunt, et per colum iunceum liquatum succum lagunculis bene picatis condunt, neque melle neque alia re ulla immixta.[5] Hic liquor non tam est durabilis, sed quamdiu sine noxa manet, utilior est ad valetudinem quam alterius myrtitis notae compositio.[6]

8   Sunt qui hunc ipsum expressum succum, si sit eius copiosior facultas, in tertiam partem decoquant, et refrigeratum picatis lagunculis condant. Sic confectum diutius permanet: sed et quod non decoxeris, poterit[7] innoxium durare biennio, si modo munde et diligenter id feceris.

XXXIX. Passum optimum sic fieri Mago praecipit, ut et ipse feci. Uvam praecoquam bene maturam

[1] sextario *ac* : -um *SA*.
[2] immiscetur *a* : miscetur *SAc*.
[3] destringent *SAa* : distinguēt *c*.
[4] tum *ac* : dum *SA*.
[5] immixta *ac* : -ae *SA*.
[6] compositio *ac* : compositiones *SA*.
[7] possit *SAac*.

BOOK XII. xxxviii. 6–xxxix. 1

juice from them to the amount of six *sextarii* with one *sextarius* of boiled-down honey and pour the mixture into a flask and seal it up. This will have to be done in the month of December, which is about the time when the myrtle-seeds are ripe, and care will have to be taken that, if possible, the weather has been fine for seven days before the berries are gathered, or, if not, for not less than three days, or at any rate that there has been no rain; also you must beware of picking them when the dew is upon them.

Many people strip off black or white myrtle-berries 7 when they are ripe and two hours later, after drying them for a time spread out in the shade, bruise them in such a way that the seeds inside may, as far as possible, remain undamaged. They then squeeze what they have bruised through a bag of flax, and having strained the juice through a rush-strainer they put it up in flasks well treated with pitch without mixing honey or anything else with it. This liquor does not last so well, but, as long as it remains without spoiling, it is more beneficial to the health than the other brand of myrtle-wine.

Some people, after this juice has been squeezed 8 out, if they have an unusually large quantity of it, boil it down to a third part of its original volume and, when it has cooled down, store it in flasks treated with pitch. Thus prepared it keeps for a longer time; but even wine which you have not boiled down may keep for two years if only you have prepared it cleanly and carefully.

XXXIX. Mago[a] gives the following instructions for making the best raisin-wine, as I myself have also made it. Gather the early grapes when they

How to make raisin-wine.

---
[a] See Book I. 1. 10 and note.

LUCIUS JUNIUS MODERATUS COLUMELLA

legere, acina mucida aut vitiosa reicere: furcas vel palos, qui cannas sustineant, inter quaternos pedes figere, et perticis iugare: tum insuper cannas ponere, et in sole pandere uvas, et noctibus tegere, ne irrorentur: cum deinde exaruerint, acina decerpere, et in dolium aut in seriam conicere, eodem mustum quam optimum, sic ut grana submersa sint, adicere: ubi combiberint uvae,[1] seque impleverit, sexto die in fiscellam conferre, et prelo premere, passumque
2 tollere: postea vinaceos calcare adiecto recentissimo musto, quod ex aliis uvis factum fuerit, quas per triduum insolaveris: tum permiscere, et subactam brisam prelo subicere, passumque secundarium statim vasis oblitis includere, ne fiat austerius: deinde post viginti vel triginta dies, cum deferbuerit, in alia vasa deliquare, et confestim opercula gypsare, et pelliculare.
3 Passum si ex uva apiana facere volueris, uvam apianam integram legito, acina corrupta purgato et secernito, postea in perticis suspendito. Perticae uti semper in sole sint facito; ubi satis corrugata erunt acina, demito et sine scopionibus in dolium conicito, pedibusque bene calcato. Ubi unum tabulatum feceris, vinum vetus spergito, postea alterum supercalcato, item vinum conspergito. Eodem modo tertium calcato, et infuso vino ita supernatare sinito dies quinque: postea pedibus

---

[1] uvas *SAac*.

---

[a] Pliny, *N.H.* XIV. § 24, says that the adjective *apianus* is applied to grapes of which bees are specially fond. However, the MSS. have the spelling *appianam*.

are quite ripe, rejecting the berries which are mouldy or damaged. Fix in the ground forks or stakes four feet apart for supporting reeds and yoke them together with poles, then put reeds on the top of them and spread out the grapes in the sun, covering them at night so that the dew may not fall on them. Then when they have dried, pluck the berries and throw them into a barrel or wine-jar and add the best possible must thereto so that the grapes are submerged. When they have absorbed the must and are saturated with it, on the sixth day put them all together in a bag and squeeze them in the wine-press and remove the raisin-wine. Next tread the wine-skins, adding very fresh must, made from other grapes which you have dried for three days in the sun; then mix together and put the whole kneaded mass under the press and immediately put this raisin-wine of the second pressing in sealed vessels so that it may not become too rough; then, twenty or thirty days later, when it has finished fermenting, strain it into other vessels and plaster down the lids immediately and cover them with skins.

If you wish to make raisin wine from "bee"[a] grapes, pick sound "bee- " grapes; clear away the rotten berries and put them aside; and afterwards hang the sound grapes on poles. Arrange that the poles shall be always in the sun, and when the berries are sufficiently withered, pull them off and throw them into a barrel without the stalks and tread them well. When you have made one layer of them, sprinkle old wine upon it, and then tread another layer of them on the top and also sprinkle wine upon it. In like manner tread a third layer and, after pouring wine upon it, let it float on the top for five days; after-

Another method.

LUCIUS JUNIUS MODERATUS COLUMELLA

4 proculcato, et in fiscina nova uvas premito. Quidam aquam caelestem veterem ad hunc usum praeparant et ad tertias decoquunt. Deinde cum uvas sicut [1] supra scriptum est, passas fecerunt, decoctam aquam pro vino adiciunt, et cetera similiter administrant. Hoc ubi lignorum copia est, vilissime constat, et est in usu vel dulcius, quam superiores notae passi.

XL. Lora optima sic fieri oportet.[2] Quantum vini uno die feceris, eius partem decimam, quot metretas efficiat, considerato, et totidem metretas aquae dulcis in vinacea, ex [3] quibus unius diei vinum expressum erit, addito: eodem et spumas defruti, sive sapae, et faecem ex lacu confundito et permisceto, eamque intritam macerari una nocte sinito, postero die pedibus proculcato, et sic permixtam prelo subicito: quod deinde fluxerit, aut doliis aut amphoris condito, et cum deferbuerit, obturato. Commodius autem servatur in amphoris. Hanc ipsam loram M.[4] Columella ex aqua vetere faciebat, et nonnunquam plus biennio innoxiam servabat.

XLI. Mulsum optimum sic facies.[5] Mustum lixivum de lacu statim tollito: hoc autem erit, quod destillaverit antequam nimium calcetur uva. Sed de arbustivo genere, quod sicco die legeris, id facito. Conicies in urnam musti mellis optimi pondo x, et diligenter permixtum recondes in lagoena, eamque

---

[1] sic *A* : sicut *Sac*.
[2] lora—oportet *SA* : *om. ac*.
[3] ex *scripsi* : sed *SAac*.
[4] M. *ac* : macrius *SA*.
[5] mulsum—facies *om. ac* : facies *SA*.

---

[a] A thin wine made for the use of labourers (Varro, *De Re Rustica*, I. 54).

wards tread with your feet and squeeze the grapes in a new wicker basket. Some people prepare old rain-water for this purpose and boil it down to a third of its original volume; then, when they have dried the grapes in the manner described above, they add the boiled-down water instead of wine and carry out the rest of the process in the same manner. Where there is plenty of wood, this wine costs very little and in use is even sweeter than the brands of raisin-wine described above.

XL. The best after-wine [a] is made as follows. Calculate to how many *metretae* the tenth part of the wine which you make in one day amounts and put that number of *metretae* of fresh water on the grape-skins out of which one day's wine has been pressed. Pour also into the same vessel the scum of the must boiled down to a third or to a half of the original volume and the dregs from the vat and mix together and let this mash soak for one night. On the next day tread it and, when it is thus thoroughly mixed, put it under the press. Then put what has flowed out either into barrels or jars and, when it has fermented, close the vessels up. It is more convenient to keep it in wine-jars. Marcus Columella used to make this same after-wine with old water and frequently kept it without its spoiling for more than two years.

How to make thin "after-wine."

XLI. The following is the way to make very good mead. Take straightway from the wine-vat must called *lixivum*—which will be that which has flowed from the grapes before they have been too much trodden—but make it with grapes from vines which grow upon trees and pick them on a dry day. You will put ten pounds of the best honey into an *urna* of must and, after carefully mixing them together, you

How to make good mead.

protinus gypsabis, iubebisque in tabulato poni; si plus volueris facere, pro portione qua supra mel adicies. Post trigesimum et alterum diem lagoenam aperire oportebit, et in aliud vas mustum eliquatum oblinire, atque in fumum reponere.

XLII. Compositio medicamenti ad tormina, quod vocatur διὰ ὀπώρας.[1] In cacabo fictili novo, vel in stagneo coquitur musti arbustivi Aminei urna, et mala cydonea grandia expurgata viginti, et integra mala dulcia granata, quae Punica vocantur, et sorba[2] non permitia divisa exemptis seminibus, quae sint instar sextariorum trium. Haec ita coquuntur, ut omnia poma deliquescant cum musto, et sit puer, qui spatha lignea vel arundine permisceat poma, ne possint aduri. Deinde cum fuerint decocta, ut non multum iuris supersit, refrigerantur et percolantur: eaque, quae in colo subsederunt, diligenter contrita levigantur, et iterum in suo sibi iure lento igni, ne adurantur, carbonibus decoquuntur, donec crassamen in modum faecis existat. Prius tamen quam de igne medicamentum tollatur, tres heminae roris Syriaci contriti et cribrati super omnia adiciuntur, et spatha[3] permiscentur, ut coeant cum ceteris. Tum refrigeratum medicamentum adicitur in vas fictile novum picatum, idque gypsatum alte suspenditur ne pallorem trahat.

[1] compositio—διὰ ὀπώρας *om. ac* : οπωρα *S*.
[2] sorba *ac* : sorva *A* : soruua *S*.
[3] sphata *A* : spathata *S* : spha *c*.

will store the must in a flagon and immediately seal it up with plaster and order it to be placed in a loft. If you wish to make more, you will add honey in the proportion mentioned above. After thirty-one days you will have to open the flagon and after straining the must into another vessel plaster it up and place it back where the smoke will reach it.

XLII. The preparation of a remedy against colic called "fruit-syrup" is made as follows. An *urna* of must made from Aminean grapes grown on trees is boiled in a new earthenware or tin cooking-pot with twenty large quinces which have been well cleaned and are sound, sweet pomegranates, which are called Carthaginian apples, and service-apples which are not very soft and have been split and had their seeds removed, a quantity weighing about three *sextarii*. These are boiled so that all the fruit dissolves in the must, and there should be a boy to stir the fruit with a wooden slice or a reed, so that it cannot burn. Then when they have been boiled down so that not much juice remains, they are allowed to cool and are strained, and what is left at the bottom of the strainer is carefully crushed and pulverized and then boiled again a second time in its own juice upon a slow charcoal fire, so that it may not burn, until a thick sediment, resembling lees, is formed. But before the preparation is removed from the fire, three *heminae* of Syrian rosemary, crushed and sifted, are added on the top of all and mixed in with a spatula, so that it may unite with the other ingredients. Then, when the preparation has cooled, it is put into a new earthenware vessel treated with pitch, and this is then plastered over and hung up high so that it may not acquire a pale colour.

How to make fruit syrup.

LUCIUS JUNIUS MODERATUS COLUMELLA

XLIII. Caseum sic condiemus.[1] Casei aridi ovilli proximi anni frusta ampla facito, et in picato vase[2] componito: tum optimi generis musto adimpleto, ita, ut superveniat, et sit ius aliquanto copiosius quam[3] caseus. Nam caseus[4] combibit, et fit vitiosus, nisi mustum semper supernatet.[5] Vas autem cum impleveris, statim gypsabis: deinde post dies viginti licet aperias, et utaris qua voles adhibita conditura. Est autem etiam per se non iniucundus.

XLIV. Uvas, ut sint virides usque ad annum, sic custodies:[6] uvas bumastos vel duracinas vel purpureas cum desecueris a vite, continuo pediculos earum impicato dura pice: deinde labellum fictile novum impleto paleis quam siccissimis cribratis, ut sine pulvere sint, et ita uvas superponito: tum labello altero adoperito, et circumlinito luto paleato, atque ita in tabulato siccissimo composita[7] labra paleis siccis obruito.

2 Omnis autem uva sine noxa servari potest, si luna decrescente et sereno caelo post horam quartam, cum iam insolata est, nec roris quicquam habet, viti detrahatur. Sed ignis in proximo decumano fiat, ut pix ferveat, in qua pediculi uvarum statim demit-

---

[1] caseum sic condiemus *om. ac.*
[2] vase *c*: vaso *SAa.*
[3] cum *ac*: quo *SA.*
[4] nam caseus *om. SAac.*
[5] et ut *post* supernatet *add. SAa*: ut *om. c.*
[6] uvas—custodies *om. ac.*: custodies *S*: -emus *A.*
[7] composita *ac*: -o *SA.*

BOOK XII. XLIII. 1–XLIV. 2

XLIII. The following is the method by which we shall preserve cheese. Cut large slices of sheep's milk cheese of last year and place them together in a vessel which has been treated with pitch; then fill it up with the best kind of must so that it covers the cheese and that there may be somewhat more liquid than cheese; for the cheese always absorbs the must and spoils, unless the latter floats above it. When you have filled the vessel, you will plaster it up immediately. Then after twenty days you may open it and use whatever seasoning you please; it is not unpleasant too by itself.

How to preserve cheese.

XLIV. In order that grapes may remain green for as much as a year, you will keep them in the following manner. When you have cut from the vine grapes in welling clusters or hard-skinned grapes or purple grapes, immediately treat their pedicles with hard pitch; then fill a new earthenware pan with the driest possible chaff, which has been sifted that it may be free from dust, and put the grapes upon it. Then cover it with another pan and daub it round with clay mixed with chaff, and then, after arranging the pans in a very dry loft, cover them up with dry chaff.

How grapes may be preserved and kept green.

Grapes of every kind can be kept without spoiling 2 if they are plucked from the vine when the moon is waning and the weather is fine, after the fourth hour and when they have already been exposed to the sun and have no dew upon them. But a fire should be made on the nearest path going from east to west,[a] so that the pitch, in which the pedicles of the grapes

---

[a] Cp. Pliny, *N.H.* XVII. § 169, *vineas limitari decumano XVIII pedum latitudinis ad contrarios vehiculorum transitus.* A similar path going north and south was called *cardo*.

tantur. In dolium bene picatum defruti amphoram
conicito, deinde transversos fustes spisse arctato, ita
ut defrutum non contingant: tum superponito
fictiles novas patinas, et in his sic uvam disponito, ut
altera alteram non contingat: tum opercula patinis
imponito et linito. Deinde alterum tabulatum, et
tertium, et quamdiu magnitudo patitur dolii,[1]
similiter superinstruito, et eadem ratione uvas componito. Deinde picatum operculum dolii defruto
large linito, et ita compositum cinere obturato.

3 Nonnulli adiecto defruto contenti sunt transversas
perticas arctare, et ex his uvas ita suspendere, ne
defrutum contingant: deinde operculum impositum
oblinire. Quidam uvas cum ita, ut supra dixi,
legerunt, doliola nova sine pice in sole siccant.
Deinde cum ea in umbra refrigeraverunt, furfures
hordeaceos adiciunt, et uvas ita superponunt, ut altera
alteram[2] non comprimat: tum generis eiusdem
furfures infundunt, et alterum tabulatum uvarum
eodem modo collocant: idque faciunt usque dum
dolium alternis furfuribus et uvis compleant. Mox
opercula imposita linunt, et uvas siccissimo frigidis-
simoque tabulato reponunt.

4 Quidam eadem ratione arida populnea vel abiegna
scobe virides uvas custodiunt. Nonnulli sicco flore

[1] dolii *a* : dolis *SA* : doliis *c*.
[2] alteram *Ac* : altera *S* : *om. a.*

may be immediately dipped, may be boiling hot. Pour an *amphora* of boiled-down must into a barrel well coated with pitch; then press in cross-pieces of wood close together in such a way as not to touch the boiled-down must; then put on the top new earthenware dishes and put the grapes on them in such a way that they do not touch one another; then put covers on the dishes and seal them up. Next construct another story similarly above, and then a third, and continue the process as long as the size of the barrel allows, and arrange the grapes in the same manner. Then, after treating the lid of the barrel with pitch, smear it generously with boiled-down must and then, when you have put it on, stop it up with ashes.

Some people, after pouring in the boiled-down must, are content with putting the cross-pieces in close together and hanging the grapes from them so as not to touch the must, and then putting on the lid and sealing it up. Some people, after gathering the grapes in the manner I have described above, dry small, new barrels in the sun without any pitch; then, when they have cooled in the shade, they put barley-bran into them and place the grapes on the top so that they do not press upon one another; then they pour in barley-bran of the same kind and arrange another layer of grapes in the same way as before. This they do until they fill the barrel with alternate strata of bran and grapes. Next they put on the lids and seal them up and store the grapes in a very dry and cool loft.

Some people, after the same method, preserve green grapes in dry saw-dust of poplar-wood or fir; others cover up the grapes, which they have picked

LUCIUS JUNIUS MODERATUS COLUMELLA

gypsi obruunt uvas, quas non nimium maturas vitibus detraxerunt. Alii cum legerunt uvam, siqua sunt in ea vitiosa grana forficibus amputant, atque ita in horreo suspendunt, in quo triticum suppositum est. Sed haec ratio rugosa facit acina, et paene tam dulcia, quam est uva passa.

5 Marcus Columella patruus meus ex ea creta qua fiunt amphorae, lata vasa in modum patinarum fieri iubebat: eaque intrinsecus [1] et exterius crasse picari: quae cum praeparaverat, tum demum purpureas et bumastos et Numisianas et duracinas uvas legi praecipiebat, pediculosque earum sine mora in ferventem picem demitti, et in praedictis patinis separatim sui cuiusque generis ita componi, ne uvae inter se contingerent: post hoc opercula superponi, et oblini 6 crasso gypso: tum demum pice dura, quae igni liquata esset, sic picari, nequis humor transire posset:[2] tota deinde vasa in aqua fontana vel cisternina ponderibus impositis mergi, nec ullam partem earum pati exstare. Sic optime servatur uva. Sed cum est exempta, nisi eo die consumitur, acescit.

7 Nihil est tamen certius, quam vasa fictilia facere, quae singulas uvas laxe recipiant. Ea debent quattuor ansas habere, quibus illigata viti dependeant:[3] itemque opercula eorum sic formari, ut media divisa sint, ut cum suspensa vasa singulas

---

[1] extrinsecus *S*. [2] possit *SAac*.
[3] dependeat *SAac*.

---

[a] This epithet, like many others applied to vines, olives, etc., is derived from the name of a Roman *gens*.

from the vines when they were not too ripe, in dry flower of gypsum. Others, when they have picked a bunch, cut off with shears any defective grapes in it, and then hang it up in the granary where there is wheat stored below them. But this method causes the grapes to become shrivelled and almost as sweet as raisins.

Marcus Columella, my uncle, used to order broad vessels like dishes to be made of clay of which wine-jars are made and to be thickly covered with pitch inside and outside. When he had got them ready, he finally gave instructions that purple and big-clustered and Numisian[a] and hard-skinned grapes should be picked and their pedicles plunged without delay into boiling pitch and that they should be arranged in the dishes mentioned above, each kind separately, in such a way that the bunches did not touch one another, and that then the covers should be put over them and sealed with thick gypsum; then finally that they should be treated with hard pitch which had been melted in the fire so that no moisture could pass through. He then ordered that the vessels should be completely submerged in spring- or cistern-water and weights placed on the top of them, no part of them being allowed to stand out of the water. On this system the grapes were excellently preserved; but, once they were taken out, they went sour unless they were eaten that day.

Nevertheless there is no more reliable method than to make earthenware vessels each with ample room to hold a single bunch of grapes. They should have four handles by which they may hang tied to the vine; also their covers should be shaped as to be divided in the middle, so that, when the suspended

uvas receperint, ex utroque latere appositi[1] operculi duae partes coëant, et contegant uvas. Et haec vasa et opercula extrinsecus et intra diligenter picata esse debebunt: deinde cum contexerunt uvas, luto paleato multo adoperiri: sed uvae dependentes a matre sic in pultarios condi debebunt, ne qua parte vasa contingant. Tempus autem quo includi debent, id fere est, quo[2] adhuc siccitatibus et sereno caelo crassa[3] variaque sint acina.

Illud in totum maxime praecipimus, ne in eodem loco mala et uvae componantur, neve in vicino,[4] unde odor malorum possit ad eas pervenire. Nam huiusmodi halitus celeriter acina corrumpunt. Eae tamen custodiendorum pomorum rationes, quas retulimus, non omnes omnibus regionibus aptae sunt, sed pro conditione locorum et natura uvarum aliae aliis conveniunt.

XLV. Antiqui plerumque Sircitulas[5] et Venuculas et maiores Amineas, et Gallicas, quaeque maioris et duri et rari acini erant, vasis condebant: nunc autem circa urbem maxime ad hunc usum Numisianae probantur. Hae sereno caelo, cum iam sol rorem sustulit, quarta vel quinta hora,[6] si modo luna decrescit et sub terris est, modice maturae rectissime leguntur: statim pediculi earum picantur: deinde in cratibus ita ponuntur, ne altera alteram collidat.

[1] at posite *SA*: apposite *a*: opposite *c*.
[2] quod *SAac*.
[3] grossa *SAac*.
[4] vicinos *SAac*.
[5] scirriculos *SAac*.
[6] horas *SAac*.

---

[a] Pliny, *N.H.* IV. § 34 says that Venuculan and Sircitulan were names applied to the same species of grape by the people of Campania; see note on Book III. 2. 2.

vessels have each received a bunch of grapes, the two halves of the lid, applied from either side, may meet and protect the grapes. Both these vessels and their covers will have to be carefully treated with pitch outside and inside. Then, when the lids have enclosed the grapes, they should be covered with a quantity of clay mixed with chaff. But the bunches hanging from the mother-vine will have to be put in pots, so that they do not touch the vessels at any point. The season when they ought to be enclosed is usually that at which it is still dry and the weather good and the grapes are big and variegated in colour.

We lay it down as a general rule that above all things apples and grapes should not be laid up in the same place or so near one another that the odour of the apples can reach the grapes. For exhalations of this kind quickly spoil the grapes. The methods of preserving apples, however, which we have mentioned, are not all adapted to every district, but different methods suit different districts according to the local conditions and the quality of the grapes.

XLV. The ancients usually stored in pots Sircitulan and Venuculan[a] and the larger Aminean[b] and Gallic grapes and those which have large, hard berries far apart; but in these days in the district round the city Numisian grapes are most commended for this purpose. These are best gathered moderately ripe, when the weather is fine and the sun has already evaporated the dew, at the fourth or fifth hour, always provided that the moon is waning and is beneath the earth. Their pedicles are immediately treated with pitch. The bunches are then placed on hurdles in such a way that they do not bruise one

[b] See note on Book II. 2. 7.

LUCIUS JUNIUS MODERATUS COLUMELLA

2 Tum demum sub tectum referuntur, et acida vel vitiosa grana forficibus amputantur: et cum paululum sub umbra refrixerint, ternae aut etiam quaternae pro capacitate[1] vasorum in ollas demittuntur et opercula diligenter pice obturantur, ne humorem transmittant. Tum vinaceorum pes bene prelo expressus proruitur, et modice separatis scopionibus resoluta intrita folliculorum in dolio substernitur, et deorsum versus spectantes ollae componuntur, ita 3 distantes, ut intercalcari possint vinacea: quae cum diligenter conspissata primum tabulatum fecerunt, aliae ollae eodem modo componuntur explentque secundum tabulatum. Deinde similiter doliis exstruuntur ollae et spisse incalcantur. Mox usque ad summum labrum vinacea condensantur, et statim operculo superposito cinere in modum gypsi temperato dolium linitur. Monendus autem erit, qui vasa empturus est, ne bibulas aut male coctas ollas emat. Nam utraque res transmisso humore vitiat uvam. Quinetiam oportebit, cum ad usum promuntur ollae, tota singula tabulata detrahi. Nam conspissata vinacea, si semel mota sunt, celeriter acescunt, et uvas corrumpunt.

XLVI. Sequuntur vindemiam rerum autumnalium compositiones, quae et ipsae curam villicae distendunt. Nec ignoro plurima in[2] hunc librum non esse collata, quae C. Matius diligentissime persecutus est. Illi enim propositum fuit urbanas

---

[1] capacitate *c*: capitate *Aa*: capietate *S*.
[2] in *om. SAac*.

---

[a] See note on Ch. 4. § 2 of this book.

BOOK XII. XLV. 2–XLVI. 1

another. Finally then they are taken indoors and the sour or defective grapes are cut off with shears; and, when they have cooled off a little in the shade, three or even four bunches, according to the capacity of the vessels, are let down into each pot and the lids are carefully closed up with pitch so that they may not let any moisture pass in. Then a foot of grape-skins well squeezed in the wine-press is emptied out and, after the stalks have been pretty well separated and the mass of the skins broken up, it is spread on the bottom of a barrel and the pots are arranged inside it with their mouths downwards at such a distance from one another that the grape-skins can be trodden between them; and when these have been carefully pressed together and have formed a first layer, other pots are arranged similarly and complete a second layer. Then in like manner more pots are piled up in the barrels and are closely trodden in. Next the grape-skins are packed in right up to the brim and immediately the cover is put on the top and the barrel is sealed with ashes mixed like plaster. The would-be purchaser, however, will have to be warned not to buy pots which absorb moisture or are badly baked; for both those defects spoil the grapes by letting in the moisture. Moreover, when the pots are brought out for use, whole layers will have to be moved together; for closely packed grape-skins, once they are moved, quickly turn sour and spoil the grapes.

XLVI. After the vintage is over comes the putting up of autumnal produce, which of itself is a great addition to the duties of the bailiff's wife. I am not ignorant that there are many subjects of which Gaius Matius[a] has carefully treated, but which have not been inserted in this book; for his purpose was to give

How to preserve pomegranates.

LUCIUS JUNIUS MODERATUS COLUMELLA

mensas et lauta convivia instruere. Libros tres edidit, quos inscripsit nominibus Coci, et Cetarii,[1] et Salgamarii.[2] Nobis tamen abunde sunt ea, quae facile rusticae simplicitati non magna impensa possunt contingere.

2 Quidam, pediculos punicorum, sicuti sunt in arbore, intorquent, ne pluviis mala rumpantur et hiantia dispereant, eaque ad maiores ramos religant, ut immota permaneant: deinde sparteis retibus arborem cludunt, ne aut corvis aut cornicibus aliisve avibus pomum laceretur. Nonnulli vascula fictilia dependentibus malis[3] aptant, et illita luto paleato arboribus haerere patiuntur: alii faeno vel culmo singula involvunt, et insuper luto paleato crasse[4] linunt, atque ita maioribus ramis illigant, ne, ut dixi, vento commoveantur.

3 Sed haec omnia, ut dixi, sereno caelo administrari sine rore debent: quae tamen aut facienda[5] non sunt, quia laeduntur arbusculae: aut certe non continuis annis usurpanda, praesertim cum liceat etiam detracta arboribus eadem innoxia[6] custodire. Nam et sub tecto fossulae[7] tripedaneae[8] siccissimo loco fiunt: eoque cum aliquantulum terrae minutae repositum est, infiguntur sambuci ramuli: deinde sereno caelo granata leguntur[9] cum

---

[1] et Cetarii *om. SAac.*
[2] salgamari *SAac.*   [3] mali *SAac.*
[4] crasse *ac* : grasse *SA.*
[5] facienda *ac* : -o *SA.*
[6] innoxiae *SAac.*
[7] fossulae *a* : possulae *SA* : passulae *c.*
[8] tripedane *S* : tripedaneae *Aa* : tripidaneae *c.*
[9] legantur *SAac.*

instructions for urban dinner-parties and sumptuous entertainments, and he produced three books on which he inscribed the titles " the Cook," " the Fishmonger " and " The Pickle-maker." We, however, are abundantly satisfied with such things as may, without great expense, fall to the lot of those living a simple, rural life.

Some people twist the pedicles of pomegranates, just as they grow upon the tree, so that the fruit may not be burst by the rains and, gaping open, utterly perish, and tie them to the larger branches that they may remain undisturbed; then they enclose the tree with nets made of broom, so that the fruit may not be torn by crows or rooks or other birds. Some people fit small earthenware vessels over the hanging fruit and, after daubing over with clay mixed with chaff, allow them to hang from the trees; others wrap up each fruit in hay or straw and smear them over thickly with clay mixed with chaff and then tie them to the larger branches, so that they may not be disturbed by the wind, as I have already said.

All these operations, as I have said, should be carried out when the weather is fine and there is no dew. But still they ought not to be carried out at all because the small trees are damaged, or at any rate they should not be put into practice year after year, particularly when one can pick this same fruit from the trees and keep it without its spoiling; for even under cover small trenches three feet across can be made in a very dry place and, after a little earth broken into small pieces has been put back, small branches of elder can be fixed in the ground, and then when the weather is calm pomegranates are picked together with their pedicles and

LUCIUS JUNIUS MODERATUS COLUMELLA

suis pediculis, et sambuco inseruntur (quoniam sambucus tam apertam et laxam medullam habet, ut
4 facile malorum pediculos recipiat). Sed cavere oportebit, ne minus quattuor digitis a terra absint, et ne inter se poma contingant. Tum factae scrobi operculum imponitur, et paleato luto circumlinitur, eaque humo, quae fuerat egesta, superaggeratur. Hoc idem etiam in dolio fieri potest, sive quis volet resolutam terram usque ad dimidium vas adicere,[1]
5 seu, quod quidam malunt, fluvialem arenam, ceteraque eadem ratione peragere. Poenus quidem Mago praecipit aquam marinam vehementer calefieri, et in ea mala granata lino vel sparto ligata paulum demitti, dum decolorentur, et exempta per triduum in sole siccari: postea loco frigido suspendi, et cum res exegerit, una nocte et postero die usque in eam horam, qua fuerit utendum, aqua frigida dulci macerari. Sed et idem auctor est creta figulari bene subacta recentia mala crasse illinire, et cum argilla exaruit, frigido loco suspendere. Mox cum exegerit usus, in aquam demittere,[2] et cretam resolvere. Haec ratio tamquam recentissimum pomum custodit.
6 Idem iubet Mago in urceo novo fictili substernere scobem populineam vel ligneam,[3] et ita disponere, ut scobis inter se calcari possit: deinde facto primo tabulato rursus scobem substernere, et similiter mala

---

[1] adicitur *S*: eicitur *Aac*.
[2] dimittere *SAac*.
[3] ilignea *SAac*.

inserted in the elder bush, for the pith of the elder is so loose and soft that it easily receives the pedicle of the fruit. But care will have to be taken that the pomegranate is not less than four inches from the ground and that the fruits do not touch one another. Next a cover is placed over the trench which you have made and is daubed all round with clay mixed with chaff and a mound is heaped over it with the earth which was dug out. The same thing can be done in a barrel, whether one chooses to fill it half-full with loose earth or, as some people prefer, with river-sand, the rest of the process being carried out as before. Mago the Carthaginian recommends that sea-water should be made exceedingly hot and that the pomegranates, tied with flax or rush, should be let down into it for a short time until they are discoloured and then taken out and dried for three days in the sun, and that afterwards they should be hung up in a cool place and, when they are required for use, they should be soaked in cold, fresh water for a night and the following day until the time when they are to be used. The same writer also suggests daubing the fruit, when it is fresh, thickly with well-kneaded potter's clay, and when the clay has dried, hanging it up in a cool place; then, when it is required for use, the fruit should be plunged in water and the clay dissolved. This process keeps the fruit as fresh as if it had only just been picked.

Mago also recommends that sawdust of poplar-wood or holm-oak wood should be spread on the bottom of a new earthenware pot and the fruit arranged so that the sawdust can be trodden in between them and that, when the first layer has been formed, sawdust should be again put down and the

disponere,[1] donec urceus [2] impleatur: qui cum fuerit repletus, operculum imponere, et crasso luto diligenter oblinire.

7 Omne autem pomum quod in vetustatem reponitur, cum pediculis [3] suis legendum est: sed, si sine arboris noxa fieri possit, etiam cum ramulis. Nam ea res plurimum ad perennitatem confert. Multi cum ramulis suis arbori detrahunt, et creta figulari cum diligenter mala obruerunt, in sole siccant: deinde si qua rimam creta fecit, luto linunt, et assiccata frigido loco suspendunt.

XLVII. Multi eadem ratione, qua granata, in scrobibus vel doliis servant cydonea. Nonnulli foliis ficulneis illigant, deinde cretam figularem cum amurca subigunt,[4] et ea linunt mala, quae cum siccata sunt, in tabulato frigido loco et sicco reponunt. Nonnulli haec eadem in patinas novas sicco gypso ita 2 obruunt, ut altera alteram non contingat. Nihil tamen certius aut melius experti sumus, quam ut cydonea maturrissima, integra, sine macula, et sereno caelo, decrescente luna, legantur, et in lagoena nova, quae sit patentissimi oris, detersa lanugine quae malis inest, componantur leviter et laxe, ne collidi possint: deinde cum ad fauces usque fuerint composita, vimineis surculis sic transversis [5] arctentur, ut modice mala comprimant, nec patiantur ea, cum acceperint liquorem, sublevari. Tum quam optimo et liquidissimo melle vas usque ad summum ita re-

---

[1] itaque sic facere *post* disponere *add. S.*
[2] urceus *ac*: urcilus *SA*.
[3] peciolis *SAac*.
[4] subiciunt *SAac*.
[5] tranversis *SAac*.

fruit similarly arranged until the pot is full, and, when it is full, the lid should be put on it and carefully sealed with thick clay.

Every fruit which is stored away to be kept for a long time ought to be picked with its pedicles; and if this can be done without harm to the tree, with a small branch also, for this contributes very greatly to making it last a long time. Many people pull the fruits from the trees together with the small boughs and, after carefully covering them with potter's clay, dry them in the sun; then they daub over with ordinary clay any crack in the potter's clay and after drying the fruits hang them up in a cool place.

XLVII. Many people keep quinces in pits or barrels in the same manner as they preserve pomegranates. Some tie them up in fig-leaves and then knead potter's clay with lees of oil and smear the quinces with it and, when they are dry, store them in a cool dry place in a loft. Others put these same fruits into new pans and cover them up with dry plaster in such a way that they do not touch one another. Nevertheless we have not experienced any more sure and satisfactory method than, when the weather is calm and the moon waning, to pick quinces which are very ripe, sound and without blemish and, after wiping off the down which is upon them, to arrange them lightly and loosely, so that they may not be bruised, in a new flagon with a very wide mouth; then, when they have been stowed in up to the neck of the vessel, they should be confined with willow-twigs laid across them in such a way that they compress the fruit slightly and do not allow them to be lifted up when they have liquid poured upon them. Then the vessel should be filled up to the top with the very best and most

3 pleatur, ut pomum submersum sit. Haec ratio non solum ipsa mala custodit, sed etiam liquorem mulsei[1] saporis praebet, qui sine noxa possit inter cibum dari febricitantibus, isque vocatur melomeli. Sed cavendum est, ne, quae in melle custodire volueris, immatura male condantur: quoniam cruda[2] si lecta 4 sunt, ita indurescunt, ut usui[3] non sint. Illud vero quod multi faciunt, ut ea dividant osseo cultro, et semina eximant, quod putent ex eis pomum vitiari, supervacuum est. Sed ratio quam nunc docui,[4] adeo quidem certa est,[5] ut etiam si vermiculus inest, non amplius tamen corrumpantur mala, cum praedictum liquorem acceperint: nam ea mellis est natura, ut coërceat vitia, nec serpere ea patiatur: qua ex causa etiam exanimum corpus hominis per annos plurimos innoxium conservat.

5 Itaque possunt etiam alia genera malorum sicut orbiculata, Cestiana,[6] melimela, Matiana, hoc liquore custodiri. Sed quia videntur in melle dulciora fieri sic condita, nec proprium saporem conservare, arculae faginae vel etiam tiliagineae, quales sunt in quibus vestimenta forensia conduntur, huic rei paulo ampliores praeparari debent, eaeque in tabulato frigidissimo et siccissimo, quo neque fumus neque taeter perveniat odor, collocantur: deinde carta[7] substrata

[1] mulsi *SAac*.   [2] grossa *SAac*.
[3] usi *SAa* : usui *c*.
[4] est—docui : esse non docui *SAac*.
[5] certa est om. *SAac*.
[6] caesiana *SA* : cesiana *ac*.
[7] cartha *S* : om. *Aac*.

---

[a] Apparently a highly esteemed kind, since *malis orbiculatis pasti* was a proverbial expression for well-fed (Cicero, *Fam.* 8. 15. 1).

liquid honey, so that the fruit is submerged. This method not only preserves the fruit itself, but also provides a liquor which has the flavour of honey-water and can without danger be given at their meals to sufferers from fever. It is called *melomeli*. But care must be taken that the fruit which you wish to preserve in honey is not stored before it is ripe; for, if it is picked when it is unripe, it becomes so hard as not to be fit for use. But the practice of many people of cutting up the fruit with a bone knife and taking out the seeds, because they think that the fruit receives harm from these, is quite superfluous, but the method which I have just described is, indeed, so sure that even if there is a small worm in the fruits, they do not deteriorate any further once they have the liquid described above added to them; for such is the nature of honey that it checks any corruption and does not allow it to spread, and this is the reason too why it preserves a dead human body for very many years without decay.

So also various kinds of apples can be kept in this liquid, such as globe-apples,[a] Cestine apples, honey-apples and Matian[b] apples; but since when thus stored they seem to become sweeter in the honey and not to keep their own particular flavour, small chests of beech or even lime wood such as those used for storing official robes but somewhat large, ought to be prepared for this purpose and placed in a very cold and dry loft, to which neither smoke nor foul odour can penetrate. Then, when a thin sheet has been spread below, the apples already men-

[b] Probably called after C. Matius; see note on Ch. 4. § 2 of this book.

LUCIUS JUNIUS MODERATUS COLUMELLA

praedicta poma[1] sic componi ut flosculi sursum pediculi deorsum spectent, quemadmodum etiam in arbore nata sunt, et ne inter se alterum ab altero contingantur. Item observandum est, ut unumquodque genus separatim propriis arculis reponatur.[2] Nam cum una clausa sunt diversa genera, inter se discordant, et celerius vitiantur. Propter quod etiam conseminalium vinearum non tam est firmum vinum, quam si per se sincerum Amineum, vel apianum, aut etiam faecinium[3] condideris. Verum sicut supra dixi, cum diligenter mala fuerint composita, operculis arcularum contegantur, et luto paleato linantur opercula, ne introire spiritus possit. Atque ea ipsa nonnulli, sicut in aliis generibus supra iam diximus, populnea, quidam etiam abiegna[4] scobe interposita, mala custodiunt. Haec tamen poma non matura, sed acerbissima legi debent.

XLVIII. Inulae curatio.[5] Cum eius radicem mense Octobri, quo maxime matura est, e terra erueris, aspero linteolo vel etiam cilicio detergito quidquid arenae inhaerebit: deinde acutissimo cultello summatim eradito, et quae plenior radicula fuerit, pro modo crassitudinis, in duas vel plures partes digiti longitudine diffindito: deinde ex aceto modice in aeneo coquito, ita ne taleolae semicrudae sint. Post haec in umbra triduo siccentur, et ita in fideliam picatam

[1] pomi *SAac*.
[2] reponantur *SAac*.
[3] fecinum *SAac*.
[4] abiegna *ac* : abiegenea *S* : abigena *A*.
[5] inulae curatio *SA* : om. *ac*.

---

[a] *I.e.* the part of the apple opposite to the pedicle, sometimes called *umbilicus*.
[b] See note on Chap. XXXIX.

tioned should be so arranged that the "floweret"[a] faces upwards and the pedicle downwards, as they grew on the tree, and so as not to touch one another. Care must also be taken that each kind is stored separately in its own chest; for when different kinds are shut up together, they disagree with one another and more quickly deteriorate. It is for this reason too that the wine from vineyards in which different kinds of vines are planted does not keep so well as when you have stored unmixed Aminean or Apian bee-wine[b] or even faecinian[c] wine by themselves. But, as I have said above, when the apples have been carefully arranged, they should be covered with the lids of the chests and these should be smeared with clay mixed with chaff, so that the air cannot enter. Furthermore, these very apples some people keep, as we have already mentioned above in speaking of other kinds, by putting poplar- or fir-tree-sawdust between them; but these apples should be picked when they are not ripe but still very bitter.

XLVIII. For the preservation of elecampane,[d] you pull the root out of the earth in the month of October, when it is at its ripest, and wipe off with a coarse linen cloth or even with a hair-cloth any sand which sticks to it, and then scrape it superficially with a very sharp knife, and split the stouter of the small roots into two or more parts, according to its thickness, of a finger's length. Then cook them slightly in a brass pot in vinegar so that the slices may not be half raw. After this they should be dried for three days in the shade and then stored in a jar treated

How to pickle elecampane.

[c] See Book III. 2. 14; the name was derived from the fact that this wine produced more dregs (faeces) than other kinds.
[d] *Inula helenium.*

LUCIUS JUNIUS MODERATUS COLUMELLA

recondantur, adiecto passo vel defruto, quod supernatet, spissamentoque cunilae imposito contectum vas pelliculetur.

2 Alia inulae conditura. Cum radices eius eraseris, taleolas ut supra facito, et in umbra triduo vel etiam quatriduo siccato: deinde siccatas in vasis sine pice, interiecta cunila conicito. Ius infundito, quod eam compositionem habeat, ut sex partibus aceti una pars sapae misceatur cum hemina salis cocti. Eo iure macerentur taleolae, donec quam minimum
3 amaritudinis resipiant. Postea exemptae [1] iterum siccentur per dies quinque in umbra: tum crassamentum [2] vini faeculenti, nec minus, si sit, mulsi, et utriusque eorum quartam partem boni defruti confundito in ollam: quae cum inferbuerit, taleolas inulae adicito, et statim ab igne removeto, ac rudicula lignea peragitato, donec perfecte refrigescant. Postea transfundito in fideliam picatam, operculo tegito, tumque pelliculato.

4 Cum radiculas diligenter eraseris, minute concisas in muria dura macerato, donec amaritudinem dimittant. Deinde effusa muria, sorba [3] quam optima et maturissima semine detracto contere, et cum inula misce. Tum sive passum seu quam optimum de-
5 frutum adicito, et vas obturato. Quidam cum

---

[1] exempta *SAac*.
[2] crassamen *S*.
[3] sorba *a* : sorva *c* : sorua *A* : soruua *S*.

with pitch, raisin-wine or boiled-down must having been added, which should cover the top of them, and, after a stuffing of marjoram has been put in, the vessel should be shut up and covered with a piece of skin.

Here is another method of preserving elecampane. 2
After scraping the roots, cut them into slices, as A second
above, and dry them for three or even four days in method.
the shade; then, when they are dry put them in
vessels without pitch, adding some marjoram to
them. Pour in a liquid composed as follows: one
part of must boiled-down to half its original volume
mixed with six parts of vinegar together with a
hemina of baked salt. Let the slices soak in this
liquid until they have the least possible taste of
bitterness. Afterwards let the slices be taken out 3
again and dried for five days in the shade; then
pour into a pot the sediment of wine which is full of
dregs and likewise of mead, if you have any, and
good boiled-down must to a fourth part of the other
two ingredients. When this mixture has boiled,
add the slices of elecampane and immediately remove
the pot from the fire and stir it well with a wooden
spoon until the slices have become perfectly cold.
Afterwards transfer them to a jar, which has been
treated with pitch, and cover it with a lid and then
put a piece of skin on the top.

After carefully scraping the small roots, cut them 4
up very small and soak them in hard brine until they A third
lose their bitterness; then, after pouring away the method.
brine, crush the very best and ripest service-apples
after taking out the seeds and mix with the elecampane. Then add either raisin-wine or the best
possible boiled-down must and stop up the vessel.
Some people, after they have preserved elecampane 5

condiverunt inulam muriaque maceraverunt, exsiccant, et malis cydoneis tritis, quae in defruto vel melle decoxerant, miscent: atque ita superfundunt [1] passum vel defrutum, et vas operculatum pelliculant.[2]

XLIX. Olivarum conditurae. Acerbam Pauseam mense Septembri, vel Octobri, dum adhuc vindemia est, contunde,[3] et aqua calida paululum maceratam exprime, faeniculique seminibus et lentisci cum cocto [4] sale modice permixtam reconde in fideliam, et mustum quam recentissimum infunde. Tum fasciculum viridis faeniculi superpositum merge, ut 2 olivae premantur, et ius superemineat. Sic curata oliva tertio die possis uti. Albam Pauseam, vel orchitem,[5] vel radiolum, vel regiam dum contundes, primum quamque, ne decoloretur, in frigidam muriam [6] demerge, cuius cum tantum paratae habueris, quantum satis fuerit implendae amphorae, faeniculi aridi fasciculum substerne in imo: deinde viridis faeniculi semina et lentisci destricta et purgata in urceolo habeto: tum exemptam [7] de muria olivam exprimito, et permixtam [8] praedictis seminibus in vas adicito: deinde cum ad fauces pervenerit eius, faeniculi aridi fasciculos superponito, et ita recentis musti duas partes et unam durae muriae permixtas 3 adicito. Hac conditura compositis olivis toto anno commode uteris.

---

[1] superfundunt *A*.
[2] pelliculant *A*.
[3] contundere *SAac*.
[4] cocto *a*: cocta *SA*: aceto *c*.
[5] orchitem *ac*: horhadem *SA*.
[6] muriam *a*: duriam *SAc*.
[7] exempta *SAac*.

and soaked it in brine, dry it and mix it with crushed quinces, which have been boiled in boiled-down must or honey; they then pour on the top raisin-wine or boiled-down must and put the lid on the vessel and cover it with a piece of skin.

XLIX. The methods of preserving olives are as follows. In the month of September or October, while the vintage is still going on, crush the bitter Pausean olive and, after soaking it for a short time in hot water, squeeze it, and after mixing it slightly for a short time with fennel and mastic seeds together with toasted salt put it in a jar and pour in the freshest possible must. Then put on the top a bunch of green fennel and sink it so that the olives may be pressed down and the liquid rise above them. Having thus treated your olive, you may use it on the third day.

How to preserve Pausean olives.

When you are going to bruise the white Pausean, or the orchites or the shuttle olive or the royal olive, first of all plunge each of them into cold brine, so that it may not lose its colour, and when you have enough ready to fill a jar, spread a bunch of dry fennel in the bottom of it; then have sprigs of green fennel and mastic ready stripped and cleaned in a small pot. Take the olives out of the brine and squeeze them, and mixing them with the said sprigs put them into the vessel. Next, when the olives have reached the neck of the vessel, put bunches of dry fennel on the top of them and then add a mixture of two parts of fresh must and one of hard brine. Olives which have been treated by this method of preservation you can perfectly well use for a whole year.

---

⁸ permixta *SAac*.

LUCIUS JUNIUS MODERATUS COLUMELLA

Quidam olivam non contundunt, sed acuta harundine insecant: idque operosius quidem, sed melius est, quia haec candidior est oliva, quam ea quae ex contusione livorem contrahit. Alii sive contuderint, sive insecuerint olivas, modico sale cocto et praedictis seminibus immiscent: deinde sapam vel passum vel, si est facultas, mellam infundunt. Mella autem quomodo fiat, paulo ante hoc ipso libro praecepimus. Cetera omnia similiter administrantur.

4 Oliva alba ex muria. Pausias olivas vel regias sine macula quam candidissimas manu destrictas [1] eligito: deinde substrato faeniculo arido in amphoram conicito [2] intermixtis seminibus lentisci nec minus faeniculi: et cum ad fauces vas [3] repleveris, adicito muriam duram: tum spissamento facto de harundinum foliis olivam premito, ut infra ius mersa sit: et iterum infundito muriam duram, dum ad summum amphorae 5 labrum perveniat. At haec oliva per se parum iucunda est; sed ad eas condituras, quae lautioribus mensis adhibentur, idonea maxime est: nam cum res exigit, de amphora promitur, et contusa recipit quamcunque volueris condituram. Plerique tamen sectivum porrum et rutam cum apio tenero et mentam minute concidant, et contusis olivis miscent: deinde exiguum aceti piperati, et plusculum mellis aut mulsi adiciunt, oleoque [4] viridi irrorant,[5] atque ita fasciculo apii viridis contegitur.

[1] tringinto S : trito et A : tritas ac.
[2] coito S : coicito Aa : conicito c.
[3] uvas SAc : vas a.   [4] oleumque SAac.
[5] inrorantur SAa : irrorantur c.

---

[a] Chap. 11.

BOOK XII. XLIX. 3-5

Some people do not bruise olives but cut them with a sharp reed; this method is indeed more laborious but it is better, because the olive thus treated is whiter than one which acquires a bluish colour from the bruising. Other people, whether they have bruised or cut their olives, mix them with a little toasted salt and the sprigs mentioned above; they then pour over them must boiled down to half its original volume or raisin-wine, or, if they can get it, bee's-wax-water: how this is made we have described a little earlier in this present book [a]; all the rest of the process is carried out in the same manner.

White olives are treated with brine as follows. Choose the whitest possible Pausean or royal olives without blemish which have been stripped by hand; then throw them into a wine-jar having put underneath them some dry fennel, mingling with them mastic-sprigs and also fennel sprigs, and when you have filled the vessel up to the neck, add hard brine. Then make a stuffing of the foliage of reeds and press the olives down with it, so that they are submerged in the liquid; pour in hard brine a second time, until it reaches the brim of the jar. However, this olive is not very pleasant by itself but is especially suitable for the preparation of preserves which are served at the more sumptuous repasts; for, when required, it is taken out of the jar and, after being crushed, blends with any other seasoning you like. Most people, however, cut up finely leeks and rue with young parsley and mint and mix them with crushed olives; then they add a little peppered vinegar and a very little honey or mead and sprinkle them with green olive-oil and then cover them with a bunch of green parsley.

LUCIUS JUNIUS MODERATUS COLUMELLA

6 Quidam sic lectae olivae in modios singulos ternas heminas salis permiscent, et adiectis[1] seminibus lentisci faeniculoque substrato amphoram usque ad fauces replent olivis: deinde aceto non acerrimo infundunt, et cum iam paene amphoram impleverunt, faeniculi spissamento deprimunt bacam, et rursus acetum usque ad summum labrum adiciunt. Postea quadragesimo die omne ius defundunt, et sapae vel defruti tres partes cum aceti una permiscent, et
7 amphoram replent. Est et illa probata compositio, ut cum muria dura Pausea alba ubi commaturuerit, omne ius defundatur, et immixtis duabus partibus defruti cum aceti una repleatur amphora. Eadem conditura possit etiam regia componi vel orchita.
8 Quidam unam partem muriae et duas aceti miscent, eoque iure olivas Pauseas colymbadas faciunt: quibus si per se quis uti velit, satis iucundas experietur, quamvis et hae[2] de muria condituram qualemcumque recipere possint.[3] Olivae Pauseae, cum iam decolorantur, antequam mitescant, cum petiolo leguntur, et in oleo quam optimo servantur. Haec maxime nota etiam post annum repraesentat viridem saporem olivarum. Nonnulli etiam cum de oleo exemerunt, trito sale aspersas pro novis apponunt.
9 Est et illud conditurae genus, quod in civitatibus Graecis plerumque usurpatur, idque vocant epityrum. Oliva Pausea vel orchita cum primum ex albo de-

[1] adiectis *ac*: adlectis *SA*.
[2] haec sic *SAa*: haec *c*.
[3] possit *SAac*.

---

[a] Pliny, *N.H.* XV. § 16: they were so called because they float on the surface of the salted water.
[b] Cato, *R.R.* 119.

BOOK XII. XLIX. 6-9

Some people mix three *heminae* of salt in each *modius* of olives picked in this manner and, having added mastic-shoots and spread fennel under them, fill the jar up to the neck with olives; they then pour upon them not very acid vinegar and, when they have almost filled the jar, press down the berries with a stuffing of fennel and again add vinegar till it reaches the very brim. Afterwards, on the fortieth day, they pour out all the liquid and mix three parts of must boiled down to a third or a half of its original volume with one part of vinegar and fill up the jar again. The following preparation is also highly approved: when the white Pausean olive has been ripened with hard brine, let all the liquid be poured out and the jar filled again with two parts of boiled-down must mixed with one part of vinegar. The royal and the orchite olive could also be treated by this method of preserving.

Some people mix one part of brine and two of vinegar and with this liquid make Pausean olives into "swimmers."[a] Anyone who cares to make use of these by themselves will find them pleasant enough, though they too, when they come out of the brine, can take on any other kind of seasoning. Pausean olives, when they are already becoming discoloured and before they grow ripe, are picked with the stem and kept in the best possible oil: this kind above all others even when a year has passed presents the fresh olive flavour. Some people also take them out of oil and after sprinkling them with pounded salt serve them up as new olives. There is also the kind of preservative which is commonly made use of in the Greek cities, and they call it *epityrum*.[b] The Pausean or orchite olive, as soon as it begins to lose its white

coloratur,[1] fitque luteola, sereno caelo manu destringitur, et in cannis uno die sub umbra expanditur: et siqui adhaerent pediculi foliaque aut surculi, leguntur. Postero die cribratur, et novo fisco inclusa prelo supponitur, vehementerque premitur, ut exsudet
10 quantulumcumque habet amurcae. Patimur autem nonnuquam tota nocte et postero die pondere pressam bacam velut exinanari,[2] tum resolutis corticulis eximimus eam, et in singulos modios olivae triti salis cocti singulos sextarios infundimus: itemque lentisci semen rutaeque et faeniculi folia sub umbra siccata, quanta satis videntur, concisa minute admiscemus, patimurque horis tribus, dum aliquatenus baca salem combibat. Tum superfundimus boni saporis oleum, ita ut obruat olivam, et faeniculi aridi fasciculum deprimimus, ita ut ius
11 supernatet. Huic autem conditurae vasa nova fictilia sine pice praeparantur: quae ne possint oleum sorbere, tamquam olivariae metretae, imbuuntur liquida gummi et assiccantur.

L. Sequitur autem frigus hiemis, per quod olivitas, sicut vindemia, curam villicae repetit. Prius itaque (quoniam incohavimus) de condituris olivarum praecipiemus, ac statim conficiendi olei rationem subiciemus. Pauseae bacae vel orchitae, nonnullis regionibus etiam Naeviae, conviviorum epulis praeparantur. Has igitur cum iam nigruerint, nec adhuc tamen per-

---

[1] decolorantur *SAac.*
[2] exinanari *scripsi*: exaniari *S*: examari *ac.*

colour and becomes yellowish, is picked by hand in fine weather and spread out for one day on reeds in the shade; any pedicles and leaves or twigs which adhere are picked off. The next day the olives are sifted, and after being enclosed in a new rush-bag, are put under the press and violently squeezed so as to exude any small quantity of lees which they have in them. But sometimes we allow the berries to remain under the weight of the press for a whole night and the next day and to be, as it were, entirely emptied, then, the rind of the shells being broken, we take the olives out and pour a *sextarius* of pounded and toasted salt on each *modius* of olives and also mix in sprigs of mastic and rue and leaves of fennel which have been dried in the shade, as much as seems enough, cut up very small, and leave them for three hours until the berries have to some extent absorbed the salt. Then we pour on the top oil of a good flavour so as to cover the olives and press down a bundle of dry fennel so that the liquid floats above them. For this kind of preserve new earthenware vessels are prepared, not treated with pitch, and in order that they may not be able to absorb the oil, they are soaked, as are olive-casks, with liquid gum and dried.

L. After this comes the cold of winter during which the olive-harvest, just as did the vintage, calls again for the attention of the bailiff's wife. We will, therefore, first of all (since we have already begun the subject) give instructions about preserving olives and immediately afterwards append an account of the way to make olive-oil. The Pausean and orchite olives and, in some districts, the Naevian also are prepared for feasts to which guests are invited; there-

How to preserve black olives.

LUCIUS JUNIUS MODERATUS COLUMELLA

maturae fuerint, sereno caelo destringere manu convenit, lectasque cribrare, et secernere, quaecumque maculosae seu vitiosae minorisve incrementi videbuntur: deinde in singulos modios olivae salis integri ternas heminas adicere, et in vimineos qualos confundere et superposito copioso sale, ita uti olivam contegat, triginta dies pati consudescere, atque omnem amurcam exstillare: postea in alveum diffundere, mundaque spongia salem, ne perveniat, detergere: tum in vas adicere, et sapa vel defruto amphoram replere superposito spissamento aridi faeniculi, quod olivam deprimat. Plerique tamen tres partes defruti aut mellis et unam miscent aceti, aliqui duas partes, et unam aceti,[1] et eo condiunt iure.

Quidam, cum olivam nigram legerunt, eadem portione,[2] qua supra, saliunt, et sic collocant in qualis, ut immixtis seminibus lentisci alterna tabulata olivarum et similiter deinde salis, tum iterum olivarum et similiter supra salis usque in summum componant: deinde post quadraginta dies, cum oliva quidquid habuit amurcae[3] exsudavit, in alveum defundunt, et cribratam separant ab seminibus lentisci, spongiaque detergent, nequid adhaereat salis: tum in amphoram confundant adiecto defruto vel sapa

---

[1] aliqui—aceti *om. S.*
[2] portionem *SAac.*
[3] amurga *S* : amurca *Aac.*

fore, when they have already turned dark but are not yet very ripe, they should be picked by hand when the weather is fine and, when picked, should be sifted and any set aside which shall seem to be spotted or spoilt or of too small growth; then to every *modius* of olives you should add three *heminae* of salt in its natural state and pour the olives into baskets of wickerwork, and after putting plenty of salt on the top of them, so that it completely covers the olives, leave them for thirty days to sweat together and let all their lees drip out; afterwards you should pour them into a tub and wipe off the salt with a clean sponge so that it may not penetrate into them. You then put them into a vessel and fill the jar with must boiled down to half or a third of its original volume, putting on the top a plug of dry fennel, so as to press the olives down. Most people, however, mix three parts, and some two parts, of boiled-down must or honey with one of vinegar and preserve the olives with this liquid.

Some people after picking black olives salt them with the same quantity of salt, as is given above, and then place them in baskets in such a way that, when mastic-sprigs have been mixed with them, they pour alternate layers first of olives and then similarly of salt and then again of olives and similarly of salt above; then after forty days, when the olives have sweated out any lees that they contained, they pour them out into a tub and after sifting them separate them from the mastic-sprigs and wipe them with a sponge so that no salt may adhere to them. Then they pour them into a jar, adding must that has been boiled down to one-third or one-half of its original volume or even honey-water, if it is

Another method.

LUCIUS JUNIUS MODERATUS COLUMELLA

vel etiam mella, si est copiosa, ceteraque similiter faciunt.

4 In singulos modios olivae singulos sextarios maturi seminis lentisci[1] et ternos cyathos seminis faeniculi; si id non est, ipsum faeniculum concisum, quantum satis videbitur, adici oportet: deinde in singulis modiis[2] olivarum salis cocti,[3] sed non moliti, ternas heminas admisceri, et ita in amphoris condi, easque fasciculis faeniculi obturari, et quotidie per terram volutari: deinde tertio quoque aut quarto 5 die quidquid amurcae inest, emitti. Post XL dies in alveum diffundi, et a sale tantummodo separari, sic ne[4] spongia detergeantur[5] olivae, sed ita ut erant exemptae,[6] massulis salis mixtis,[7] in amphoram condantur, et spissamentis impositis ad usus in cellam reponantur.

Maturam olivam in salsura[8] factam colymbadem de muria tollito, spongia tergito: deinde canna viridi scindito duobus vel tribus locis, et triduo in aceto habeto: quarto die spongia extergito, in vas, id est urceum aut cacabum novum, mittito substrato apio et modica ruta. Conchis deinde pleno vase olivis immitte defrutum usque ad os. Lauri

[1] lentisque *SAac*.
[2] modiis *ac* : modios *A* : modis *S*.
[3] cocti *ac* : -is *SA*.
[4] sic ut non *Sa* : sīc ñ *A* : sic nõ *c*.
[5] detergatur *SAac*.
[6] exempta *SAac*.
[7] maxime salis micis *SA* : maximi salis miscis *Ac*.
[8] salsura *Gesner* : stativam *SAac*.

---

a I.e. bundles, probably of fennel, to keep the olives below the surface of the liquid.

BOOK XII. L. 3-5

abundant, and carry out the rest of the process as before.

To each *modius* of olives must be added a single *sextarius* of ripe mastic-seed and three *cyathi* of fennel-seed; if the latter is unobtainable, as much fennel itself cut up small as seems enough should be added. Then with each *modius* of olives three 4 *heminae* of toasted, but not ground salt should be mixed and then the olives should be put up in jars and closed in with bundles of fennel and daily rolled along the ground. Then on every third or fourth day any lees that there is in them should be let out. After forty days the olives should be poured into a 5 tub and merely separated from the salt, but without being wiped with a sponge, put back into the jar just as they were when taken out, with small lumps of salt mixed with them, and after stuffings [a] have been put on them they should be laid up for future use in the store-room.

Another method.

Ripe olives which have become "swimmers" [b] in the process of salting, you should take out of the brine and wipe with a sponge; then cut them in two or three places with a green reed and keep in vinegar for three days: on the fourth day wipe clean with a sponge and put them into a vessel, that is, a new pot or jar, after laying parsley and a little rue underneath. When the vessel is full of olives pour in by means of shells [c] must boiled down to a third of its original volume till it reaches the mouth of the vessel, and put in young shoots of bay with the

[b] See Ch. 49. 7.
[c] For the use of shells for ladling liquid see Horace, *Sat.* I. 3. 14, and Wickham's note. *Cf.* iron ladles, Ch. LII. 8.

LUCIUS JUNIUS MODERATUS COLUMELLA

turiones in hoc usu mittito, ut [1] olivas deprimant. Post dies viginti utere.

LI. Oliva nigra maturissima sereno caelo legitur, eaque sub umbra uno die in cannis porrigitur, et quaecumque est vitiosa baca, separatur. Item siqui adhaeserunt pediculi, adimuntur, foliaque et surculi, quicumque sunt intermixti, eliguntur. Postero die diligenter cribratur, ut siquid inest stercoris separetur: deinde intrita [2] oliva novo fisco includitur, et prelo subicitur, ut tota nocte exprimatur.
2 Postero die inicitur quam mundissimis molis suspensis, ne nucleus frangatur. Et cum est in sampsam redacta, tunc sal coctus tritusque manu permiscetur cum ceteris aridis condimentis. Haec autem sunt, careum, cyminum, semen faeniculi, anisum Aegyptium. Sat erit autem totidem heminas salis adicere, quot sunt modii [3] olivarum, et oleum superfundere, ne exarescat: idque fieri debebit, quotiescumque videbitur assiccari.

3 Nec dubium est, quin optimi saporis sit, quae ex oliva Pausia [4] facta est. Ceterum supra duos menses sapor eius non permanet integer. Videntur autem alia genera huic rei magis esse idonea, sicut Liciniae et Culmineae. Verumtamen habetur praecipua in hos usus olea Calabrica, quam quidam propter similitudinem oleastellum vocant.

[1] et *SAac*. [2] intra *Aac* : int̄ *S*.
[3] modii *ac* : modi *SA*. [4] postea *S*.

---

[a] *I.e.* the upper mill-stone is so fixed as not to grind very hard on the lower.

BOOK XII. L. 5–LI. 3

object of pressing the olives down. After twenty days you can use them.

LI. Black olives are gathered very ripe when the weather is fair and are spread in the shade on reeds for one day, and any damaged berries are set aside. Also any pedicles which have adhered to the fruit are removed, and any leaves and twigs which are intermingled with them are picked out. On the following day the olives are carefully sifted in order that any dirt which is among them may be separated: then the unbruised olives are enclosed in a new frail and put under the press that they may be squeezed for a whole night. Next day they are thrown into mills which are as clean as possible and are suspended [a] so that the kernels may not be broken, and when they are reduced to a pulp, salt which has been toasted and rubbed in the hand is mixed with the other dry seasonings, which are caraway, cumin, fennel seed and Egyptian anise-seed. It will, however, be enough to add the same number of *heminae* of salt as there are *modii* of olives and to pour oil upon the top so that they may not become dry; this will have to be done whenever they seem to be getting dry.

There is no doubt that the preparation which is made from the Pausean olive has the best flavour, but it does not remain unimpaired for more than two months. Other kinds of olives seem more suitable for this purpose, such as the Licinian and Culminian; however the Calabrian olive, which some people call the "little wild olive," because of its resemblance to the wild olive, is regarded as particularly good for these purposes.

LUCIUS JUNIUS MODERATUS COLUMELLA

LII. Media est olivitas plerumque initium mensis Decembris. Nam et ante hoc tempus acerbum oleum conficitur, quod vocatur aestivum, et circa hunc mensem viride premitur, deinde postea maturum. Sed acerbum oleum facere patrisfamilias rationibus non conducit, quoniam exiguum fluit, nisi baca tempestatibus in terram decidit, et necesse est eam sublegere, ne a domesticis pecudibus 2 ferisve consumatur. Viridis autem notae conficere vel maxime expedit, quoniam et satis fluit, et pretio paene duplicat domini reditum. Sed si vasta sunt oliveta,[1] necesse est ut[2] aliqua pars eorum maturo fructui reservetur.

Locus autem in quo confici oleum debet, etiam descriptus est priore volumine; pauca tamen ad rem pertinentia commemoranda sunt, quae prius omi-3 seram. Tabulatum, quo inferatur olea, necessarium est, quamvis praeceptum habeamus, ut[3] uniuscuiusque diei fructus molis et prelo statim subiciatur. Verumtamen quia interdum multitudo bacae torculariorum[4] vincit laborem,[5] esse oportet pensile horreum, quo importentur fructus: idque tabulatum simile esse debet granario, et habere lacusculos tam multos, quam postulabit modus olivae, ut separetur et seorsum reponatur uniuscuiusque diei coactura.

[1] oliveta *ac* : -o *SA*.
[2] ut *om. SAac*.
[3] ut *om. SAac*.
[4] torculariorum *c* : troclariorum *Aa* : trocariorum *S*.
[5] laborem sit laborem *SA* : labore sit *a* : laborem sic labor est *c*.

BOOK XII. LII. 1–3

LII. The beginning of the month of December generally falls in the middle of the olive-harvest, for, earlier than this, the oil that is made, the so-called "summer" oil, is bitter, and this month is about the time when the green oil is pressed and then afterwards the ripe oil. But, because the yield is small, it is not to the interest of the master of the household to make bitter oil, unless the berries have fallen to the ground owing to storms and it is necessary to gather them up, so that they may not be consumed by domestic or wild animals. But it is 2 most expedient that the master should make some of the green kind, because the yield of it is satisfactory and it almost doubles his profit by the price which it fetches; but if the olive-groves are extensive, it is essential that some part of them should be reserved for the ripe fruit.

The place in which the oil should be made has been described in an earlier volume;[a] there are, however, a few points connected with this subject which I had omitted before and must mention now. A loft to 3 which the olives may be carried is necessary, though we have already given instructions that each day's picking should be immediately placed under millstones and the press. However, since the large quantity of berries sometimes defeats the efforts of the men who work the presses, there ought to be a storeroom supported on arches, to which the fruit may be conveyed. This loft ought to be like a granary and contain as many bins as the quantity of olives will require, so that each day's picking may be kept apart

How to make olive-oil.

---

[a] The culture of olives is treated in Book V. Ch. 8 and 9, but Columella says nothing there about the place where the oil is made.

4 Horum lacusculorum solum lapide vel tegulis oportet consterni, et ita declive fieri, ut celeriter omnis humor per canales aut fistulas defluat. Nam est inimicissima oleo amurca, quae si remansit in baca, saporem olei corrumpit. Itaque cum lacus, quemadmodum diximus, exstruxeris, asserculos inter se distantes semipedalibus spatiis supra solum ponito, et cannas diligenter spisse textas inicito, ita ut ne bacam transmittere queant, et olivae pondus 5 possint sustinere. Iuxta omnes autem lacusculos, ea parte qua defluet amurca, sub ipsis fistulis in modum fossularum concavum pavimentum, vel canalem lapideum esse oportebit, in quo consistat,[1] et unde exhauriri possit quidquid defluxerit. Praeterea lacus vel dolia praeparata sub tecto haberi oportebit, quae seorsum recipiant sui cuiusque generis amurcam, sive quae sincera defluxerit, sive etiam quae salem receperit. Nam utraque usibus plurimis idonea est.

6 Oleo autem conficiendo molae utiliores sunt, quam trapetum; trapetum, quam canalis et solea.[2] Molae quam [3] facillimam patiuntur administrationem, cum [4] pro magnitudine bacarum vel submitti vel etiam elevari possunt, ne nucleus, qui saporem olei vitiat, confringatur. Rursus trapetum plus operis facilius-

7 que quam solea et canalis efficit. Est et organum erectae tribulae simile, quod tudicula vocatur:

[1] consistant *SAac.*
[2] est olea *SAac.*
[3] quia *S.*   [4] tum *SAac.*

---

[a] Cato, *R.R.* 22. 1, describes a press (*trapetum*) but his account is difficult to follow; Varro, *R.R.* I. 55. 5, says that it was fitted with hard, rough mill-stones.

and stored separately. The floor of these bins must 4 be paved with stones or tiles and made to slope so that all moisture may flow quickly away by gutters or pipes; for lees are most harmful to the oil, and if they remain in the berries, spoil the flavour of the oil. So, when you have constructed the receptacles in the manner which we have described, place on the floor small boards half a foot distant from one another and put reeds on the top carefully, closely woven so as not to allow the berries to pass through and yet be able to support the weight of the olives. Adjoining all the 5 bins on the side where the lees flow down from them, under the pipes themselves, there will have to be a concave pavement in the form of little trenches or else a stone channel, in which any liquid which has flowed down may stand and from which it can be drawn off. Moreover it will be necessary that vats or barrels should be kept in readiness indoors to receive the lees of each kind of olive separately, whether they be such as flow down without admixture with anything else or have been salted; for both kinds are suitable for very many purposes.

Mills are more practicable for making oil than an 6 oil-press,[a] and an oil-press than a *canalis* or a *solea*.[b] Mills are very easily managed, since they can be either lowered or raised according to the size of the berries, so that the kernel, which spoils the flavour of the oil, is not broken. Moreover the oil-press does more work and with greater ease than the *solea* or *canalis*. There is also a machine which resembles an 7 upright threshing-sledge, which is called a *tudicula*;[c]

[b] Nothing is known of these two machines, which are not mentioned by any other author.

[c] Literally "a little hammer."

LUCIUS JUNIUS MODERATUS COLUMELLA

idque non incommode opus efficit, nisi quod frequenter vitiatur, et si bacae plusculum ingesseris, impeditur. Pro conditione tamen et regionum consuetudine praedictae machinae exercentur. Sed et optimum [1] molarum opus est,[2] tum etiam trapeti. Haec ante quam de oleo conficiendo dissererem, praefari necesse habui.

8 Nunc ad ipsam rem veniendum est, quamquam multa omissa sunt, quae sicut ante vindemiam, sic et ante olivitatem praeparanda sunt, tamquam lignorum copia, quae multo ante apportanda est, ne cum res desideraverit, operae avocentur; tum scalae, corbulae, decemmodiae, trimodiae satoriae, quibus destricta baca suscipitur,[3] fisci, funes cannabini,[4] spartei, conchae ferreae, quibus depletur oleum, opercula, quibus vasa olearia conteguntur, spongiae maiores et minores, urcei, quibus oleum progeritur, cannae, tegetes, quibus oliva excipitur, et siqua sunt 9 alia, quae nunc memoriam meam refugiunt. Haec omnia multo plura esse debent: quoniam in usu depereunt, et pauciora fiunt; quorum siquid suo tempore defuerit, opus intermittitur. Sed iam quod pollicitus sum exequar.

Cum primum bacae variare coeperint, et iam quaedam nigrae fuerint, plures tamen albae, sereno caelo manibus destringi olivam oportebit, et sub-
10 stratis [5] tegetibus aut cannis cribrari et purgari. Tum diligenter emundatam protinus in torcular deferri, et

---

[1] optimum *scripsi*: optima *SAac*.
[2] opus est *om. SAac*.
[3] suspicitur *SAc*: subicitur *a*.
[4] cannavini veli *S*: canna veli *A*: cannales *a*: canali *c*.
[5] substractis *SA*: substratis *a*: subtractis *c*.

this performs its function without much trouble except that it frequently gets out of order, and, if you put a little too many berries into it, it becomes clogged. The above machines are used, however, according to conditions and local custom; but the performance of the mill is best and that of the oil-press the next best. I have thought it necessary to mention these points before treating of the actual making of oil.

We must now come to our real subject, though many details have been omitted of the preparations which have to be made before the olive harvest just as before the vintage; for example, there should be plenty of wood, which should be brought a long time beforehand, so that the workmen may not be called away when circumstance demands wood. Then there must be ladders, little baskets, sowers' baskets holding ten and three *modii*, to contain the berries when they are stripped from the trees, frails, ropes of hemp and of broom, iron ladles for emptying out the oil, lids for covering the oil vessels, large and small sponges, pitchers in which the oil is carried out, rushes and mats in which the olives are caught, and various other things which escape my memory at the moment. All these things ought to be much more numerous than are required, because they perish with the using and become fewer and fewer, and if anything is wanting when it must be used, the work is interrupted. But I will now proceed to carry out my promise.

As soon as the berries begin to show different colours and some are already black but more of them white, the olives will have to be stripped by hand in fine weather and be sifted and cleansed upon mats and reeds spread under them. Then when they have been carefully cleaned they must be immediately

LUCIUS JUNIUS MODERATUS COLUMELLA

integram in fiscis novis includi, prelisque subici, ut quantum possit paulisper exprimatur. Postea resolutis corticulis et emollitis [1] debebunt, adiectis [2] binis sextariis integri salis in [3] singulos modios, et aut regulis, si consuetudo erit regionis, aut certe novis fiscis sampsae exprimi. Quod deinde primum defluxerit in rotundum labrum, nam id melius est, quam plumbeum quadratum, vel structile gemellar, protinus capulator depleat et in fictilia labra huic usui praeparata defundat.

11 Sint [4] autem in cella olearia tres labrorum ordines, ut unus primae notae, id est primae pressurae oleum recipiat, alter secundae, tertius tertiae. Nam plurimum refert non miscere iterationem, multoque minus tertiationem cum prima pressura: quoniam longe melioris saporis est, quod minore vi preli, quasi lixivum [5] defluxerit. Cum deinde paululum [6] in labris primis constiterit oleum, eliquare id capulator in secunda labra debebit, et deinde in sequentia usque ad ultima. Nam quanto saepius translatione ipsa ventilatur, et quasi exercetur, tanto fit liquidius,
12 et amurca liberatur. Sat erit autem in singulis ordinibus tricena componi labra, nisi si vasta fuerint oliveta, et maiorem numerum [7] desideraverint.

---

[1] torculis emoli *SA* : torculis et moli *ac*.
[2] adiectis *c* : adlectis *SA*.
[3] in *om. SAa*.
[4] sin *SAac*.
[5] lexivium *SAac*.
[6] paulum *S*.
[7] malorum munerum *S*.

---

a This seems to be the meaning of *regula* in the only other passage outside this work where it is used in connexion with the pressing of olives (*Digesta*, 19. 2. 19. § 2).

BOOK XII. LII. 10–12

taken down to the press and be enclosed, while still whole, in new frails and put under the presses to be squeezed for the shortest possible time. Next, when their skins have been opened and softened, after the addition of two *sextarii* of natural salt to each *modius*, the pulp will have to be squeezed out either with the disks of the press,[a] if it is the custom of the locality, or at any rate in new frails. Then immediately the man who is using the ladle must empty out what first has flowed into the round pan, for that is better than a square leaden vessel or a built-up receiver with two divisions in it, and pour it into the earthenware tubs prepared for this purpose.

In the oil-cellar there should be three rows of pans, one to hold oil of the first quality, that is of the first pressing, another of the second and a third of the third; for it is very important not to mix the second pressing, much less the third with the first pressing, because the first pressing has a far better flavour, because it has flowed from a less violent use of the press as though it had not been treated.[b] Then when the oil has stood for a very short time in the first[c] pans, the man who holds the ladle will have to strain it into the second and then into the next until he comes to the last; for the more often it is aerated by being transferred from one pan to another and, as it were, kept moving, the more transparent it becomes and the freer from dregs. It will be enough for thirty pans to be placed in each row, unless the olive-groves are extensive and require a larger number.

[b] *Cf.* Cato, *R.R.* 23. 2, where *lixivum mustum* is used of must made from untrodden grapes.
[c] *I.e.* the first in each of the rows.

LUCIUS JUNIUS MODERATUS COLUMELLA

Quod si frigoribus oleum cum amurca congelabitur, plusculo sale cocto utique utendum erit. Ea res resolvit oleum, et separat ab omni vitio. Neque verendum est, ne salsum fiat. Nam quantumcunque adieceris salis, nihilominus saporem non recipit oleum. Solet autem ne sic [1] quidem resolvi, cum maiora frigora incesserunt: itaque nitrum torretur, et contritum inspergitur et commiscetur; ea res 13 eliquat amurcam. Quidam quamvis diligentes olearii bacam integram prelo non subiciunt, quod existimant aliquid olei deperire. Nam cum preli pondus accepit, non sola exprimitur amurca, sed et aliquid secum pinguitudinis attrahit.

Illud autem in totum praecipiendum habeo, ut neque fumus neque fuligo, quamdiu viride oleum conficitur, in torcular [2] admittatur, aut in cellam oleariam. Nam est utraque res inimica huic rei; peritissimique olearii vix patiuntur ad unam lucernam opus fieri. Quapropter ad eum statum caeli et torcular et cella olearia constituenda est, qui maxime a frigidis ventis aversus est, ut quam minime vapor 14 ignis desideretur. Dolia autem et seriae, in quibus oleum reponitur, non tantum eo tempore curanda sunt cum fructus necessitas cogit, sed ubi fuerint a mercatore vacuata, confestim villica debet adhibere curam, ut si quae faeces aut amurcae in fundis vasorum subsederint, statim emundentur, et non calidissima

---

[1] si *SAac*.
[2] torcular *c* : trocular *A* : troclar *a* : trocla *S*.

But if in cold weather the oil freezes together with the lees, you will certainly have to use a little more toasted salt; this melts the oil and frees it from anything harmful. You need have no fear that it may become salty; for, however much salt you may have added, the oil nevertheless does not acquire the flavour of it. Even thus, however, the frozen oil does not generally melt when unusually sharp frosts have occurred, so nitre is baked and pounded up and sprinkled and mixed with the oil. This has the effect 13 of clarifying the lees. There are some producers of oil who, although they are careful, do not put the berry in its original state under the press, because they think that some part of the oil is then lost. For when the olive has been subjected to the weight of the press, the lees are not the only thing which is squeezed out but they also take some of the fatness with them. Here is a general precept which I have to give you, that no smoke or soot should be admitted into the room where the press is, during all the time when green oil is being made, nor into the oil-cellar, for both are injurious to the process, and the most experienced makers of oil scarcely allow the work to be done by the light of a single lamp. Wherefore both the press-room and the oil-cellar should be built facing the quarter of the sky which is away from the cold winds, so that the warmth of a fire may be as little needed as possible. The barrels and jars in which the oil is 14 stored should be carefully looked after not only when the requirements of the fruit season make this necessary; but, after they have been emptied by the merchant, the bailiff's wife should immediately make it her business to see that any sediment or lees that have settled at the bottom of the vessels is

LUCIUS JUNIUS MODERATUS COLUMELLA

lixivia, ne vasa ceram remittant, semel atque iterum eluantur:[1] deinde aqua tepida leviter manibus defricentur, et saepius eluantur, atque ita spongia omnis humor assiccetur. Sunt qui cretam figularem in modum liquidae faecis aqua resolvant, et cum vasa laverint, hoc quasi iure intrinsecus oblinant, et patiantur arescere: postea cum res exigit, pura eluunt[2] aqua. Nonnulli prius amurca, deinde aqua vas perluunt, et assiccant. Tum considerant, numquid ceram novam dolia desiderent. Nam fere sexta quaque[3] olivitate cerari oportere antiqui dixerunt. Quod fieri posse non intellego. Nam quemadmodum nova vasa si calefiant, liquidam ceram facile recipiunt,[4] sic vetera non crediderim propter olei succum ceraturam pati. Quam tamen et ipsam ceraturam nostrorum temporum agricolae repudiaverunt, existimaveruntque satius esse nova dolia liquida gummi perluere, siccataque suffumigare alba cera, ne pallorem aut malum odorem capiant. Eamque suffitionem semper faciendam iudicant quotiescunque[5] vel nova vel vetera vasa curantur et oleo novo praeparantur. Multi cum semel nova dolia vel serias crassa[6] gummi[7] linierunt,[8] una in perpetuum gummitione contenti sunt. Et sane quae semel oleum testa combibit,[9] alteram gummitionem non recipit. Respuit enim olei pinguitudo talem materiam, qualis est gummis.

[1] eluantur *A* : elevantur *Sa* : eleventur *c*.
[2] eluunt *om. SAac*.
[3] quaeque *S*.   [4] recipiant *SAac*.
[5] quotiensque *SAac*.
[6] crasse *SAac*.
[7] gummi *ac* : cummi *SA*.
[8] linierunt *scripsi* : linentur *ac* : lientur *SA*.
[9] conbibitam *SAac*.

promptly cleaned and washed out once and again with lye, which should not be very hot, lest they should lose their wax, and then they should be gently rubbed with the hands in tepid water and rinsed out several times and all the moisture dried out with a sponge. Some people dissolve potter's clay to form a kind of liquid sediment and, after they have washed the vessels, daub them inside with this soup-like preparation and let it dry; afterwards, when they require them for use, they wash them out with pure water. Some people wash the vessels thoroughly first with lees, then with water, and dry them; they then examine the jars to see if they require fresh wax; for the ancients declared that they ought to be waxed about every sixth oil-making. I do not understand how this can be done; for, while new vessels, if they are heated, readily admit the wax, yet I cannot believe that old vessels, being saturated with oil-juice, can bear the application of wax. However, this very process of waxing has been abandoned by the husbandmen of our own day and they have come to the opinion that it is enough to wash new jars thoroughly with liquid gum and, when they are dry, to fumigate them from below with white wax, that they may not take on a pallid colour or an evil odour. In their opinion this fumigation ought always to be carried out whenever new or old vessels are being treated and prepared for new oil. Many people, when once they have daubed their new jars and barrels with thick gum, are satisfied forever with one gumming, and certainly a vessel which has once absorbed oil does not admit of a second gumming; for the fatty nature of the oil rejects such matter as gum.

LUCIUS JUNIUS MODERATUS COLUMELLA

Post mensem Decembrem circa calendas Ianuarias eadem ratione, qua superius, destringenda erit olea, et statim exprimenda. Nam si reposita in tabulatum fuerit, celeriter concalescet: quoniam hiemalibus pluviis amurcae plus concipit, quae est contraria huic rei. Cavendum est itaque,[1] ne fiat oleum cibarium. Quod uno modo vitari poterit, si protinus illata de agro baca commolita et expressa erit,[2] quae sic administrata fuerit, ut supra diximus. Plerique agricolarum crediderunt, si sub tecto baca deponatur, oleum in tabulato crescere: quod tam falsum est, quam in area frumenta grandescere;[3] idque mendacium vetus ille Porcius Cato sic refellit. Ait enim in tabulato corrugari olivam, minoremque fieri. Propter quod cum facti unius mensuram rusticus sub tecto reposuerit, et post multos dies eam molere voluerit, oblitus prioris mensurae quam intulerat, ex alio acervo similiter seposito quantumcunque mensurae defuit supplet, eoque facto videtur plus olei[4] requieta[5] quam recens baca[6] reddere, cum longe plures modios acceperit. Attamen ut maxime id verum esset,[7] nihilominus ex pretio viridis olei plus quam multitudine[8] mali[9] nummorum contrahitur. Sed et Cato dixit: Et sic quidem quicquam[10] ponderis aut mensurae oleo accedit, si

---

[1] itaque *SAac.*
[2] cera *SAac.*
[3] grandescere *S* : crescere *Aac.*
[4] olei *om. SAac.*
[5] sequiete *SAa* : esse quietem *c.*
[6] bace *SAac.*
[7] esse *SAac.*
[8] multitudinem *SAac.*
[9] alii *SAac.*
[10] quicquam *ac* : quiquam *SA.*

BOOK XII. LII. 17–20

After the month of December, about the first of January, the olives will have to be stripped from the trees in the manner described above and immediately pressed; for if they have been stored in a loft they quickly grow heated, since, when the winter rains come, they engender more lees, which are a hindrance to the making of oil; you must, therefore, be particularly on your guard against making oil for eating from such olives: this can be avoided in one way only, namely, if the berries are crushed and pressed immediately when they have been brought in from the field, being treated in the manner which I have described above. Most husbandmen have believed that, if the berries are stored under cover, the yield of oil is increased by putting the olives in a loft. This is quite as untrue as to say that grain increases on the threshing-floor. This falsehood the ancient writer Porcius Cato [a] refuted by saying that the olive becomes wrinkled in the loft and grows smaller; for which reason the husbandman, when he has stored under cover the quantity for one oil-making and some days later wishes to crush it, having forgotten the previous amount which he had brought in, makes up any quantity which was lacking from another heap which was similarly set apart, and, when he has done this, the berries which have remained undisturbed seem to yield more oil than the newly gathered, whereas he has taken many more *modii* of olives to make it. If, however, this idea were ever so true, nevertheless a larger profit is made from the price which green oil fetches than from a huge quantity of inferior oil. Cato also said: Thus indeed whatever addition of weight or measure accrues to the oil, if you care to

[a] *R.R.* LXIV.

portiones velis in factum adiectae bacae computare, non proventum, sed detrimentum senties.[1] Quapropter dubitare non debemus lectam olivam primo quoque tempore commolere, preloque subicere.

21 Nec ignoro etiam cibarium oleum esse faciendum. Nam ubi vel exesa vermiculis oliva decidit, vel tempestatibus et pluviis in lutum defluxit, ad praesidium aquae calidae decurritur, ahenumque[2] calefieri debet, ut immundae bacae eluantur. Sed id non ferventissima fieri oportet, verum modice calida, quo commodior gustus olei fiat: nam si excoctus est, etiam vermiculorum ceterarumque immunditiarum saporem trahit. Sed cum fuerit oliva elauta, reliqua, sicut supra praecepimus, fieri debebunt.

22 Fiscis autem non iisdem probum et cibarium oleum premi oportebit. Nam veteres ad caducam olivam, novi autem ordinario aptari oleo, semperque cum expresserint facta, statim ferventissima debent aqua bis aut ter elui: deinde si sit profluens, impositis lapidibus, ut pondere pressi detineantur, immergi: vel si nec flumen est, in lacu, aut in piscina quam purissimae aquae macerari, et postea virgis verberari, ut sordes faecesque decidant, et iterum elui, siccarique.

[1] non—senties *om. SAac.*
[2] ęnumque *S*: aenumque *A*: ahenumque *a*: aenumque *c*.

---

[a] The reading of this sentence is doubtful; it is not a quotation from Cato, *De Re Rustica*.

reckon the quantity of berries added for the oil-making, you will find that there has not been a profit but a loss.[a] Therefore, we ought not to hesitate to crush the olives, when they have been picked, at the first possible opportunity and put them under the press.

I know perfectly well too that oil for ordinary eating has to be made. For when the olive drops from the tree because it has been eaten by small worms or has fallen down into the mud owing to storms or rain, immediate recourse is had to warm water and the cauldron should be heated up so that the dirty berries may be washed. This, however, ought not to be done with very hot but with moderately warm water, so that the taste of the oil may be more agreeable; for if the agreeable taste is boiled out, the heat draws out the flavour of the worms and the other impurities. When the olives have been washed, the rest of the process is as described above.

The superior oil and the ordinary eating-oil must not be pressed in the same frails; for old frails ought to be fitted for the fallen olives but new ones for the first-class oil, and when the batches of olives have been pressed, the frails ought to be washed out immediately two or three times in very hot water; then, if there is running water at hand, they should be sunk in it by having heavy stones put on the top of them, so that they may be kept under water by the pressure of the weight; or if there is no stream, they should be soaked in a lake or pond of the purest possible water and afterwards beaten with rods that the dirt and lees may fall off them, and then they must be washed again, and finally dried.

LUCIUS JUNIUS MODERATUS COLUMELLA

LIII. Quamvis non erat huius temporis olei gleucini compositio, tamen huic parti voluminis reservata est: ne parum opportune vini conditionibus interponeretur. Hac autem ratione confici debet. Vas oleare [1] quam maximum, et aut novum aut certe bene solidum praeparari oportet: deinde per vindemiam musti quam optimi generis et quam recentissimi sextarios sexaginta cum olei pondo octoginta in id confundi: tum aromata non cribrata, sed ne minute quidem contusa, verum leviter confracta in reticulum iunceum aut linteum adici, et ita in olei 2 atque musti partem [2] demitti. Sint autem iis portionibus pensata,[3] quas [4] infra subicimus, calami, schoeni, cardamomi, xylobalsami,[5] corticis de palma, faeni Graeci vetere vino [6] macerati et postea siccati atque etiam torrefacti, iunci [7] radicis, tum etiam iridis Graecae, nec minus anisi Aegyptii pari pondere, id est, uniuscuiusque [8] libram et quadrantem, ut supra diximus, reticulo inclusa demittito,[9] et metretam linito. Post septimum diem aut nonum apertae metretae siquid faecis aut spurcitiae faucibus inhaerebit, manu eximito, et detergito: deinde oleum 3 eliquato, novisque vasis recondito. Mox reticulum eximito, et aromata in pila quam mundissime contundito, tritaque in eandem metretam reponito, et tantundem olei quantum prius infundito, et obturato, in sole ponito. Post septimum diem oleum depleto, et quod reliquum est musti picato cado recondito.

---

[1] oleare S: olea Aac.
[2] pre SAc: pondo a.
[3] pensatae SA: pensate ac.
[4] atque SAac.
[5] cardamo mixto balsami SAa: balsamo c.
[6] primo SAac.   [7] iuncti SA: iunci ac.
[8] uniusquusque S.   [9] dimittito SAac.

LIII. Although the preparation of "gleucine" oil[a] did not fall in this period of the year, I have nevertheless kept it for this part of my book, so that it might not be inserted rather inconveniently among the recipes for making wine. It ought to be made in the following manner: an oil-vessel, as large as possible and either new or at any rate very solid, should be got ready; then at vintage-time sixty *sextarii* of must, of the best quality and as fresh as possible, should be poured into it together with eighty pounds weight of oil; then spices, which have not been sifted and not even pounded into small pieces, but slightly broken, should be put into a small net made of rush or linen and let down into part of the must and oil. The spices should be weighed out in the proportions which we append below—reed, sweet-rush, cardamum, balsam-wood, palm-tree-bark, fenugreek soaked in old wine and afterwards dried and also toasted, bulrush root, Greek iris, and likewise Egyptian anise, the same weight of each, namely, a pound and a quarter; these you should let down enclosed in a small net, as we have said above, and you should then seal up the cask. After the seventh or the ninth day, open the cask and, if there are any lees or impurity adhering to the neck of it, remove them by hand and clear them away; then strain the oil and store it in new vessels. Next take out the small net and pound up the spices as cleanly as possible in a mortar and, when they are ground up, replace them in the same cask and pour in the same quantity of oil as before, stop up the vessel and put it in the sun. After the seventh day empty the oil out and store what remains of the must in a barrel treated

[a] *Cf.* Pliny, *N.H.* XV. § 29.

LUCIUS JUNIUS MODERATUS COLUMELLA

Nam id si non exacuerit,[1] medicamentum dabitur potandum imbecillis bubus et cetero pecori. Oleum autem secundarium non insuavis odoris quotidianam unctionem praebere poterit dolore nervorum laborantibus.

LIV. Oleum ad unguenta sic facito. Ante quam oliva nigrescat, cum primum decolorari coeperit, nec tamen adhuc varia fuerit, maxime Liciniam, si erit, si minus, Sergiam;[2] si nec haec fuerit, tunc Culminiam bacam manu stringito, et statim purgatam prelo integram subicito, et amurcam exprimito: 2 deinde suspensa mola olivam frangito, eamque vel in regulas, vel in novo fisco adicito, subiectamque prelo sic premito, ne vasa intorqueas, sed tantum ipsius preli pondere quantulumcumque exprimi patiaris; deinde cum sic fluxerit, protinus capulator amurca separet, et diligenter seorsum in nova labra transferat, quousque[3] eliquet. Reliquum olei, quod postea fuerit expressum, poterit ad escam vel cum alia nota mixtum vel per se approbari.

LV. Hactenus de oleo dixisse abunde est; nunc ad minora redeamus.

De sucidia et salsura facienda. Omne pecus et praecipue suem pridie quam occidatur, potione prohiberi oportet, quo sit caro siccior. Nam si biberit, plus humoris salsura habebit. Ergo sitientem cum

[1] exaugerit *S* : exsaucerit *A* : exacesceant *a* : exhauserit *c*.
[2] sergiam *SA* : regiam *ac*.
[3] eiusque *SAac*.

---

[a] See note on Ch. 52. 10.

with pitch. For if this must does not go sour, it will be given as a medicine for sick oxen and the other cattle to drink. But oil of this second quality, which has not an unpleasant odour, may serve as a daily ointment for those suffering from pain in the sinews.

LIV. Oil for ointment should be made in the following manner. Before the olive turns black, as soon as it begins to change colour but when it is not yet mottled, pick by hand preferably Licinian olives, if not, Sergian, or, if these are not available, Culminian, and immediately clean them and put them whole beneath the press and squeeze out the lees; then break up the olives in a hanging mill, and put them either upon the disks of the press[a] or in a new frail, and having placed them under the press and squeezed them in such a way that you do not strain the vessels, but allow only ever so little at a time to be squeezed out by the weight of the press; then, when the oil has flowed out in this manner, the man who holds the ladle should immediately separate the oil from the lees and carefully transfer it into new pans by itself, until it is clarified. The rest of the oil, which has been pressed out subsequently, may be acceptable for food, either mixed with some other kind or by itself.

How to make oil for ointment.

LV. What we have said about oil up to this point is enough: let us now return to less important subjects.

The salting of pork.

As to the killing and salting of pigs, every farmyard animal and especially the pig ought to be prevented from drinking on the day before it is to be killed, so that its flesh may be drier; for, if it has drunk anything, the flesh which is to be salted will have more moisture in it. Therefore, having killed

LUCIUS JUNIUS MODERATUS COLUMELLA

occideris, bene exossato : nam ea res minus ungui-
2 nosam [1] et magis durabilem salsuram facit. Deinde
cum exossaveris, cocto sale nec nimium minuto, sed
suspensa mola infracto diligenter salito, et maxime
in eas partes, quibus ossa relicta sunt, largum salem
infarcito, compositisque supra tabulatum tergoribus
aut frustis vasta pondera imponito, ut exsanietur.[2]
Tertio die pondera removeto, et manibus diligenter
salsuram fricato; eamque cum voles reponere,
minuto et trito sale aspergito, atque ita reponito:
nec desieris [3] eius quotidie salsuram fricare, donec
3 matura sit. Quod si serenitas fuerit iis diebus,
quibus perfricatur caro, patieris eam sale conspersam
esse novem diebus : at si nubilum aut pluviae, un-
decima vel duodecima die ad lacum salsuram deferri
oportebit, et salem prius excuti, deinde aqua dulci
diligenter elui, necubi sal inhaereat, et paululum [4]
assiccatam in carnario suspendi, quo modicus fumus
perveniat qui, siquid humoris adhuc continetur,
siccare eum possit. Haec salsura luna decrescente
maxime per brumam, sed etiam mense Februario
ante idus commode fiet.
4 Est et alia salsura, quae etiam locis calidis omni
tempore anni potest usurpari. Cum ab aqua pridie
sues prohibitae sunt, postero die mactantur, et vel
aqua candente, vel ex tenuibus lignis flammula facta
glabrantur,[5] nam utroque modo pili detrahuntur

[1] unguinosam *Gronovius* : incynosam *SAac.*
[2] exanietur *SAac.*
[3] desieris *ac* : desit *S* : desi A.
[4] paulum *S.*
[5] glabrata *SAac.*

---

[a] The reading is doubtful here.

BOOK XII. LV. 1-4

your pig when it is thirsty, bone it thoroughly; for this makes the salted flesh less oily *a* and causes it to keep better. Then when you have boned it, salt 2 it carefully with salt that has been toasted and is not too fine but has been broken up in a hanging mill, and in particular, stuff a large quantity of salt into the parts where there are bones still left, and, after arranging the flitches or pieces on boards, put huge weights on the top of them so that the blood may be pressed out of them. On the third day remove the weights and carefully rub the salted meat with your hands, and when you are minded to put it back, sprinkle it with salt crushed small and then replace it, and do not cease to rub the salt into it daily until it is completely ready. If there is fine weather on 3 the days on which the flesh is rubbed, you will let it remain with the salt sprinkled upon it for nine days; but if there is cloud or rain, it will be advisable on the eleventh or twelfth day to carry the salted flesh down to a pond and first shake off the salt and then wash the meat in fresh water, so that there is no salt adhering to it anywhere, and when it has dried a little, it should be hung up in the larder, where a moderate amount of smoke can reach it, which can dry up any moisture still retained by it. This salting will be suitably carried out when the moon is waning, especially at mid-winter, but also in February before the fifteenth day of that month.

There is another method of salting, which can be 4 employed even in hot climates at any time of year. After the pigs have been kept from drinking the day before, next day they are killed and their hairs removed either with boiling water or a small flame from thin strips of wood, for their bristles are got rid of by

LUCIUS JUNIUS MODERATUS COLUMELLA

caro in libraria frusta conciditur: deinde in seria substernitur sal coctus, sed modice ut supra diximus infractus: deinde offulae carnis spisse componuntur, et alternis sal ingeritur. Sed cum ad fauces seriae perventum est, sale reliqua pars repletur, et impositis ponderibus in vas [1] comprimitur: [2] eaque caro semper conservatur,[3] et tanquam salsamentum in muria sua permanet.

LVI. Rapa quam rotundissima sumito, eaque, si sunt lutosa, detergito, et summam cutem novacula decerpito: deinde sicut consueverunt salgamarii decussatim ferramento lunato incidito. Sed caveto, ne usque ad imum praecidas rapa. Tum salem [4] inter incisuras raporum non nimium minutum aspergito, et rapa in alveo aut seria componito, et sale plusculo aspersa triduo sinito, dum exudent.
2 Post tertiam diem mediam fibram rapi gustato,[5] si receperit salem. Deinde cum videbitur satis recepisse, exemptis omnibus, singula suo sibi iure eluito: vel si non multum liquoris fuerit, muriam duram adicito, et ita eluito: et postea in quadratam cistam vimineam, quae neque spisse, solide tamen et crassis viminibus contexta [6] sit, rapa componito: deinde sic aptatam [7] tabulam superponito, ut usque ad fundum, si res exigat, intra cistam deprimi possit.
3 Cum autem eam tabulam sic aptaveris, gravia pondera superponito, et sinito nocte tota et uno die siccari. Tum in dolio picato fictili, vel in vitreo

[1] ovas *S*: uvas *A*: vas *a*: in vas *c*.
[2] conteritur *SAac*.
[3] consumatur *SAac*.
[4] sale *SAac*.
[5] gusto *SAac*.
[6] contexta *ac*: contesta *SA*.
[7] aptatam *Ac*: aptata *a*: apta *S*.

both methods, and the flesh is cut up into pieces weighing a pound each. Then salt, toasted but only slightly broken up, as I have described above, is spread at the bottom of a tub; next the pieces of meat are packed closely together in layers and salt is put in between each layer. When the brim of the barrel is almost reached, the rest is filled with salt and it is pressed down into the vessel by weights put on the top. This meat always keeps good and, like salt fish, remains in its own brine.

LVI. Choose the roundest turnips you can find and wipe them if they are dirty, and pare off the outer skin with a sharp knife. Then, as pickle-makers usually did, make an incision in the form of an X with a crescent-shaped iron instrument; but take care not to cut down to the bottom of the turnip. Then sprinkle salt, which should not be very fine, amongst the incisions in the turnips and arrange them in a tray or tub and after sprinkling them with a little more salt leave them alone for three days until they exude their moisture. After the third day you should taste 2 the inner fibre of the turnip to see if it has absorbed the salt. Then, when it seems to have absorbed enough, take all the turnips out and wash them each in its own juice, or if there is not much liquid, add some hard brine and then wash them. After that arrange them in a square wickerwork basket, which is not closely woven but still solidly and made of thick withies. Then put on the top a board so fitted that it can, if necessary, be pressed down inside the crate right to the bottom. Having thus fitted the board, 3 put heavy weights on the top of it and leave the turnips to dry for a whole night and a day; then put them into an earthenware jar treated with pitch, or

componito, et sic infundito sinapi et aceto, ut a iure contegantur.

4 Napi quoque, sed integri, si minuti sunt, maiores autem insecti, eodem iure, quo rapa, condiri possunt: sed curandum est, ut haec utraque antequam caulem agant et cymam faciant, dum sunt tenera, com-
5 ponantur. Napos minutos integros, aut rursus amplos in tres aut quattuor partes divisos in vas conicito, et aceto infundito, salis quoque cocti unum sextarium in congium aceti adicito. Post trigesimum diem uti poteris.

LVII. Semen sinapis diligenter purgato, et cribrato: deinde aqua frigida eluito, et cum fuerit bene lautum, duabus horis in aqua sinito. Postea tollito, et manibus expressum in mortarium novum aut bene emundatum conicito, et pistillis conterito. Cum contritum fuerit, totam intritam ad medium mortarium contrahito, et comprimito manu plana. Deinde cum compresseris, scarifato, et impositis paucis carbonibus vivis aquam nitratam suffundito, ut omnem amaritudinem eius et pallorem exsaniet.[1] Deinde statim mortarium erigito, ut omnis humor eliquetur. Post hoc album acre acetum adicito, et pistillo permisceto, colatoque. Hoc ius ad rapa
2 condienda optime facit. Ceterum si velis ad usum conviviorum praeparare, cum exsaniaveris sinapi, nucleos pineos quam recentissimos et amygdalam [2] adicito, diligenterque conterito infuso aceto. Ce-

---

[1] exaniet *SAac*.
[2] amylum *SAa*: amilium *c*.

BOOK XII. LVI. 3–LVII. 2

a glass vessel, and pour mustard and vinegar over them so that they are covered with liquid.

Navews also may be preserved in the same liquid as turnips, whole if they are small, but cut up if they are rather large. Care, however, must be taken that, in either case, they are put in store while they are still tender, before they form a stalk and put forth shoots. The small navews you should throw whole into the vessel, the larger ones, on the other hand, after they have been cut into three or four parts; and you should pour vinegar on them and also add one *sextarius* of toasted salt to a *congius* of vinegar. You can use them after the thirtieth day.

LVII. Carefully cleanse and sift mustard-seed, then wash it in cold water and, when it has been well cleaned, leave it in the water for two hours. Next take it out and, after it has been squeezed in the hands, throw it into a new or thoroughly cleaned mortar and pound it with pestles. When it has been pounded, collect the whole mash in the middle of the mortar and compress it with the flat of the hand. After you have compressed it, scarify it and, after placing a few live coals upon it, pour water mixed with nitre on it in order to eliminate any bitterness and paleness from it. Then immediately lift up the mortar, so that all moisture may be drained away, and after this add sharp white vinegar and mix it with the pestle and strain it. This liquid serves very well for preserving turnips. If, however, you should wish to prepare mustard for the use of your guests, when you have squeezed out the mustard, add pine-kernels, which should be as fresh as possible, and almonds, and carefully crush them together after pouring in vinegar. Finish the process in the manner

How to prepare mustard.

tera, ut supra dixi, facito. Hoc sinapi ad embamma non solum idoneo, sed etiam specioso uteris: nam est candoris eximii, si sit curiose factum.

LVIII. Priusquam olusatrum coliculum agat, radicem eius eruito mense Ianuario vel etiam Februario, et diligenter defricatam [1] nequid terreni habeat, in aceto et sale componito: deinde post diem trigesimum eximito, et corticem eius delibratum abicito. Ceterum medullam eius concisam in fideliam vitream vel novam fictilem conicito, et adicito ius, quod sicut 2 infra scriptum est fieri debebit: sumito mentam, et uvam passam, et exiguam cepam aridam, eamque cum torrido farre et exiguo melle subterito: quae cum fuerit bene trita, sapae vel defruti duas partes et aceti unam permisceto: atque ita in eandem fideliam confundito, eamque operculo contectam pelliculato. Cum deinde uti voles, cum suo iure concisas radiculas promito, et oleum adicito. Hoc ipso tempore siseris radicem poteris eadem ratione, qua supra, condire: sed cum exegerit usus, eximes de fidelia, et oxymeli cum exiguo oleo superfunde.

LIX. Addito in mortarium satureiam, mentam, rutam, coriandrum, apium, porrum sectivum, aut si id non erit viridem cepam, folia lactucae, folia erucae, thymum viride vel nepetam, tum etiam viride puleium, et caseum recentem [2] et salsum: ea omnia pariter conterito, acetique piperati exiguum, permisceto. Hanc mixturam cum in catillo composueris, oleum superfundito.

[1] diffricatam *S* : defricatam *Aac*.
[2] recentem *c* : recens *SAa*.

I have described above. You will find this mustard not only suitable as a sauce but also pleasing to the eye; for, if it is carefully made, it is of an exquisite brilliance.

LVIII. Before the alexanders[a] puts forth its stalk, pull up its root in January or even February, and carefully rub it, so that it has no soil upon it, and place it in vinegar and salt; then after the thirtieth day take it out and peel off the skin and throw it away, but cut up the inner part and put it into a glass pot or a new earthenware vessel and add a liquid, which will have to be made according to the directions given below. Take mint and raisins and a small dried onion which you should pound up with parched wheat and a little honey. When it has been well bruised, mix with it two parts of must boiled down to half or a third of its original volume and one part of vinegar, and then pour them all together into the same jar and put the lid on it and cover it with a cap of skin. Then when you wish to use it, bring out the chopped-up roots with their own juice and add oil. You will be able to pickle the skirwort-root[b] at the same time and by the same method as that given above, and when there is occasion to use it, you will take it out of the jar and pour oxymel with a little oil over it.

How to pickle alexanders and skirwort.

LIX. (1) Put into a mortar savory, mint, rue, coriander, parsley, leeks or, if you have no leeks, a green onion, leaves of lettuce and of rocket, green thyme or calamint. Also green pennyroyal and fresh and salted cheese: pound them all together and mix a little peppered vinegar with them. When you have put this mixture in a bowl, pour oil over it.

Recipes for salads.

[a] *Smyrnium olusatrum.* [b] *Sium sisarum.*

2 Aliter. Cum viridia, quae supra dicta sunt, contriveris, nuces iuglandes [1] purgatas, quantum satis videbitur, interito, acetique piperati exiguum permisceto, et oleum infundito.

Aliter. Sesamum leviter torrefactum cum iis viridibus, quae supra dicta sunt, conterito. Item aceti piperati exiguum permisceto,[2] tum supra oleum superfundito.

3 Aliter. Caseum Gallicum vel cuiuscunque notae volueris minutatim concidito et conterito, nucleosque pineos, si eorum copia fuerit, si minus, nuces Avellanas torrefactas adempta [3] cute, vel amygdalas [4] aeque supra condimenta pariter misceto, acetique piperati exiguum adicito et permisceto, compositum-
4 que oleo superfundito. Si condimenta viridia non erunt, puleium aridum vel thymum vel origanum aut aridam satureiam cum caseo conterito, acetumque piperatum et oleum adicito. Possunt tamen haec arida, si reliquorum non sit potestas, etiam singula caseo misceri.

Oxypori compositio. Piperis albi, si sit, si minus, nigri unciae tres, apii seminis unciae duae, laseris radicis, quod silphium Graeci vocant, sescunciam, casei sextantem: haec contusa et cribrata melli permisceto, et in olla nova servato: deinde cum exegerit usus, quantulumcumque ex eo videbitur, aceto et garo diluito.

5 Aliter, Ligustici unciam, passae uvae detractis vinaceis sextantem, mentae aridae sextantem, piperis albi vel nigri quadrantem: haec, si maiorem impensam vitabis, possunt melli admisceri, et ita

[1] grandes *SAac*.   [2] permiscetum *S*.
[3] ademptas *S* : adempta *Aa* : adepta *c*.
[4] amigdas *S*.

BOOK XII. LIX. 2–5

(2) When you have crushed the green stuffs detailed above, rub into them walnuts, well cleaned, in what seems a sufficient quantity, and mix in a little peppered vinegar and pour oil on the top.

(3) Crush up some slightly parched sesame with the green stuffs detailed above; also mix in a little peppered vinegar and then pour oil on the top.

(4) Cut Gallic cheese, or any other sort you like, in minute pieces and pound it up; take pine-cones, if you have plenty of them, but, if not, toasted hazel-nuts after taking off their shells, or almonds and mix them in the same quantity over the herbs used for seasoning and add a little peppered vinegar and mix it in, and pour oil over the compound thus formed. If there are no green seasonings crush dry pennyroyal or thyme or marjoram or dried savory with the cheese and add peppered vinegar and oil; but any one of these herbs when dry, if the rest are not available, can also be mixed by itself with cheese.

Quick digestives are prepared thus:

How to make quick digestives.

(1) Take three ounces of white pepper, if you have it, or, if not, of black, two ounces of parsley seed, an ounce and a half of laser root, which the Greeks call *silphium*, and a *sextans* of cheese. Crush and sift them and mix them with honey and keep them in a new pot. Then, when you have occasion to use the mixture, dilute whatever slight quantity you think fit with vinegar and fish-sauce.

(2) Take an ounce of lovage, a *sextans* of raisins from which the skins have been removed, a *sextans* of dry mint, and a *quadrans* of white or black pepper; these, if you wish to avoid further expense, can be mixed with honey and so preserved. But if you wish

servari. At si pretiosius oxyporum facere voles, haec eadem cum superiore compositione miscebis, et ita in usum repones: quod si etiam Syriacum laser habueris pro silphio, melius adicies pondo semunciam.

Clausulam peracti operis mei P. Silvine non alienum puto indicem lecturis, si modo fuerint qui dignentur ista cognoscere, nihil dubitasse me paene infinita esse, quae potuerint huic inseri materiae: verum ea quae maxime videbantur necessaria, memoriae tradenda censuisse. Nec tamen canis natura[1] dedit cunctarum rerum prudentiam. Nam etiam quicumque sunt habiti mortalium sapientissimi, multa scisse dicuntur, non omnia.

[1] naturam *SAac.*

to make a more expensive dish for quick digestion, you will mix these same ingredients with the preparation described above and then lay it by for future use. But if you have Syrian laser[a] instead of *silphium*, you will do better to add half an ounce of it.

I think that it is not out of place, Publius Silvinus, as a conclusion to my task which is now finished, to declare to my future readers, if there be any who deign to take cognizance of such things, that I have never doubted that the number of subjects which might have found a place in the material of my treatise is almost infinite, but I have deemed it proper to set forth only such things as seemed most essential. However, nature has not bestowed even upon the grey-headed a practical knowledge of all things; for even those who are held to be the wisest of mortals are said to have known many things, but not all.

[a] Assafoetida. See page 340.

ADDITIONAL NOTE

Laser denoted the juice of Cyrenian silphium until this plant became extinct about the middle of the first century A.D. It was a species of the asafoetida-producing group, similar in appearance to *Narthex asa foetida* and closely related to *Scorodosma foetida*. The laser from Cyrene referred to in *De Arboribus* 23.1. was silphium juice, but the Syrian laser suggested as a substitute for silphium (XII. LIX. 5) was asa foetida, used as a substitute when the Cyrenian variety had become extinct (Pliny, *N.H.* XIX. 38 ff., XXII. 100 ff.).

ON TREES

DE ARBORIBUS

I. Quoniam de cultu agrorum abunde primo volumine praecepisse videmur, non intempestiva erit arborum virgultorumque cura, quae vel maxima pars habetur rei rusticae. Placet igitur, sicuti Vergilio, nobis quoque duo esse genera surculorum: quorum alterum sua sponte gignitur, alterum cura mortalium procedit. Illud, quod non ope humana provenit, materiae est magis aptum: hoc cui labor adhibetur, 2 idoneum fructibus. De hoc [1] itaque praecipiendum est, atque id ipsum genus tripartito dividitur: nam ex surculo vel arbor procedit, ut olea, ficus, pirus; vel frutex, ut violae,[2] rosae, harundines; vel tertium quiddam, quod neque arborem neque fruticem proprie dixerimus, sicuti est vitis. Arborum et fruticum docebimus cultum, si prius de vitibus praeceperimus.

3 Qui vineam vel arbustum constituere volet, seminaria prius facere debebit: sic enim sciet cuius generis vitem positurus sit. Nam quae pretio parata

---

[1] fructibus. De hoc *Lundström*: fructibunde S: fructib, unde hoc A.
[2] violę S: olivę Aac.

ON TREES

I. Since we appear to have given ample instruction on the cultivation of fields in our first book,[a] it will be not out of place to deal with the care of trees and shrubs, which is considered to hold a very important place, indeed, in husbandry. We too, like Vergil, think proper, then, to divide growing trees into two classes, those which come into being of their own accord and those which are the result of human care. The former class, which does not come up by the help of man, is better suited for timber, the latter, on which labour is expended, is adapted to the production of fruit. We must, therefore, give instructions about the latter class. And this falls of itself into three parts; for from the shoot proceeds either a tree, such as the olive, fig and pear, or a shrub, such as the violet, rose or reed, or a third class which cannot properly be called either a tree or a shrub, such as the vine. We shall be teaching the cultivation both of trees and of shrubs if we first give directions about vines.

He who wishes to plant a vine or a tree on which to train a vine will have first to make a nursery; for in that way he will know what kind of vine he is going to plant. For a vine which is purchased and

Preface.

The choice of suitable land for growing vines.

[a] Probably the first book of an earlier and shorter edition of the *De Re Rustica* of which only the *De Arboribus* has survived.

LUCIUS JUNIUS MODERATUS COLUMELLA

disponitur, certam generositatis fidem non habet: quoniam dubium est, an is qui vendidit, legendis seminibus adhibuerit diligentiam: tum etiam quod ex longinquo petitur, parum familiariter nostro solo venit, propter quod difficilius convalescit alienum 4 exterae regionis. Optimum est itaque eodem agro, quo vitem dispositurus es, vel certe vicino facere seminarium: idque multum refert loci natura. Nam si colles vineis vel arbustis occupaturus es,[1] providendum est, ut siccissimo loco fiat seminarium, et iam quasi ab incunabulis vitis exiguo assuescat humori: aliter cum transtuleris de humido in aridum locum, 5 viduata pristino[2] alimento deficiet. At si campestres et uliginosos agros possidebis, proderit[3] quoque seminarium simili loco facere, et vitem largo consuescere humori. Nam quae ex sicco[4] in aquosum agrum transfertur, utique[5] putrescit. Ipsum autem agrum, quem seminario destinaveris[6] planum et succosum, sat erit bipalio vertere quod vocant[7] rustici sestertium. Ea repastinatio[8] altitudinis habet plus sesquipedem, minus tamen quam duos pedes. Iugerum agri vertitur operis quin-6 quaginta. Collem autem et clivosum modum iugeri, sed ne minus duobus pedibus alte, repastinabis[9] operis sexaginta: vel si eodem loco, quo vineam

[1] est *SA*.
[2] viduata pristino *ed. pr.*: videat apristino *SA*.
[3] prodiderit *SA*.
[4] nam quae ex sicco in *Lundström*: nāquae exis quo in *S*: namq; ex his co in *A*.
[5] utique *Schneider*: utrimque *SA*: utrumque *a*.
[6] designaberis *Aa*.
[7] vocant *S*: vertunt *Aac*.
[8] ea res pastinatio *SAac*.
[9] repastinaberis *Sc*: repastenaberis *A*: repastinabis *a*.

put into the ground cannot be guaranteed to possess noble qualities, since it is doubtful whether the vendor took pains in the selection of seeds. Again, a plant which is brought from a distance arrives not properly acclimatized to our soil, and, therefore, being an alien from a foreign land, thrives only with difficulty. It is, therefore, best to make a nursery 4 in the same ground in which you are going to plant your vine or at any rate in the close neighbourhood. The nature of the ground is of great importance. For if you are going to cover hill-sides with vines or trees to support vines, you must take care that your nursery is placed in a very dry spot and that the vine may then become accustomed, as it were, from infancy, to only a small amount of moisture; otherwise, when you transplant it from a moist to a dry place, being deprived of its former nourishment it will fade away. On the other hand, if you possess flat, marshy lands, 5 it will be best to make your nursery too on similar ground and accustom the vine to a generous amount of moisture; for when it is transplanted from dry to watery ground, it is certain to rot away. It will be enough if the actual flat and juicy land which you choose for your nursery is turned over with a trenching-spade,[a] which the country people call a "two-foot-and-a-half." Trenching of this kind goes to a depth of more than a foot and a half but less than two feet. A *iugerum* of land can be turned up in one day by fifty labourers: but you will need sixty 6 labourers to trench a hill and sloping ground which measures a *iugerum* if you trench it to a depth of not less than two feet; or, if you wish to make your nursery in the actual place where you intend to plant

[a] See III. 5. 3.

LUCIUS JUNIUS MODERATUS COLUMELLA

ordinaturus es, facere voles seminarium, tribus pedibus alte [1] repastinabis iugerum operis octoginta : ita tamen si neque lapis, neque tophus aut alia materia difficilior intervenerit : quae res, quot operas absumat, parum certum est. Nos autem de terreno loquimur.

II. Peracta repastinatione, mense Februario vel prima parte Martii semina legito. Sunt autem optima, quae de vitibus notatis leguntur. Nam cui cordi est bona seminaria facere, circa vindemiam vites, quae et magnum et incorruptum fructum ad maturitatem perduxerint, rubrica cum [2] aceto, ne pluviis abluatur,[3] permixta denotat, nec hoc [4] uno tantummodo anno facit, sed continuis tribus vel pluribus vindemiis easdem vites inspicit, an perseverent esse fecundae. Sic enim manifestum est generositate vitium, non anni ubertate fructum provenire. Si compluribus vindemiis eundem tenorem servarint, ex eiusmodi vitibus lecta semina multum bonumque vinum praebebunt. Namque qualiscumque generis uvae, quae integrae et incorruptae ad maturitatem perveniunt, longe melioris saporis vinum faciunt, quam quae praecipientur aestu, aut alia de causa.

III. Semina autem eligito grandi acino, tenui folliculo, paucis minutisque vinaceis, dulci sapore. Optima habentur a lumbis; secunda ab humeris; tertia a summo vitis vertice [5] lecta, quae celerrime comprehendunt, et sunt feraciora, sed aeque [6] celeriter

---

[1] altere *SA*. [2] rubrica cum *vett. edd.* : rubricatū *SA*.
[3] abluit *SAac*. [4] hoc *om. Sa*.
[5] a summo vitis vertice *Lundström* : a sūme vite vita *S* : sūme vitae vita *A*.
[6] aeque *Lundström* : ea quā *S* : et qua' *A*.

---

[a] *Semen* denotes scion, grafting, shoot, sprig, as well as seed. The distinction can generally be inferred from the context.

your vineyard, you will need eighty labourers to trench a *iugerum* to the depth of three feet, provided, however, that no stones or tufa or other more difficult material causes obstruction. For the latter task it is quite uncertain how many labourers the work may require; but we are now speaking of ground which consists of nothing but soil.

II. When the trenching is finished, in the month of February or the early part of March, you must choose your plants. The best are those which are gathered from vines which have been marked; for he who is anxious to make good nurseries, about the time of the vintage, marks those vines which have brought large, sound fruit to maturity, with red ochre mixed with vinegar, so that it may not be washed off by the rain. This he does not merely one year, but inspects the same vines at three or even more successive vintages to see if they continue to be fruitful; for it is thus clear that the fruit is the result of the generous quality of the vines and not due to the productivity of any particular year. If the 2 vines follow the same course for several vintages, plants taken from such will produce an abundance of good wine. For grapes of whatever kind they are which come to maturity sound and unspoiled, provide wine of a far better flavour than those which are forced on by the heat or some other cause.

On the choice of vine-seeds and plants.

III. Choose plants [a] which have large berries, thin skins, few and small pips and a sweet flavour. Those taken from the "loins" of the vine are considered the best, next those taken from the "shoulders," and third those taken from the top of the vine. These take root very quickly and are more prolific, but they also grow old more rapidly. I do not hold

On the choice of vine-shoot and the condition of the soil.

LUCIUS JUNIUS MODERATUS COLUMELLA

senescunt. Pampinaria sarmenta deponi non placet, quia sterilia sunt. Locis pinguibus et planis et humidis praecoques vites serito, raris [1] acinis, brevibus nodis, imbecillas: nam tali generi [2] vitium
2 eiusmodi ager aptus est. Locis aridis et macris et siccis vitem sero maturantem [3] et validam, crebrisque acinis. Quod si pingui agro validas vites deposueris, pampinis magis eluxuriabuntur, et qualemcumque fructum tulerint, ad maturitatem non perducent: rursus imbecillae exili agro celeriter deficient, exiguumque fructum dabunt. Unumquodque genus vitium separatim serito: ita suo quodque tempore
3 putabis et vindemiabis. Semina cum novello sarmento [4] deposita cito comprehendunt et valenter crescunt sed celeriter senescunt: at quae vetere sarmento panguntur, tardius convalescunt, sed tardius deficiunt. Semina quam recentissima terrae mandare convenit. Si tamen mora intervenerit, quo minus statim serantur, quam diligentissime obrui tota oportet eo loco, unde neque pluvias neque ventos sentire possint. Plantaria facito ab exoriente ad decimam lunam,[5] et a vigesima ad tricesimam.
4 Haec melior est vitibus satio. Sed cum seris, frigidos ventos vitato.

Malleolum sic deponito. Virgam malleolarem non amplius quam sex gemmarum esse convenit, ita tamen sunt, si brevia internodia habent. Eius imam partem, quam in terram demissurus [6] es, acutissima

[1] raris *S* : paris *Aa*.
[2] tali generi *vett. edd.* : talis generis *SAac*.
[3] sero maturantem *Lundström* : sere maturam *S* : sera matura *A*.
[4] cum novello sarmento *Lundström* : cū novello vetere (veterae *A*) sarm̄to *SA*.
[5] a decima luna *SAac*.

with setting tendril-shoots, since they are unproductive. For rich, flat and moist ground you should plant early vines which have few berries and joints close to one another, and are weak: for that kind of soil suits this type of vine. In arid, lean and dry places you should plant a vine which comes late to maturity and is strong and has abundant berries. But if you plant strong vines in rich soil, they produce a greater luxuriance of tendrils and, whatever kind of fruit they bear, they will not bring it to maturity; on the other hand, weak vines will quickly fail in poor soil and will produce little fruit. You should plant each kind of vine separately; thus you will prune and harvest each sort at its proper season. Plants put into the ground with a young shoot take root quickly and grow up strongly, but they rapidly grow old; but those which are planted with an old shoot thrive more slowly but deteriorate more slowly. It is advisable that the plants should be as fresh as possible when they are committed to the soil; if, however, any delay has occurred to prevent their being planted immediately, they should be completely covered up as carefully as possible in a spot where they can feel neither the rain nor the winds. You should make your plantations from the new moon to the tenth day and from the twentieth to the thirtieth day. The latter is the better time for planting; but when you are planting, avoid the cold winds.

Set a mallet-shoot in the following manner. It is advisable that the rod of the mallet shoot should not have more than six " eyes," provided that the spaces between the joints are short. The lower part of it which you are going to let down into the ground you

---

⁶ dimissurus *S* : dimisurus *A*.

LUCIUS JUNIUS MODERATUS COLUMELLA

falce iuxta nodum, sic ne gemmam laedas,[1] rotunda plaga amputato, et statim fimo bubulo linito: tum in terram bene pastinatam et stercoratam rectum sarmentum defigito, ita ut ne minus quattuor gemmae abscondantur. Pedale quoquoversus spatium sat erit inter semina relinqui: cum comprehenderint, identidem pampinentur, ne plura sarmenta quam debent, nutriant.[2] Item quam saepissime fodiantur: ferro ne tangantur. Vicensimo et quarto mense resecentur:[3] post tricesimum et sextum mensem transferantur.

In agro requieto vineam ponito.[4] Nam ubi vinea fuit, quod citius decimo anno severis, aegrius comprehendet, nec unquam roborabitur. Agrum antequam vineis obseras, explorato qualis saporis sit: talem enim etiam gustum[5] vini praebebit. Sapor autem sicuti primo docuimus volumine comprehendetur, si terram aqua diluas, et cum consederit[6] tum demum aquam degustes. Aptissima vitibus terra est arenosa, sub qua consistit dulcis humor: probus consimilis ager, cui subest tophus: aeque utilis congesta et mota terra. Sabulum quoque, cui subest dulcis argilla, vitibus convenit. Omnis autem qui per aestatem finditur ager, vitibus[7] arboribusque inutilis. Terra inferior alit vitem et arbo-

---

[1] ledas S : ridas Ac : radas a.
[2] nutriant om. Aac.
[3] resecantur SAac.
[4] posito SAR : ponito c.
[5] gustus SAac.
[6] conderit SAac.
[7] vitib; S : vitio A.

should cut with a very sharp pruning-knife near a joint, so that the section has a rounded surface, in such a way as not to damage the " eye," and you should immediately smear it with ox-dung. You should then fix the shoot upright in earth which has been well trenched and manured, in such a way that not less than four " eyes " are hidden. It will be enough 5 to leave the distance of one foot each way between the plants. When they have taken root, they should be repeatedly trimmed back, so that they do not have to give nourishment to more shoots than they ought. Also they should be dug round as often as possible and they should not be touched by any iron tool. In the twenty-fourth month they should be cut back, and they should be transplanted after the thirty-sixth month.

You should plant your vineyard on ground which has lain fallow; for where there has been a vineyard, anything which you plant sooner than the tenth year will only take root with difficulty and will never attain to any strength. Before you plant a piece of ground 6 with vines, you should examine what sort of flavour it has; for it will give the wine a similar taste. The flavour can be ascertained, as we described in our first book, if you soak the earth in water and taste the water when the earth has gone to the bottom. Sandy soil under which there is sweet moisture is the most suitable for vines. Similar soil under which there is tufa is excellent; earth which has been heaped up after being brought from elsewhere is equally beneficial. Gravel, too, which has sweet clay underneath it, suits vines; but any soil which is split during the summer is useless for vines and trees. The soil under the surface feeds the vine

LUCIUS JUNIUS MODERATUS COLUMELLA

7 rem; superior custodit. Saxa summa parte terrae et vites et arbores laedunt, ima parte refrigerant. Et mediocri raritudine optima est vitibus terra: sed ea quae [1] transmittit imbres, aut rursus in summo diu retinet, vitanda est. Utilissima [2] autem est superior modice rara, circa radices densa. Montibus clivisque difficulter vineae convalescunt, sed firmum probumque saporem vini praebent. Humidis et planis locis robustissimae, sed infirmi saporis vinum nec perenne faciunt. Et quoniam de seminibus atque habitu soli praecepimus, nunc de genere vinearum disputabimus.

IV. Vites maxime gaudent arboribus, quia naturaliter in sublime procedunt, tum et materias ampliores creant,[3] et fructum aequaliter percoquunt. Hoc genus vitium arbustivum vocamus, de quo pluribus suo loco dicemus. Vinearum autem fere genera in usu tria sunt, iugata, humi proiecta, et deinde tertia, quae est a Poenis usurpata,[4] more arborum in se
2 consistens. Id genus [5] comparatum iugata quadam parte deficitur, quadam superat: iugata plus aëris [6] recipit, et altius fructum fert, et aequalius concoquit, sed difficilior est eius cultus: at haec ita constituta est, ut etiam arari possit; eoque ubertatem maiorem consequitur, quod saepius et minore impensa excolitur. At quae protinus in terram porrecta [7] est,

---

[1] quā (quam) *SAac.*
[2] *post* utilissima *add.* vitanda *SA.*
[3] creant *S* : careant *Aac.*
[4] a Poenis usurpata *Lundström* : aponis urpata *S* : aponi surpata *A.*
[5] consistens . id genus *ed. Ald.* : consistent ... ingens *S* : consistent In genus *Aac* : iugatae *ed. Ald.* : iugatum *SAac.*
[6] aeris *ac* : eris *SA.*
[7] porrecta *S* : precta *A* : proiecta *c* : poiecta *a.*

and the tree, while the top soil protects them. Stones on the upper part of the earth are harmful both to vines and to trees; in the lower soil they keep them cool. Moderately loose soil is best for vines; but soil which lets the rain through or, on the other hand, retains it for a long time near the surface, must be avoided. The most beneficial is that which is moderately loose near the surface and dense round the roots. On mountains and slopes vines only thrive with difficulty but produce wine of a lasting and excellent flavour. Vines which grow on damp, flat ground are very strong but produce wine of a weak flavour which does not keep. Since we have now given instructions about vine-plants and conditions of soil, we will now discuss the different kinds of vines.

IV. Vines take a great delight in trees, because they naturally have an upward tendency; also in that position they grow more abundant firm-wood and ripen their fruit evenly. This kind of vine we call a "tree-vine" and we will speak of it at greater length in its proper place. There are generally three types of vine in use; that which is trained on a trellis, that which spreads along the ground, and thirdly that which is in use among the Carthaginians and stands by itself like a tree. The last type as compared with the trellised vine is in some respects inferior, in others superior. The trellised vine is more open to the air and bears its fruit higher up and ripens it more evenly, but its culture is more difficult; its arrangement is such that the plough can be used round it and it attains a greater degree of fruitfulness, because it can be given more frequent attention and at less cost. However, the type which spreads directly along the ground produces a large quantity

On sites for vineyards.

LUCIUS JUNIUS MODERATUS COLUMELLA

multum sed non bonae notae vinum facit. Vinea optime repastinato agro ponitur, nonnumquam tamen vel melius quibusdam locis sulcis committitur: interdum etiam scrobibus deponitur. Sed, ut dixi, repastinatur iugerum in altitudinem pedum trium operis [1] octoginta; sulcum autem terrenum pedum duorum altum,[2] et longum septuaginta una opera [3] effodit; scrobes ternarios, id est quoquoversus pedum trium, una opera facit xviii. Vel si cui cordi est laxius vites ponere, scrobes quaternarios, id est quoquoversus pedum quaternum, una opera xii facit; vel bipedaneos quoquoversus una opera xx effodit. Curandum autem est, ut locis aridis et clivosis altius vites deponantur quam si humidis et planis. Item si scrobibus aut sulcis vineam posituri erimus, optimum erit ante annum scrobes vel sulcos facere.

Vinea, quae angustissime conseritur, quoquoversus quinque pedum spatio interposito ponitur; laxius vero inter pedes vii vel viii; sed quae rarissime ut etiam facile arari possit, inter denos pedes constituitur. Haec positio vinearum modum sine dubio agri maiorem occupat, sed valentissima et fructuosissima est. Cum semina depones, imum scrobem vel sulcum bidentibus [4] fodito, mollemque reddito. Vitem quam ponis, fac ut ad Orientem spectet adminiculo religata. In imo scrobe lapides circa pondo quina ita ponito, ne vitem premant,[5] sed iuxta radices

---

[1] operis *S* : optos *A*.   [2] altum *om. A*.
[3] opera *c* : opere *SAa*.
[4] videntibus *S* : videntib; *A*.
[5] premat *Sa* : p̄mat *A*.

of wine but not of a good quality. It is best to plant vines in well-trenched ground, but it is sometimes and in some localities better to set them in furrows; sometimes also they are put into planting-holes. But, as I have already said, a *iugerum* can be trenched in a day to a depth of three feet by eighty labourers, whereas 3 one labourer in the same time digs a furrow in the earth two feet deep and seventy feet long. One labourer in a day makes eighteen three-foot planting-holes, that is, measuring three feet in each dimension; or, if one wishes to plant the vines less closely, a single labourer can make twelve four-foot planting-holes, that is to say, measuring four feet each way, or twenty measuring two feet each way. Care, however, must be taken that in dry and sloping ground the vines are planted deeper than on damp, level ground. Also, if we are going to plant out a vineyard in planting-holes or furrows, it will be best to make our planting-holes or furrows a year beforehand.

In a vineyard which is planted very closely, the 4 vines are planted with five feet between them in every direction; but if more widely, at intervals of seven or eight feet. The widest space ever allowed, in order that a plough may be easily used, is fixed at ten feet. This arrangement of the vines certainly takes up a greater extent of ground but it produces the strongest and most abundant fruit. When you are putting in the seedlings, you should dig the bottom of the planting-hole or furrow with hoes and make it soft. You should see that the vine which you are planting faces east when it is fastened to the support. In the bottom of the planting-hole you should place stones about five pounds in weight in such a way as not to press upon the vine but so as to

5 sint. Praeterea post haec vinaceae heminam uvae albae in nigra, uvae nigrae in alba ponito, atque ita scrobem vel sulcum cum stercorata terra ad medium completo. Triennio deinde proximo paulatim scrobem vel sulcum usque in summum completo: sic vites consuescent radices deorsum agere. Spatium autem radicibus, qua repant, lapides praebent, et hieme aquam repellunt, aestate humorem praebent; vinaceae [1] radices agere cogunt. Quoniam praecepimus quemadmodum vites ponendae sint, nunc culturam earum docebimus.

V. Vineam novellam omnes gemmas agere sinito: simul atque pampinus instar quattuor digitorum erit, tum demum pampinato, et duas materias relinquito: alteram quam vitis constituendae causa submittas, alteram subsidio habeas, si forte illa ordinaria interierit: hanc rustici custodem vocant. Proximo deinde anno, cum putabis vitem, meliorem unam virgam relinquito, alteras tollito.[2] Tertio anno vitem, in quam formam [3] voles, dum tenera est, com-
2 ponito. Si iugatam [4] eris facturus,[5] unam materiam [6] submittito, ita ut duas gemmas, quae sunt proximae a terra, falce acuta [7] radas, ne [8] possint germinare: deinde tres sequentes relinquas, reliquam partem virgae amputes. Sin autem vitem in se consistere voles,

---

[1] vineaceis *SAac*.
[2] alteras tollito *om. Aac*.
[3] in quā formā *S* : in qua forme *A*.
[4] iugata *SAac*.
[5] facturus *Pontedera* : facturis *S* : facturi *Aac*.
[6] una materia *S* : uma materia *A*.
[7] acute *SAac*.   [8] ne *om. SAac*.

be near the roots. Further, after doing this, you 5
should put in a *hemina* of the skins of white grapes
for a black vine and of black grapes for a white vine
and then half fill the planting-hole or furrow with
well-manured earth; then during the next three
years you should gradually fill the planting-hole or
furrow right up to the top. By this method the
vines will become accustomed to strike their roots
downwards. The stones give the roots space into
which they can creep and in winter they keep away
the water while they provide moisture in the
summer; the grape-skins force the vines to put
forth their roots. Since we have given directions
about the way in which the vines should be planted,
we will now give instructions about their cultivation.

V. You should allow the newly-planted vineyard to
put forth all its " eyes " and, as soon as the tendrils are   On the
about four inches long, you should then trim them back,   cultivation
and you should leave two of them for firm-wood, one   of vines.
of which you should allow to grow in order to form
the vine, while you should keep the other in reserve
in case the one destined for a place in the row should
die. The country folk call this second shoot the
" guardian." Then next year, when you prune the
vine, you should leave the better rod on each plant
and remove the others. In the third year you
should arrange the vine, while it is still tender, in the
shape which you want. If you are going to train it 2
over a trellis, you should allow one firm-wood shoot
to grow, pruning away with a sharp pruning-hook
the two " eyes " which are nearest to the ground, so
that they cannot bud; then you should leave the
next three " eyes " and cut off the rest of the rod.
If, however, you wish the vine to stand by itself, you

LUCIUS JUNIUS MODERATUS COLUMELLA

sicuti arbori bracchia submitti patieris, et dabis operam, ut in orbem quam rotundissime formetur. Nam praeterquam quod speciem habet sic composita, tum etiam minus laborat, cum undique velut aequilibrio stabilita in se requiescit. Sat erit autem cum primo bracchia submittuntur, singulas gemmas singulis sarmentis relinqui, ne protinus onere gravetur.
3 Post hanc putationem lectis [1] sarmentis, bidentibus alte et aequaliter [2] vineam fodito: vel, si ita late disposita erit, arato.

Ab idibus Octobribus ablaqueare incipito, ante brumam ablaqueatam habeto. Per brumam vitem ne colito, nisi si voles eas radices, quae in ablaqueatione apparebunt, persequi. Tum demum optime amputabis, sed ita ne codicem laedas, sed potius instar digiti unius a matre relinquas, et ita radicem reseces. Nam quae propius abraditur, praeterquam quod vulnus viti praebet eoque [3] nocet, tum etiam de ipsa cicatrice plures radices prorepunt.
4 Itaque optimum est exiguam partem relinqui, atque ita summas partes, quas aestivas rustici [4] appellant, resecare: quae sic resectae inarescunt,[5] nec ultra vitibus obsunt. Possunt etiam soboles per brumam caedi, eo magis quod frigoribus extirpatae minus recreantur. Peracta ablaqueatione ante brumam

[1] lecti *SAac*.
[2] qualiter *SAac*.
[3] eoque *vett. edd.*: eo q̄d *Sc*: eo quod *Aa*.
[4] rusti *S*: fusti *A*.
[5] inarescunt *Lundström*: inalescunt *SAa*.

will allow the branches to grow freely, as you do with a tree, and be careful that it takes a circular shape with as round a form as possible; for, besides having a pleasing appearance when so arranged, it is also subjected to less strain, since, being stabilized by being equally balanced on all sides, it rests upon itself. It will be enough, if, when first the branches are allowed to grow, one " eye " is left on each shoot, so that the vine may not be overweighted at first. After this pruning, when you have chosen which 3 shoots you are going to leave, you should dig up the vineyard deeply and uniformly with mattocks, or if it has been laid out with sufficiently wide spaces, you should plough it.

From the 15th of October you should begin to trench the vineyard and so have it trenched before mid-winter. During the winter do not cultivate the vines, unless you wish to trace the roots which make their appearance during the trenching; for that will be the best time to cut them back, but in such a way as not to damage the stem of the vine but preferably leave about an inch from the mother-tree and so cut back the root. If the root is pared away more closely, besides the fact that it wounds and therefore harms the vine, more roots also make their way out from the actual scar. Thus it is best 4 that a small portion of the root should be left and that thus the furthest parts, which the country folk call the " summer roots ", should be cut back. These, when thus cut back, dry up and do not hamper the vines any more. The shoots can also be cut during the winter, with the additional advantage that, having been destroyed by the cold, they are less likely to grow again. When the trenching is finished,

tertio quoque anno macerati stercoris, ne minus sextarios binos ad radices vitium posuisse [1] conveniet, praeterquam columbinum; quod si quo amplius quam heminam posueris, viti nocebit. Post brumam deinde ablaqueationem circumfodito. Ante aequinoctium vernum, quod est octavo [2] cal. April., ablaqueationem adaequato. Post idus Aprilis terram ad vitem aggerato. Aestate deinde quam potes saepissime occato. Iugerum vineae quinque operis ablaqueatur, quinque foditur, tribus occatur. Iugerum valentis et iam constitutae vineae quattuor operis putatur, sex alligatur. Arbusto nihil eiusmodi potest ante [3] finiri, quia inaequalitas arborum non patitur operis iusta comprehendi. Quibusdam placet vitem proximo anno translatam non putare, sequenti deinde anno purgare, et unam virgam, quam submittamus, ad tertiam gemmam resecare: tertio deinde si vitis recte convaluerit, una plus gemma submittere: quarto duas gemmas proximae putationi adicere, atque ita quinto demum anno vitem iugare. Hunc eundem ordinem culturae nos quoque experti comprobavimus.

VI. Veterem vineam, si in summo radices habebit, resecare nolito: alioquin [4] etiam novellam vineam, quae ex resectione enata fuerit, inutilem habebis,[5] summa parte terrae natantibus radicibus, neque

---

[1] posuisse *ac*: posuisset *SA*.
[2] octava *Aac*.
[3] potest ante *Lundström*: potestate *SAac*.
[4] aliqui *SAac*.
[5] habebit *SAac*.

every third year before the winter comes, it will be advisable to have put not less than two *sextarii* of soaked manure, but not pigeon-dung, round the roots of the vines; but if you put more than a *hemina*, it will harm the vines. When winter is over, you should dig round the ground which you have trenched. Before the spring equinox, which occurs on March 25th, you should level the soil which has been trenched. After April 13th you should heap the soil round the vine; then in the summer you should hoe the ground as often as possible. Five labourers can trench a *iugerum* of vineyard in a day, and the same number can dig it up, and three can hoe it. Four labourers can prune a *iugerum* of a thriving and well-established vineyard, and six can tie up the vines in a day. When the vines grow upon trees no definite calculation of this kind can be made beforehand, because the varying size of the trees does not allow a fixed calculation in terms of labourers. Some people are of opinion that a vine which has been transplanted should not be pruned during the next year but should be cleared in the following year and one of the rods, which we allow to grow, cut back to the third " eye ": then in the third year, if the vine has thrived properly, it should be allowed to grow at one more " eye," and in the fourth year two " eyes " should be added at the next pruning, and finally in the fifth year the vine should be placed upon the trellis. This same method of culture we have ourselves tried and approved.

VI. Do not cut back an old vine, if it has roots on the surface; otherwise you will find even the new vine, which will grow from the cut, of no use, since its roots float, as it were, on the surface of the ground, and you

LUCIUS JUNIUS MODERATUS COLUMELLA

fructum uberem percipies, et nihilominus celeriter consenescet. Eiusmodi itaque vinea, si non peraridos[1] habet[2] truncos, et flecti potest, factis sulcis optime sternitur, atque ita renovellatur. Sin autem usque eo exaruit, ut curvari[3] non possit, primo anno summatim, ita ne radices eruas[4] aut laedas, ablaqueato, et stercus ad radices addito, atque ita putato, ut[5] paucas et certas materias relinquas, et fodias diligenter, et saepius pampines, ne omnino supervacua sarmenta nutriat. Sic exculta longas et firmas materias creabit, quas sequenti anno scrobibus inter ordines factis propaginabis:[6] ac deinde triennio post, cum convalescat, saepius fodies, matremque ipsam onerabis, nihil in posterum prospiciens ei quam sublaturus es. In[7] posterum annum matrem radicitus tolles, atque ita novellam vineam ordinabis. Sin autem vetus vinea, dumtaxat[8] generis boni, radices alte positas habebit, ita ut ablaqueatae non conspiciantur, eam vineam circa calend. Martias, antequam reseces, ablaqueato, et protinus, cum alte ablaqueaveris, sic resecato. Quattuor digitos ab radicibus truncum relinquito, et si fieri poterit, iuxta aliquem nodum serrula desecato, et plagam acutissimo ferro delevato. Deinde minutam terram mediocriter stercoratam ita superponito, ut adobruto trunco ne minus tres digiti terrae super plaga sint. Hoc idcirco, ne sole inarescat, et ut melius

---

[1] vinea si non peraridˆ *S* : vineas inoperari *A*.
[2] habent *SAac*.
[3] curari nonpossint *SAac*.
[4] eruḅas *SA*.   [5] aut *SAac*.
[6] propaginabis *Pontedera* : propaginib; *S* : propaginibus *Aac*.
[7] in *add. Lundström*.
[8] vineam taxat *SA*.

will not reap an abundance of fruit, and the vine will, nevertheless, quickly grow old. Therefore, this kind of vine, if its stems are not dried up and it can be bent, is best laid in furrows made for the purpose and is thus planted anew. But if it has dried up 2 to such a degree that it cannot be curved, you should in the first year trench the ground slightly in such a way as not to tear up or damage the roots, and put some manrue round them, and prune the vine in such a way as to leave a few dependable firm-wood shoots, and carefully dig the soil and rather frequently trim the tendrils, so that the vine may not nourish entirely useless shoots. If it is cultivated in this way it will 3 produce long, firm-wood shoots, which in the following year you will propagate in planting-holes made between the rows of vines; and then for three years, while it is gaining strength, you will rather often dig round it, and you will suppress the mother-plant, making no future provision for it, since you intend to get rid of it. The next year you will remove the mother-vine root and branch and then place the young vine in position. But if an old vine, which is, never- 4 theless, of a good kind, has its roots situated so deep that they are not visible after the trenching, you should trench this vine about March 1st, before you cut it back, and when you have trenched it deeply, you should cut it back immediately. You should leave the stock four inches from the roots and, if possible, cut it off with a small saw near some knot and smooth off the cut with a very sharp knife; then put a little earth moderately well manured on the top, so that, when the stock is buried, there are not less than three inches of soil above the place where it was cut. The object of this is that it may not be 5

materias citet¹ percepto humore, quem terra praebet. At quae mali generis et infructuosa vinea est, summasque partes et eiuncidas² et exesas habet,³ si radices eius satis alte positae sunt, optime inseretur ita ut ablaqueata et nudata pars ima secundum terram sic amputetur, ne cum aggerata fuerit, supra terram exstet.

VII. Propagationum genera tria sunt in usu maxime: unum quo virga edita a matre sulco committitur: alterum, quo ipsa mater prosternitur, atque in plures palos per suas virgas dividitur: tertium, quo vitis finditur in duas vel tres partes, si diversis ordinibus diducenda est. Hoc genus tardissime convalescit, quia vitis divisa medullam amittit. Et quoniam genera proposuimus, unumquodque qualiter faciendum sit, demonstrabimus.⁴

2 Virgam cum a matre in terram deprimere voles, scrobem quoquoversus quattuor pedum facito ita, ut⁵ propago⁶ non laedatur alterius radicibus. Deinde quattuor gemmas, quae in imum scrobem perveniant, relinquito, ut ex iis radices citentur. Reliquam partem, quae continens matri est, adradito, ne 3 supervacua sarmenta procreet. Diversae autem quae supra terram exstare debent,⁷ ne passus fueris plus quam duas aut ut maxime tres gemmas habere.

---

¹ materia scit & *S* : materiascit et *Aa*.
² eiuncidas *Schneider* : et iuncidas *SAac*.
³ habent *SAac*.
⁴ demonstravimus *SAac*.
⁵ ut *om. SAac*.
⁶ ppala *SA* : propala *ac*.
⁷ debet *SAac*.

dried up by the sun and that by taking up the moisture which the earth supplies, it may in a better manner send out firm-wood shoots. But a vine which is of a bad kind and unfruitful and of which the upper parts are meagre and rotten, if its roots are put sufficiently deep in the ground, will undergo engrafting very successfully, provided that the lower part, after having been dug round and stripped bare, is cut off next to the ground so that it does not stand out above it when soil has been heaped upon it.

VII. There are three kinds of propagation chiefly in use: one in which a rod which has sprung from the mother-vine is planted in a furrow, a second in which the mother-vine itself is laid down flat and distributed branch by branch over a number of props, and a third in which the vine is cleft into two or three parts according to the number of different rows over which it has to be trained. The last method is very slow in thriving, because the vine, being divided up, loses its inner pith. Since we have suggested several methods, we will show how each should be carried out.

On the propagation of vines.

When you wish to press down a rod from the mother-vine into the ground, you should make a planting-hole measuring four feet each way so that the layer is not damaged by the roots of another vine. Then you should leave four " eyes," extending to the bottom of the planting-hole, that roots may spring from them; scrape the rest of the rod, that is, the part nearest to the mother-vine, so that it may not produce superfluous shoots. The parts which lie in the other direction, and should project above the ground, you should not allow to have more than two or at most three " eyes." Pare

LUCIUS JUNIUS MODERATUS COLUMELLA

Reliquas, quae in terram absconduntur, exceptis quattuor imis, fac adradas, ne in summo radices vitis citet.[1] Hoc modo propagata celeriter convalescet: et tertio anno a matre separabitur. Sin autem ipsam vitem sternere voles, iuxta radicem [2] ita, ne ipsam laedas, curiose fodito, et vitem ita supplantato, ne radicem abrumpas. Cum eam statueris[3] et videris quousque possit pertingere, sulcum facies unum, in quem vitem integram demittas:[4] deinde ex eo sulco quasi ramos fossarum facies, per quos, uti quaeque virga postulavit, propaletur, atque ita terra adoperies. Sin autem vitis exiguam materiam habebit, et in diversos ordines diducenda erit, neque aliter potuerit palos, ad quos perducitur, pertingere, quam ut diffluvietur, curabis ut quam acutissima falce ab ea parte, qua bifurca est, findas eam, et item[5] ferro acuto plagam emendes, sicubi inaequaliter findi videbitur. Sic diducta poterit in plures ordines dividi.

Non inutilis est etiam illa propagatio, quam nos reperimus, si quando in ordinem vitis deest, neque est tam procera virga quae cum in imum scrobem demissa[6] fuerit, retorqueri et erigi supra terram possit. Brevitatem ne reformidaveris, sed qualemcumque virgam, cuius cacumen in imum scrobem pervenit, deprimito et obruito: deinde gemmas, quae secundum ipsam matrem sunt, submittito, ut

---

[1] radice vites cit & *S* : radicae vitis cit et *A*.
[2] radicē *S* : radice *A*.
[3] quousque *post* statueris *add. SA*.
[4] dimittas *SAac*.
[5] idem *SAac*.
[6] dimissa *SAac*.

away the other "eyes" which are hidden in the ground except the four lowest, so that the vine may not put forth roots on the surface. Propagated in this manner the vine will quickly thrive, and in the third year will be separated from the mother-vine. But if you wish to lay the vine itself on the ground, 4 carefully dig round it near the root, so as not to damage it, and train it along the ground without breaking off the root. When you have put it in position and have seen how far it can reach, you will make a single furrow into which to lower the entire vine; you will then make branches, as it were, leading out of the furrow along which the vine may be spread out as each rod has required, and then you will cover the whole over with earth. But if the vine 5 has very little firm-wood and will have to be trained to several different rows and cannot reach the stakes to which it is being conducted without being divided, you will be careful to cleave it with the sharpest possible pruning-knife at the point where it forks and also with a sharp instrument smooth off the cut, if it seems to be unevenly cleft at any point. Thus separated it will be possible to distribute it over several rows.

The following method of propagation too which we 6 have discovered has its advantages, if at any time a vine is missing in a row and the rod which has been pressed into the bottom of the planting-hole is not tall enough to be pressed back and set up above ground. Do not be concerned at its shortness but press down and cover up the rod, of whatever kind it be, the top of which reaches the bottom of the planting-hole; then allow the "eyes" which are nearest to the mother-vine itself to grow freely, so

LUCIUS JUNIUS MODERATUS COLUMELLA

materias a superiore parte citent. Tum demum post triennium a matre amputato, et ad suum palum eam partem, quam a matre [1] praecideris, reducito, et caput vitis facito. Propaginis scrobem minime triennio paulatim completo: summas radices praecidito: crebro fodito.

VIII. Cum vitem inserere voles, optimi generis sarmenta fructuaria tum, cum gemmas agere incipient, vento Austro a matre praecidito. Sarmentum, quod inseris, de summa vite [2] sit rotundum, bonis crebrisque nodis. Tres deinde nodos integerrimos relinquito: infra tertiam gemmam ex utraque parte duorum digitorum spatium in modum cunei tenuissimo scalpello acuto,[3] ita ne medullam laedas.
2 Vitem deinde, quam insiturus es, resecato, et plagam levato, atque ita findito, et paratos surculos in fissuram demittito,[4] eatenus qua adrasi sunt, ita ut cortex surculi corticem vitis aequaliter contingat. Quidquid inserueris, ulmi diligenter [5] libro vel vimine ligato luto subacto paleato oblinito plagam, et alligato, ne aqua ventusve penetrare possit: deinde supra lutum muscum imponito, et ita religato: ea
3 res praebet humorem, nec inarescere sinit. Infra insitionem et alligaturam falce acuta leviter vitem vulnerato ex utraque parte, ut ex his potius plagis

[1] *post* matre *add.* ei *S* : et *A*.
[2] summa vite *ac* : sūmā vitem *S* : sūmā vite' *A*.
[3] acuto *SAac*.
[4] dimittito *SAac*.
[5] ulmi diligentur libro vel vimine *Lundström* : uimide uluidini legenter libro *SA*.

that they may put forth firm-wood from the upper part of the rod. Then eventually after three years lop the rod away from the mother-vine and train to its own stake the part which you have cut away from the mother-vine and form it into a separate vine. Gradually fill up in a minimum of three years the planting-hole in which the layer is; cut away the surface roots and dig the ground frequently.

VIII. When you wish to engraft a vine, cut off fruit-bearing shoots of the best quality from a mother-vine at the time when they begin to put forth " eyes " and when the wind is in the south. A shoot which you are using as a graft should be taken from the top of the vine and be round with a large number of good knots in it. Then leave the three soundest knots and below the third " eye " pare away a space of two inches with a very fine knife on both sides without damaging the pith so as to form a wedge. Then make a cut in the vine which 2 you are going to engraft and smooth off the wound, and then make clefts and insert in them the shoots which you have got ready up to the point at which they have been pared down, in such a way that the bark of the slip meets the bark of the vine evenly all round. Any graft which you have put in, you should carefully bind with a withy or the bark of an elm, and you should smear the cut with well-kneaded clay mixed with chaff and bind it, so that neither water nor wind can penetrate into it; then put moss outside the clay and bind it round. This 3 provides moisture and does not allow it to dry up. Below the insertion and the binding make a slight wound with a sharp pruning-knife on both sides of the vine, so that the moisture may flow out from these

How to make vines yield plentifully by grafting.

LUCIUS JUNIUS MODERATUS COLUMELLA

humor[1] defluat, quam ex insitione ipsa abundet; nocet enim nimius humor, nec patitur surculos insertos comprehendere. Quibusdam antiquorum terebrari[2] vitem placet, atque ita leviter adrasos surculos demitti: sed nos meliore ratione hoc idem fecimus. Nam antiqua terebra scobem facit, et propter hoc urit eam partem quam perforat: praeusta autem perraro unquam comprehendit insertos surculos.
4 Nos rursus terebram,[3] quam gallicanam[4] dicimus, huic insitioni aptavimus: ea excavat,[5] nec urit, quia non scobem, sed ramenta facit. Itaque cavatum foramen cum purgavimus, undique adrasos surculos inserimus, atque ita circumlinimus. Talis insitio facillime convalescit. Igitur secundum aequinoctium perfectam vitium insitionem[6] habeto: humida loca de uva alba: sicca de nigra inserito.
5 Vites quae exiguum dant fructum, aceto acri cum cinere irrigato, ipsumque codicem eodem cinere linito. At si fructum quem[7] ostendunt, ad maturitatem non perducunt, sed priusquam mitescant, uvae inarescunt, hoc modo emendabuntur. Cum instar ad ervi[8] amplitudinem acini habuerint, radice tenus vitem praecidito, plagam acri aceto pariter ac lotio veteri permixta terra linito, eodemque radices rigato, saepe fodito. Haec materias citant, eaeque fructum perferunt.

[1] umorē *SA*.
[2] tenerebrari *S* : tenebrari *A*.
[3] terebrā *c* : terebras *SAa*.
[4] gallicanam *c* : gallinacam *SAa*.
[5] excabat *SA*.
[6] insitione *A*.
[7] quem *c* : quae *SA*.
[8] instar ader vi *S* : insitione *Aac*.

cuts rather than overflow from the actual graft; for too much moisture is harmful and does not allow the slips which have been engrafted to take hold. Some of the ancients think that a hole ought to be bored in the vine and then the slips put in after being smoothly pared away; but we have carried out the same process by a better method. For the old-fashioned auger creates sawdust and therefore burns the part which it perforates and, being burnt, it rarely ever takes hold of the slips which are inserted. We, on the other hand, have adapted what we call the Gaulish auger for this kind of grafting. This makes a hollow without causing burning, because it produces shavings instead of saw-dust. So when we have cleaned out the hole which has been bored, we insert the slips after they have been pared on every side and then daub them round. This kind of graft thrives very readily. You should, then, have the grafting of your vines completed about the equinox. Engraft moist places from a white grape, dry places from a black.

You should water vines which produce very little fruit with sharp vinegar mixed with ashes and daub the stock itself with the same ashes. But if the vines do not bring to maturity the fruit which they display, but the grapes dry up before they grow mellow, they will be corrected by the following method: when the berries have reached the size of pulse, cut down the vine to the very root and daub the cut with sharp vinegar and similarly with stale urine mixed with earth, and water the roots with the same mixture, and dig round frequently. This method will cause the growth of firm-wood branches, and it is these which bear the fruit.

LUCIUS JUNIUS MODERATUS COLUMELLA

IX. Est etiam genus insitionis, quod uvas tales creat, in quibus varii generis ac saporis[1] colorisque reperiuntur acini. Hoc autem ratione tali efficitur: Quattuor vel quinque sive etiam plures voles virgas diversi generis sumito, easque diligenter et aequaliter compositas colligato, deinde in tubulum fictilem vel cornu arcte inserito, ita ut aliquantum exstent ab utraque parte, easque partes, quae exstabunt, resolvito, in scrobem deinde ponito, et terra stercо-
2 rata obruito, ac rigato, donec gemmas agant. Cum inter se virgae cohaeserint, post biennium aut triennium facta iam unitate, dissolves tubulum, et circa medium fere crus, ubi maxime videbuntur coisse, vitem serra praecidito, et plagam levato, terramque minutam aggerato, ita ut tribus digitis alte plagam operiat: ex eo codice cum egerit coles, duos optimos submittito, reliquos deicito: sic uvae nascentur, quales proposuimus.

3 Malleolum findito ita, ne gemmae laedantur, medullamque omnem eradito, tum demum in se compositum colligato, sic ne gemmas allidas,[2] atque ita terra stercorata deponito, et rigato. Cum coles agere coeperit, saepe et alte refodito. Adulta vitis tales uvas sine vinaceis creabit.

X. Vindemia facta, statim putare incipito ferramentis quam optimis et acutissimis: ita plagae[3] leves fient, neque in vite aqua consistere poterit:

[1] ac saporis *om. Aac.*
[2] allides *c*: alliges *SAa.*
[3] acutissimis ita plage *S*: acutissima sit aplage *A.*

ON TREES IX. 1–X. 1

IX. There is another kind of grafting which produces bunches of grapes in which berries of different kinds and flavours and colours are found. It is carried out in the following manner: Take four or five or, if you like, even more rods of different kinds and, after arranging them carefully and uniformly in a bunch, tie them together and then insert them, packed closely together, in an earthenware pipe or a horn, so that they project a little way at both ends; loosen the parts which project and then place them in a planting-hole and cover them with well-manured soil, and water them until they produce " eyes." After two or three years, when the rods have cohered and formed a unity, you will break up the pipe and cut the vine with a saw about the middle of the stem, where it shall seem to have formed the closest mass, and smooth off the cut; then heap over it fine soil, so as to cover the cut to a depth of three inches. From the stock thus formed, when it has put forth stalks, allow the two best to grow and discard the rest; in this manner bunches of grapes will grow of the kind which we have suggested.

How bunches of grapes may have berries of different kinds.

2

You should split a mallet-shoot in such a way that the " eyes " may not be hurt, and scrape away all the pith; then bind it closely together as it was before in such a way as not to damage the " eyes," and then plant it in well-manured earth and water it. When it begins to put forth stalks, dig round it frequently and deeply. When the vine has come to full growth it will produce grapes without pips.

3

X. When the vintage is finished, you should immediately begin pruning with the best and sharpest instruments: thus the cuts which are made will be smooth and water will not be able to remain in the

How to prune vines after the vintage.

373

quae simulatque immorata[1] est, corrumpit vitem, vermesque et alia creat animalia, quae materiam exedunt. Plagas autem rotundas[2] facito: nam
2 celerius cicatricem ducunt. Sarmenta lata, vetera, male nata, contorta, omnia haec praecidito: novella et fructuaria, et interdum sobolem idoneam, si iam superficies parum valebit, submittito bracchiaque conservato. Quam celerrime poteris putationem perficito. Arida et vetera, falce quae amputari non possunt, acuta dolabra abradito. In agro macro et sicco vineam imbecillam ante brumam[3] putato: quam partem non deputaveris, circa calend. Febr. repetito: ab. idibus Decemb. ad idus Ianuarias ferro
3 tangi vitem et arborem non convenit. Cum vitem putabis, inter duas gemmas secato: nam si iuxta ipsam gemmam secueris, laborabit,[4] nec materiam citabit. Cicatrix autem semper deorsum spectet, ita neque aqua sole laborabit, humoremque recte capiet. In agro crasso validaque vinea plures gemmas et palmas relinquito, in exili pauciores. Sicubi in vite bracchium desiderabis,[5] falce acuta semel aut bis eo loco alte instar digiti mucrone ferito. Bracchium quamvis longum cave totum tollas, nisi si totum aruerit.
4 Vineam novellam ante brumam ablaqueatam habeto, ut omnes imbres limumque concipiat. Vites arboresque quo citius ablaqueaveris, erunt

---

[1] inmortale *SAac*.   [2] rotundos *SA*.
[3] ante brumam *a* : ad brumā *S* : an brumān *A*.
[4] laborabit *a* : laboravit *SAc*.
[5] desideravit *SAac*.

ON TREES x. 1-4

vine; for water, as soon as it is allowed to stagnate, causes the vine to decay and breeds worms and other creatures which eat into the firm-wood. But make your cuts round, for then they scar over more quickly. Cut away all shoots which are too broad, 2 or old, or badly grown, or twisted; but allow those to grow which are young and fruitful and sometimes a suitable off-shoot, if the part of the vine which is above ground is not thriving well, and conserve the branches. Finish the pruning as quickly as possible. Shoots which are old and dry and cannot be cut away with a pruning-knife, you should pare away with a sharp axe. On land which is lean and dry you should prune a weak vine before the shortest day, and come back about Feb. 1st to any part of the vineyard which you have not pruned. Between Dec. 13th and Jan. 13th it is not advisable that a vine or a tree should be touched with the knife. When you prune a vine, 3 make the cut between two " eyes "; for if you make it near the " eye " itself, it will suffer and will not put forth firm-wood. The scar should always face downwards; it will then not suffer from water or the sun's heat and will catch the moisture in the right way. On rich soil and in a thriving vineyard you should leave more " eyes " and top branches, but fewer on poor soil. Whenever you find a branch lacking on a vine, take a sharp pruning-knife and strike deeply on the spot with the point once or twice to the depth of about an inch. However long a branch may be, take care not to remove it entirely unless it is quite dried up.

See that a young vine is trenched round before the 4 winter, so that it may take up all the rain and mud. The sooner you trench round vines and trees, the

valentiores. Sed quaecumque in clivis erunt positae, ita ablaqueandae sunt, ut a superiore parte secundum codicem lacusculi fiant, ab inferiore autem pulvilli [1] altiores excitentur, quo plus aquae limique contineant. Vinea vetus neque ablaqueanda est, ne radices, quas in summo habet, inarescant, neque aranda, ne radices abrumpantur. Bidentibus saepe et alte fodito aequaliter, et stercore vel palea conspergito solum ante brumam, vel cum circum ipsam vitem summatim ablaqueaveris, stercorato.

XI. Vineam quam putare tam bene pampinare utile est: nam et materiae quae fructum habent, melius convalescunt, et putatio sequentis anni expeditior,[2] tum etiam vitis minus cicatricosa fit: quoniam quod viride et tenerum decerpitur, protinus convalescit. Super haec quoque melius uvae [3] maturescunt. Ante dies decem quam vinea florere incipit, pampinatam habeto. Quidquid supervacui enatum fuerit, tollito. Quod in cacumine aut in bracchiis natum erit, decerpito, dumtaxat quae uvam non habebunt. Cacumina virgarum, ne luxurientur, demutilato.[4] Uvas, quae meridiem aut occidentem spectabunt, ne praeurantur, suo [5] sibi pampino tegito.

XII. Simulatque uva variari coeperit, fodito tertiam fossuram: et cum iam maturescet, ante meridiem, priusquam calere incipiet; cum desierit, post meridiem fodito pulveremque excitato: ea res et a sole et a nebula maxime uvam defendit. Lutulentam terram neque arare neque fodere oportet, quia

[1] pulvilli altiores *Lundström* : pullialiliores *SA*.
[2] expeditior *a* : expeditur *SA* : expeditius *c*.
[3] uvae *add. Schneider.*
[4] dimutilato *SAac.*   [5] sua *SAac.*

quicker they will thrive. But any vines which are planted on slopes must be trenched in such a way that pools may be formed on the higher ground next the stem, and ridges raised to a greater height on the lower ground, so as to contain more water and mud. An old vineyard must not be trenched, lest the roots which it has on the surface may dry up, nor must it be ploughed, lest the roots be broken off. You should dig it frequently and deeply to a uniform depth and strew the ground with dung or chaff alone before the winter, or else manure it after trenching round the vine itself on the surface only.

XI. It is just as beneficial to trim a vine well as to prune it, for then both the firm-wood branches, which bear the fruit, thrive better, and the next year's pruning is more quickly done; also the vine shows fewer scars, for, when that which is removed is young and tender, the vine immediately recovers; moreover the grapes ripen better. See that your vineyard is trimmed ten days before it begins to blossom, and remove anything superfluous which has grown. Remove anything which has grown on the top and on the branches, provided that they are not going to bear grapes. Cut back the tops of the rods, so that they may not run riot. Cover the bunches of grapes which face south or west with their own tendrils, so that they may not be burnt by the sun. *Of the trimming of vines.*

XII. As soon as the grapes begin to turn colour, give the vineyard a third digging, and when they are already ripening, do the digging in the forenoon, before it begins to be hot; when the heat has ceased, dig and stir up the dust after midday. This is the best way of protecting the grapes both from sun and from fog. Clayey ground ought not to be ploughed *On the digging of vineyards.*

2 valde durescit et finditur. Bidentibus terram vertere utilius est quam aratro. Bidens aequaliter totam terram vertit: aratrum praeterquam quod scamna facit, tum etiam boves, qui arant, aliquantum virgarum et interdum totas vites frangunt. Finis autem fodiendi vineam nullus est: nam quanto saepius foderis, tanto uberiorem fructum reperies.[1]

XIII. Palearum acervos inter ordines verno tempore positos habeto in vinea.[2] Cum frigus contra temporis consuetudinem intellexeris, omnes acervos incendito, ita fumus nebulam et rubiginem removebit.

XIV. Lupinum terito, et cum fracibus[3] misceto, eoque imam vineam circumlinito: vel bitumen cum oleo coquito, eo quoque imas vites tangito, formicae[4] non excedent.[5]

XV. Vites, quae secundum aedificia sunt, a soricibus aut muribus infestantur. Id ne fiat plenam lunam observabimus, cum erit in signo Leonis vel Scorpionis vel Sagittarii vel Tauri, et noctu ad lunam putabimus.[6] Genus est animalis, volucra appellatur;[7] id fere praerodit teneros[8] adhuc pampinos et uvas: quod ne fiat, falces, quibus vineam putaveris, peracta putatione, sanguine ursino linito: vel si pellem fibri habueris, in ipsa putatione, quoties falcem acueris, ea pelle aciem detergito, atque ita

---

[1] repperies *Aa*. [2] vineam *SAac*.
[3] fragibus *SAa* : frugibus *c*.
[4] formicam *SAac*.
[5] exedent *Ursinus*.
[6] putavimus *SA*.
[7] appellantur *SAac*. [8] teneras *SA*.

or dug, because it grows very hard and splits. It is more profitable to turn up the soil with mattocks than with the plough; the mattock turns up all the soil in a uniform manner; when the plough is used, besides the fact that it forms ridges, and, besides, the oxen used for ploughing break off some of the rods and sometimes whole vines. There is never an end of digging a vineyard; for the more often you dig it, the greater will you find the abundance of fruit.

XIII. In the spring see that you have heaps of chaff placed between the rows in the vineyard. When you notice that there is cold which you would not usually expect at the season, set fire to all the heaps; the smoke then will get rid of the fog and mildew. *How to avoid the effects of fog and mildew.*

XIV. Crush lupines and mix them with dregs of oil and smear the lowest part of the vines with this; or else boil bitumen with oil and touch the lowest parts of the vines with this; ants will then not crawl beyond. *How to keep off ants.*

XV. Vines which are near buildings are infested by shrew mice and rats and mice. To get rid of them we shall look out for a full moon when it is in the constellation of the Lion or the Scorpion or the Archer or the Bull, and prune the vines at night by moonlight. There is a kind of animal called the leaf-roller [a] which commonly gnaws off the tendrils and grapes while they are still tender. To prevent this, when the pruning is finished, smear the pruning-knives with which you have pruned the vineyard with bear's blood; or, if you have a beaver's skin, during the actual pruning, whenever you sharpen your pruning-knife, wipe the edge of it on this skin, and then recommence your pruning. Having *How to keep off mice.*

[a] Also called *convolvulus* (Cato, *R.R.* 95).

LUCIUS JUNIUS MODERATUS COLUMELLA

putare incipito. Quoniam de vineis abunde diximus, de arbustis praecipiamus.

XVI. Vitem maxime populus alit, deinde ulmus, deinde fraxinus. Opulus[1] quoniam frondem non[2] idoneam habet, a plerisque improbatur. Ulmus autem quam Atiniam vocant rustici, generosissima est et laetissima, multamque frondem habet: eaque maxime serenda est locis pinguibus vel etiam mediocribus: sed si aspera et siticulosa loca arboribus obserenda[3] erunt, neque populus neque ulmus tam idoneae sunt quam orni: eae autem silvestres[4] fraxini[5] sunt, paulo latioribus tamen foliis quam ceterae fraxini, nec deteriorem frondem, quam 2 ulmi praestant. Caprae quidem et oves vel libentius etiam hanc frondem appetunt.

Igitur qui arbustum constituere volet, ante annum quam deponantur[6] arbores, scrobes faciat quattuor quoquoversus pedum.[7] Deinde circa calen. Mart. in eandem[8] scrobem ulmum et populum, vel fraxinum deponat, ut si ulmus defecerit, populus vel fraxinus locum obtineat. Si autem utraque vixerint, altera eximatur,[9] et alio loco deponatur. Arbustum inter quadragenos pedes dispositum esse convenit: sic enim et ipsae arbores, et appositae vites melius convalescent, fructumque meliorem dabunt. Segetes etiam, quae in eo erunt, minus umbra laborabunt. 3 Arborem quam deposueris saepius circumfodito, quo

[1] populus *codd.*   [2] non *om. codd.*
[3] obserenda *vett. edd.*: observanda *SAac.*
[4] orni: eae autem silvestres *Lundström*: orniae silvestres *S*: orniae āt silvestre *A*.
[5] fraxini *ac*: fraxi *SA*.   [6] deponant *SAac*.
[7] quattuor quoquo annum *S*: quattuor q̄ annū *A*.
[8] eadē *SAa*.
[9] exima *ante* eximatur *add. SA*.

dealt at length with vineyards, let us now give directions about trees used for supporting them.

XVI. The poplar is the tree chiefly used for supporting the vine, and next to it the elm, and then the ash-tree. The maple is not approved of by most people, because it has not a suitable leaf. The kind of elm which the country folk call Atinian,[a] is the noblest and most luxuriant and has an abundance of foliage. It is chiefly to be planted in rich soil but can also be planted in moderately rich ground. If places which are rough and waterless have to be planted with trees, neither the poplars nor the elms are so suitable as mountain-ashes; these are wild ash-trees but with slightly, however, broader leaves than the other ash-trees, and their leaves are not inferior to those produced by elms. Goats and sheep, indeed, have a greater liking for their leaves than any other.

He, therefore, who wishes to establish a plantation of trees for supporting vines, should make planting-holes measuring four feet each way a year before the trees are planted. Then about March 1st he should plant an elm, and a poplar or a mountain-ash in the same planting-hole, so that if the elm should fail, the poplar or the mountain-ash may take its place; if, however, both trees have lived, one should be taken up and planted elsewhere. Trees for supporting vines are best arranged at intervals of forty feet; for then both the trees themselves and the vine trained upon them will thrive better and produce better fruit ; crops, too, which are grown in the plantation will suffer less harm from the shade. Dig frequently round the supporting tree which you have planted, so that it may reach maturity more

[a] Called after the name of a Roman *gens*.

LUCIUS JUNIUS MODERATUS COLUMELLA

celerius adolescat:[1] et triennium ferro ne tetigeris. Completis sex et triginta mensibus, ad recipiendam vitem formabis, supervacuous ramos amputabis, alterna bracchia in modum scalarum relinques, alternisque annis putabis.[2] Sexto anno, si iam firma videbitur, maritabis hoc modo. Ab ipso arboris crure pedale spatium intermittito, deinde sulcum in [3] quattuor pedes longum, in tres altum, in dupondium semissem latum cum feceris, patiere minime duobus mensibus eum tempestatibus verberari. Tum demum circa cal. Martias vitem de seminario ne minus [4] decem pedum sternito, et adminiculato, arborique iungito: eam proximo anno ne putaveris: tertio vero ad unam virgam redigito, paucasque gemmas [5] relinquito, ne antequam invaluerit, in altitudinem repat: deinde ubi amplum incrementum habuerit,[6] per omnia arboris tabulata disponito materias, ita tamen ne vitem oneres, sed certa et robustissima flagella submittas. Arbustivam vitem quam putare, tam alligare [7] diligenter oportet. Nam in eo fructus maxime vis consistit, diutiusque perennat,[8] quae firmis toris et idoneis locis religata est. Itaque omnibus annis convenit subsequi putationem, ita ut tori renoventur, et vitis per idoneos ramos disponatur.

XVII. Olea maxime collibus siccis et argillosis gaudet: at humidis campis et pinguibus laetas frondes sine fructu affert. Melius autem truncis

---

[1] adolescant *SAac*.
[2] putabis *om. Aac*.
[3] inter *SAac*.   [4] deminus *SA*.
[5] gemmasque *SA*.
[6] haberet *SAc*: habet *a*.
[7] putare tam alligare *vett. edd.*: putar & amalis gera *S*: putaret amallis gera *Aac*.
[8] pannat at *S*: pannat *A*: perennat quae *ac*.

quickly and do not touch it with the knife for three years. When thirty-six months are completed, you will shape the tree for receiving the vine and lop off the superfluous branches and leave every alternate bough so as to form a sort of ladder, and you will prune the tree every other year. In the sixth year, if the tree seems now to be firmly established, you will " wed " the vine to it in the following way. Leave a space of one foot from the base of the tree, then, having made a furrow four feet long and three feet deep and two and a half feet wide, you will allow it to be buffeted by the weather for at least two months. Then about March 1st take from the nursery a vine not less than ten feet long and spread it out flat and prop it up and attach it to the tree. In the following year do not prune it, but in the third year reduce it to a single rod and leave only a few " eyes," so it may not creep up to a great height before it has gained strength. Then when it has attained an ample growth, distribute the firm-wood shoots over every " story " of the tree, in such a way, however, as not to burden the vine, but so as to allow the shoots which are reliable and very strong to grow freely. You must be just as careful in binding up a vine which is supported on a tree as in pruning it; for it is on this that the strength of the fruit chiefly depends, and a vine which has been attached to a tree with firm ties and in suitable places lasts longer. Therefore, every year you should supplement the pruning by seeing that the ties are renewed and the vine trained over suitable branches.

XVII. The olive-tree takes most delight in hills which are dry and clayey; but in damp, rich plains it produces luxuriant foliage without fruit. It is better

How to form a plantation of olive-trees.

LUCIUS JUNIUS MODERATUS COLUMELLA

quam plantis olivetum constituitur. Magoni placet siccis locis olivam autumno post aequinoctium seri ante brumam. Nostrae aetatis agricolae fere 2 vernum tempus circa calend. Mart. servant. Oportet autem scrobem oleae quoquoversus pedes quaternos patere, in imum scrobem lapidem glareamque abicere, deinde super terram quattuor digitorum inicere,[1] tum arbusculam deponere ita rectam, ut quod scrobe exstiterit, in medio sit. Arbusculam autem a tempestatibus tueri diligenter oportet adminiculando, et terram, quae in scrobe reponitur, 3 stercore immiscere. Oleam decet inter sexagenos pedes disponi, ut spatium in latitudinem crescendi habeat: nam quae in proceritatem extenduntur, evanidae fiunt, parumque fructus ferunt. Optima est oleo Liciniana, Pausia[2] secunda oleo, escae Orcita. Sunt et regiae, et radii non sine specie, neque oleo nec esui tam[3] gratae, quam quas supra diximus. Si oleam posueris eo loco, unde quercus effossa est, emorietur, ideo quod quidam vermes, qui raucae dicuntur,[4] in radice quercus nascuntur, eique maxime semina oleae consumunt. Si in olea unus ramus aliquanto ceteris laetior est, nisi eum recideris, arbor tota 4 fiet retorrida. Omnes arbusculas[5] prius quam transferas,[6] rubrica notare convenit, ut cum serentur, easdem caeli partes aspiciant, quas etiam in seminario

[1] inigere *SA*.
[2] posita *SAac*.
[3] esui tam *Schneider*: sui tā *S*: suuitā *A*.
[4] qui raucae dicuntur *Lundström*: quirunt educunt *S*: quirt ẹducuntur *A*.
[5] arbusculas *S*: arbusculus *Aac*.
[6] transferat *SAac*.

---

[a] Pliny, *N.H.* 17. 18. 30, § 130.

to form an olive-grove with stocks than with slips. Mago is of opinion that the olive-tree should be planted in dry ground during the autumn after the equinox before mid-winter. The husbandmen of today generally observe the beginning of the spring season about March 1st. The planting-hole for an olive-tree ought to be four feet each way and you should throw into the bottom of it stones and gravel and then cast on the top four inches of soil; next, you should set the small tree in an upright position, so that the part which stands out from the planting-hole is in the middle. You must protect the small tree carefully from storms by propping it up and you should mix dung with the earth which is put back in the planting-hole. Olive-trees should be arranged with sixty feet between them so that they may have room to grow in breadth; for those which run to height become feeble and bear little fruit. The best tree for oil is the Licinian. The Pausean is the next best for oil and the orcita for eating. The royal and the shuttle-shaped are not without a beauty of their own, but are not so acceptable for making oil or for eating as those which we have mentioned above. If you plant an olive-tree in a place from which an oak has been dug up, it will die, for the reason that there is a kind of worm, which is called *rauca*,[a] born in the root of the oak and it is particularly liable to eat up olive seedlings. If one branch of an olive-tree thrives somewhat better than the rest, unless you cut it off, the whole tree will become burnt up. It is well to mark all the small trees with ochre before you transplant them, in order that, when they are planted, they may face the same quarter of the sky towards which they had looked in the nursery-bed; otherwise they

conspexerant: alioquin frigore vel calore laborabunt ab iis partibus, quas praeter consuetudinem sub alio aëre [1] positas habuerint.

XVIII. Priusquam pomarium constituas, quam magnum habere voles circummunito maceria, aut fossa, ita ut non solum pecori, sed ne homini quidem transitus sit, nisi per ostium,[2] dum adolescant semina. Nam si saepius cacumina manu praefracta aut a pecore praerosa fuerint, in perpetuum corrumpuntur. 2 Generatim autem arbores disponere utilius,[3] maxime ne imbecillae a valentioribus premantur, quia [4] nec viribus nec magnitudine sunt pares, neque pariter crescunt. Terra quae vitibus apta est, eadem quoque utilis est arboribus.

XIX. Ante annum quam pomaria disponere voles, scrobes fodito: ita sole pluviaque macerabuntur, et quod posueris cito comprehendet. Sed si quo anno scrobes feceris, etiam semina ponere voles, minime ante duos menses fodito scrobes, postea stramentis eos impleto, et incendito. Quo latiores patentioresque scrobes feceris, eo laetiores erunt uberioresque [5] 2 fructus. Scrobis clibano similis esto,[6] imus quam summus patentior, ut laxius radices vagentur, ac minus frigoris hieme, minusque aestate vaporis per angustum ostium intret:[7] tum etiam clivosis locis terra, quae in eum congesta est, pluviis non abluitur. 3 Arbores raris [8] intervallis serito,[9] ut cum creverint,

---

[1] alio aëre positas *Lundström* : alio aepositas *SA*.
[2] odium *SAac*.   [3] utilis *SAac*.
[4] qua *SAa* : quae *c*.
[5] uberioresque *S* : superioresque *Aac*.
[6] esto *Lundström* : est *SAac*.
[7] ostium intret *Lundström* : ostentarit *SAac*.
[8] raris *Ursinus* : pares *SAac*.
[9] sepito *SAac*.

will suffer from cold or heat in those parts which are exposed to an atmosphere to which they are not accustomed.

XVIII. Before you establish an orchard, enclose the extent of land which you wish to have, either with a wall or with a ditch, so that there may be no access to it until the seedlings are reaching maturity, not only to cattle but also to human beings, except through the entrance. For if the tops be frequently broken off by hand or gnawed off by cattle, they are spoilt forever. It is more practical to arrange the trees according to their different kinds, chiefly in order to prevent the weak being overwhelmed by the stronger, since trees are not of the same strength or size and do not grow uniformly. Land which is suitable for vines is also suitable for fruit-trees.[a]

How to form an orchard.

XIX. A year before you wish to establish an orchard, dig planting-holes; then they will be softened by sun and rain and what you plant will quickly take root. If, however, you wish to plant the seedlings in the same year as you have made the planting-holes, dig the latter at least two months beforehand and afterwards fill them with straw and set fire to it. The broader and more open you make the planting-holes, the more luxuriant and abundant will be the fruit. The planting-holes ought to resemble an oven, being wider at the bottom than at the top, in order that the roots may spread more widely and that less cold in winter and less heat in summer may enter through the narrow aperture; also on sloping ground the earth which is heaped into the hole is not washed away by the rains. Plant the trees at wide intervals,

How to make planting-holes.

---

[a] The rest of the *De Arboribus* is slightly shorter, but on a few occasions longer than *De Re Rustica*, V. 10 to the end.

LUCIUS JUNIUS MODERATUS COLUMELLA

spatium habeant, quo ramos extendant. Nam si spisse posueris,[1] neque infra quid serere poteris, nec sic[2] ipsae fructuosae erunt, nisi eas intervulseris.[3] Itaque placet inter ordines quadragenos pedes, minimumque tricenos, relinqui.

XX. Semina lege, ne minus crassa, quam manubrium est bidentis, recta, levia, procera, sine ulceribus, integro libro. Ea bene et celeriter comprehendunt. Semina si ex arboribus sumes, de iis potissimum sumito, quae omnibus annis bonos et uberes ferunt fructus. Observabis autem ab humeris, qui sunt contra solem orientem, ut eosdem decerpas.[4] Sed si cum radice plantam[5] posueris, incrementum ei maius[6] futurum[7] quam ceteris senties. 2 Arbos insita fructuosior est, quam quae insita non est, id est, quam quae ramis[8] aut plantis[9] ponitur. Priusquam arbusculas transferas, rubrica vel alia qualibet[10] re signato, ut iisdem ventis, quibus ante steterunt, constituas eas: curamque adhibeto, ut ab superiore et sicciore et exiliore in planiorem, humidiorem, pinguiorem agrum transferas. Semina trifurca maxime ponito: ea exstent supra terram tribus pedibus. In eodem scrobe si[11] duas aut tres arbusculas ponere voles, curato ne inter se contingant, nam 3 ita[12] vermibus interibunt.[13] Cum semina depones,

---

[1] poteris S : potieris A : potueris *ac.* [2] si $SAac.$
[3] intervulseris S : intervaseris $Aac.$
[4] decerras s& S : decertasset $Aac.$
[5] planta $SAac.$
[6] maius *Lundström* : eius $SAac.$
[7] futurum quam *Lundström* : fruitq. S : fruit quae A : feruitque a : fueritque c.
[8] ramus Sac : ram' A.
[9] plantis c : planis SAa.
[10] alia qualibent S : aliqualibent A : aliqualibet *ac.*

so that, when they have grown, they may have space to spread their branches. For if you place them close to one another, you will not be able to plant anything underneath them, nor will the trees themselves be fruitful thus if you do not thin them out. The approved method, therefore, is that forty or at least thirty feet should be left between the rows.

XX. Choose seedlings which are not less thick than the handle of a fork, straight, smooth, tall, without excrescences and with the bark intact. Such plants take root well and quickly. If you take slips from the trees, take them preferably from those which bear good and abundant fruit each year. You will look for slips on the shoulders of the trees which face the rising sun and pluck these. If, however, you have set a plant which has a root, you will perceive the growth to be greater than in the other plants. A tree which is engrafted is more fruitful than one which is not, that is, one which is planted in the form of a branch or of a seedling. Before you transplant the small trees, mark them with ochre or anything else you please, in order that you may plant them facing the same wind as they faced before, and be careful to transfer them from higher, drier and poorer soil to ground which is flatter, moister and richer. Set preferably seedlings which have three shoots, and let them project three feet above the ground. If you wish to put two or three small trees in the same planting-hole, take care that they do not touch one another; for in that way they will be destroyed by worms. When you plant seedlings in the ground,

---

[11] si *om. Aac.*
[12] ita *ed. Ald.*: ut *SAac.*
[13] interibunt *Lundström*: interbunt *S*: interimunt *Aac.*

LUCIUS JUNIUS MODERATUS COLUMELLA

dextra sinistraque usque[1] in imum scrobem fasciculos sarmentorum bracchii[2] humani crassitudine deponito, ita ut supra terram paululum exstent, per quos aestate parvo labore aquam radicibus subministres. Arbores aut semina cum[3] radicibus autumno serito circa id. Octobres. Taleas[4] et ramos vere, ante quam germinare arbores incipiant, deponito. Sed ne tinea molesta sit seminibus ficulneis, in imum scrobem taleam lentisci, ita ut cacumen eius deorsum spectet, obruito.

XXI. Ficum frigoribus ne serito: loca aprica, calculosa, glareosa,[5] interdum et saxosa amat. Eiusmodi agro cito convalescit, si scrobes amplos et idoneos feceris. Ficorum genera etiam si sapore et habitu differunt, tamen uno modo, sed dispari loco pro differentia agri,[6] seruntur. Locis frigidis et autumni temporibus aquosis praecoques serito, ut ante pluviam fructum deligas: locis calidis hibernas serotinas serito. At si voles ficum, quamvis non natura, seram facere, cum grossuli minuti erunt, fructum decutito: ita alterum edet fructum, et in 2 hiemem seram[7] differet[8] maturitatem. Nonnumquam etiam cum frondere coeperint arbores, cacumina fici acutissimo ferramento summa amputare prodest. Sic firmiores arbores et feraciores fiunt. Semper proderit simulac folia agere coeperit ficus,

---

[1] dextra sinistraque usque *Pontedera*: dextra sinistra sinis quaeusque *S*: dextra sinistra sinis quousq; *Aac*.
[2] brachiũ *S*: brachi *A*: bracchi *c*: bracchii *a*.
[3] cum *om. SAac*.
[4] taleas et *vett. edd.*: talaer *S*: tala et *Aac*.
[5] glareosũ *Aac*.
[6] sed dispari loco pro differentia agri *Lundström*: sed dispro differentia agri pari loco *SA*: sed dispari differentia agri pari loco *ac*.   [7] serũ *SAac*.   [8] differat *SAac*.

deposit small bundles of twigs of the thickness of a man's arm on the right and left of them reaching to the bottom of the planting-hole, so as to project a little above the ground, so that you may supply the root with water without much trouble. Plant trees and seedlings with roots in the autumn about October 15th; but set cuttings or branches in the spring before the trees begin to bud. But to prevent the caterpillar from causing trouble to fig-tree plants, bury in the bottom of the planting-hole a cutting from a mastic-tree so that the top of it faces downwards.

XXI. Do not plant a fig-tree in cold weather; it likes places exposed to the sun, pebbly, gravelly and sometimes even stony. In this sort of ground it quickly thrives, if you have made roomy and suitable planting-holes. The several kinds of fig-trees, even if they differ in flavour and habit, yet are planted in one way but in dissimilar positions according to the difference of soil. In cold districts and where the weather is wet in autumn you should plant early fig-trees, so that you may gather the fruit before the rain comes; in warm places you should plant late winter fig-trees. But if you wish to make a fig-tree ripen late, though it does not naturally do so, shake down the fruit while the little figs are small; it will then produce a second crop and put off its ripening until late winter. Sometimes, too, when the trees 2 begin to bear leaves, it is beneficial to cut off the uppermost tops of the fig-tree with a very sharp knife; the trees will then become stronger and more prolific. It will always be beneficial, as soon as the fig-tree begins to put forth leaves, to dilute ochre

When the fig-tree should be planted.

LUCIUS JUNIUS MODERATUS COLUMELLA

rubricam amurca diluere, et cum stercore humano ad radicem infundere: ea res efficit uberiorem fructum, et farctum fici speciosius et plenius.

XXII. Nucem Graecam serito Arcturi signo, vel circa calend. Febr. quia prima gemmascit. Agrum calidum durum et siccum desiderat. Nam in locis diversis eiusmodi natura si posueris nucem, protinus putrescet. Antequam nucem deponas, in aqua mulsa, nec nimis dulci macerato: ita iucundioris fructum, cum adoleverit, praebebit, et interim melius 2 atque celerius nascetur. Ternas nuces in trigonum statuito, parsque acutior inferior sit, quia inde radices mittit, nuxque a nuce minime palmo absit, et apex[1] ad Favonium spectet. Omnis nux unam radicem mittit, et simplici stilo prorepit. Cum ad scrobis solum radix pervenit, duritia humi coërcita recurvatur, et extensa in modum ramorum alias radices emittit. Nucem Graecam et Avellanam 3 Tarentinas hoc modo facere poteris. In quo scrobe destinaveris nucem serere, terram minutam in modum semipedis ponito, ibique semen ferulae iacito.[2] Cum ferula fuerit enata, eam findito, et in medullam eius sine putamine nucem Graecam vel Avellanam abscondito, et ita adobruito. Hoc ante cal. Mart. facito, vel etiam inter nonas et idus Martias. Hoc eodem tempore iuglandem et pineam et castaneam serere oportet.

[1] apex *scripsi*: anceps *codd.*
[2] iacito *S*: facito *Aac.*

---

[a] The reading here is doubtful: the MS. reading *anceps* gives no sort of sense.

with lees of oil and pour it over the roots together with human ordure: this makes the fruit more abundant and the inner part of the fig more attractive in appearance and fuller.

XXII. You should plant the almond-tree, since it is the first nut-tree to blossom, when the constellation of the Bear rises on about Feb. 1st. It requires warm, hard and dry ground; for if you plant it in places which differ from this kind of nature, it immediately rots. Before you set a nut-tree, you should soak it in honey-water which should not be too sweet; it will then, when it comes to maturity, produce fruit of a pleasanter flavour, and meanwhile it will grow better and quicker. Place three nuts so as to form a triangle and make the sharper end point downwards, because it is from there that it puts out its roots, and let the nuts be at least a hand's breadth away from one another, and an apex[a] of the triangle face towards the west. Every nut sends out one root which creeps forth in a single stem; when the root reaches the bottom of the planting-hole, it is checked by the hardness of the ground and is bent back and, extending like the branches of a tree, it puts forth other roots.

You can make an almond and a hazel into Tarentine nuts in the following manner. In the planting-hole in which you intend to sow the nut place fine soil to a depth of half a foot and throw in there fennel-seed. When the fennel has grown up, split it and secrete in the pith of it an almond or a filbert without its shell, and then cover it over with earth. Do this before March 1st or even between March 7th and 15th. You should plant the walnut, pine-nut and chestnut at the same time of year.

LUCIUS JUNIUS MODERATUS COLUMELLA

XXIII. Malum Punicum vere usque in cal. Apriles recte seritur. Quod si acidum aut minus dulcem fructum feret, hoc modo emendabitur. Stercore suillo et humano et lotio humano veteri radices rigato. Ea res et fertilem arborem reddet, et primitivos annos fructum vinosum, postea vero etiam dulcem et apyrinum facit. Nos exiguum admodum laser Cyrenaicum [1] vino diluimus, et ita cacumina arboris summa oblevimus: ea res emendavit acorem ma-
2 lorum. Mala Punica ne rumpantur in arbore, remedio placuit lapides tres, si, cum seres arborem, ad radicem ipsam collocaveris. At si iam arborem satam habueris, scillam secundum radices arboris serito. Alio modo, cum iam matura mala fuerint, antequam rumpantur, petiolos, quibus pendent, intorqueto. Eo modo servabuntur etiam [2] anno toto.

XXIV. Piros [3] autumno ante brumam serito, ita ut minime dies quinque et viginti ad brumam supersint. Quae ut sint feraces, cum iam adoleverint, alte ablaqueato, et iuxta ipsam radicem truncum findito, et in fissuram cuneum pineum tedae adicito, et ibi relinquito: deinde obruta ablaqueatione cinerem supra terram spargito.

XXV. Mala aestiva, cydonea,[4] sorba, pruna, post mediam hiemem usque in idus Febr. serito. Morum [5] ab idibus Febr. usque in aequinoctium vernum recte seres. Siliquam Graecam, quam quidam ceratium vocant, item Persicum ante brumam per autumnum

---

[1] Lacer cirenaicum *a* : Lacire nai cū *S* : laci renai cū *A*.
[2] et cū *A* : et cum *ac*.   [3] pirus *S* : prius *Aa*.
[4] cythonea *SAa* : citonea *c*.
[5] morum *vett. edd.* : marum *S* : mabrum *Aac*.

---

a *Cf.* p. 340.
b See note on Book V. 10. 20.

XXIII. It is right to plant the pomegranate in the spring up to April 1st. If it bears fruit which is bitter and not sweet enough this will be remedied by the following method: moisten the roots with sow-dung and human ordure and stale human urine. This will both render the tree fertile and during the first years cause the fruit to have a vinous taste, and afterwards, too, makes it sweet and its kernels soft. We ourselves have mixed just a little laser from Cyrene [a] with wine and thus smeared the uppermost tops of the trees. This has remedied the tartness of the 2 fruit. To prevent pomegranates from bursting on the trees, the approved remedy is to place three stones at the very root of the tree when you plant it; if however, you have already planted it, sow squill near the roots of the tree. According to another method, when the fruit is already ripe and before it bursts, you should twist the stalks on which it hangs; by the use of this device they will also keep for a whole year.

How to plant pomegranate-trees.

XXIV. Plant pear-trees in the autumn before mid-winter, so that at least twenty-five days remain before mid-winter. In order that the trees may be fruitful when they have now come to maturity, trench deeply round them and split the trunk close to the very root and in the fissure insert a wedge of pitch-pine and leave it there; then, when the trenching has been filled in, scatter ashes over the soil.

How to plant pear-trees.

XXV. Plant summer-apples, quinces, service-apples and plums from the middle of winter to February 13th. You will be right in planting the mulberry from February 13th to the vernal equinox. You should plant the carob-tree, which some people call *ceratium*,[b] and also the peach during the autumn

How to plant apple trees, etc.

serito. Amygdala si parum feracia erunt, perforata arbore lapidem adigito: ita librum arboris inolescere sinito.

2   Omnium autem generum ramos[1] circa calend. Martias in hortis, ubi et subacta et stercorata terra est, per pulvinos arearum disponere convenit: deinde cum tenuerint, danda est opera, ut dum teneros ramulos habent, veluti pampinentur, et ad unum stilum primo anno semina redigantur: et cum autumnus incesserit, ante quam frigus cacumina adurat, omnia folia decerpere expediet, et ita crassis harundinibus, quae ab una parte nodos integros habent, quasi pileolos induere, atque ita a frigore et gelicidiis teneras adhuc virgas tueri. Post quartum et vicesimum deinde mensem[2] sive transferre et disponere in ordinem voles, seu inserere, satis tuto utrumque facies.

XXVI. Omnis surculus inseri potest, si non est ei arbori, cui inseritur, dissimilis cortice: si vero fructum etiam eodem tempore fert, sine ullo scrupulo optime inseritur. Tria autem genera insitionum antiqui tradiderunt: unum, cum resecta et fissa arbor recipit insertos surculos: alterum, quo resecta inter librum et materiam admittit semina: quae utraque genera verni temporis sunt: tertium, cum ipsas gemmas cum exiguo cortice in partem sui delibratam recipit, quam vocant agricolae emplastrationem. Hoc genus aestatis est. Quarum[3] insitionum rationem cum

[1] ramos *S*: ramus *Aac*.
[2] messem *Aac*.
[3] quorum *SAac*.

before winter comes. If almond-trees are not productive enough, make a hole in the tree and drive in a stone, and then allow the bark of the tree to cover it.

It is proper to plant out branches of all kinds of fruit-trees about March 1st in gardens in raised beds after the soil has been well worked and manured. Then, when they have taken root, care must be taken to trim them like vines while the little branches are young and tender, and in the first year the plants should be reduced to a single stem. When autumn has come on, before the cold nips the tops, it will be well to strip off all the foliage, and then cover the trees with caps, as it were, of thick reeds which have their knots intact on one side, and thus protect the still tender rods from cold and frosts. Then, after the twenty-fourth month, you will quite safely do whichever you wish of two things—either transplant them and arrange them in rows or engraft them.

XXVI. Any kind of scion can be grafted on any tree if it is not dissimilar in respect of bark to the tree in which it is grafted; indeed, if it also bears fruit at the same season, it can perfectly well be grafted without any scruple. The ancients have handed down to us three kinds of grafting; one in which the tree, having been cut and cleft, receives the scions inserted in it; the second in which the tree having been cut admits seedlings between the bark and the hard-wood, both these methods belong to the season of spring; and the third, when the tree receives actual buds together with a little bark into a part of it which has been stripped of the bark. The last kind the husbandmen call "emplastration"; it takes place in summer. When we have imparted the

tradiderimus, a nobis quoque repertam aliam docebimus.

2 Omnes arbores simulatque gemmas agere coeperint, luna crescente, inserito, olivam autem circa aequinoctium vernum usque in idus Apriles. Ex qua arbore inserere voles et surculos ad insitionem sumpturus es, videto ut sit tenera et ferax nodisque crebris: et cum primum germina tumebunt, de ramulis anniculis[1] qui solis ortum spectabunt et integri erunt, eos legito, crassitudine minimi digiti: 3 surculi sint bisulci. Arborem, quam inserere voles, serra diligenter exsecato ea parte, quae maxime nitida et sine cicatrice est, dabisque operam, ne librum laedas. Cum deinde truncam recideris, acuto ferramento plagam levato: deinde quasi cuneum ferreum vel osseum inter corticem et materiem, ne minus digitos tres, sed lente demittito, ne 4 laedas aut rumpas corticem. Postea surculos, quos inserere voles, falce acuta ab ima parte eradito tam alte quam cuneum demisisti, sed ita ne medullam neve alterius partis corticem laedas: ubi surculos paratos habueris, cuneum vellito, statimque surculos demittito in ea foramina, quae cuneo adacto inter corticem et materiem facta sunt. Ea autem fine,[2] qua adraseris, surculos demittito ita, ut sex digitis de arbore exstent. In una autem arbore duos aut tres ramulos figito, dum ne minus quaternum digi-

---

[1] anni circulis *Aac*. [2] fine *S* : fines *Aac*.

method of these graftings, we will also set forth another which we have discovered.

You should engraft all other trees as soon as they begin to put forth buds and when the moon is waxing, but the olive-tree about the spring equinox and until April 13th. See that the tree from which you intend to graft and are going to take scions for grafting is young and fruitful and has frequent knots and, as soon as the buds begin to swell, choose from among the small branches which are a year old those which face the sun's rising and are sound and have the thickness of the little finger. The scions should be forked. You should cut the tree, into which you wish to insert the scion, carefully with a saw in the part which is most healthy and free from scars, and you will take care not to damage the bark. Then, when you have cut away part of the tree, smooth over the wound with a sharp iron instrument; then put in a kind of wedge of iron or bone between the bark and firm-wood to a depth of not less than three inches, but do so gently, so as not to damage or break the bark. Afterwards with a sharp pruning-knife pare down the scions, which you wish to insert, at the bottom end of them to the depth to which you have inserted the wedge but in such a way as not to damage the pith or the bark on the other side. When you have got the scions ready, pull out the wedge and immediately push down the scions into the holes which have been formed by driving in the wedge between the bark and the firm-wood. Put in the scions by inserting the ends where you have pared them down and in such a way that they stand out six inches from the tree. Fix two or three little branches as grafts in one tree, provided that the space between them is not less

LUCIUS JUNIUS MODERATUS COLUMELLA

torum inter eos sit spatium. Pro arboris magnitudine et corticis bonitate haec facito. Cum omnes surculos, quos arbor patietur, demiseris, libro ulmi vel vimine arborem astringito: postea paleato luto bene subacto oblinito totam plagam, et spatium quod est inter surculos usque eo, ut duobus digitis insita exstent: supra lutum muscum imponito, et ita alligato, ne pluvia dilabatur. Si pusillam arborem inserere voles, iuxta terram abscindito, ita ut sesquipedem a terra exstet. Cum deinde absciderIs, plagam diligenter levato, et medium truncum acuto scalpro modice findito, ita ut fissura trium digitorum sit. In eam deinde cuneum, quoad patietur,[1] inserito, et surculos ex utraque parte adrasos demittito, ita ut librum seminis libro arboris aequalem facias. Cum surculos diligenter aptaveris, cuneum vellito: deinde arborem, ut supra dixi, alligato et oblinito: dein terram[2] circa arborem aggerato usque ad ipsum insitum. Ea res a vento et calore maxime tuebitur.

Tertium genus insitionis, cum sit subtilissimum, non omni generi arborum[3] idoneum est: et fere eae recipiunt talem insitionem, quae humidum succosumque et validum librum habent, sicuti ficus. Nam et lactis plurimum remittit, et corticem robustum habet. Optime itaque ea inseritur tali ratione. Ex qua arbore inserere voles, in ea quaerito novellos et nitidos ramos. In his deinde observato gemmam, quae bene apparebit, certamque spem germinis

---

[1] quoad patietur *Schneider*: qd̄ is patetur *SA*: quod is patietur *a*: quod his patietur *c*.
[2] terra *S*: tr̃a *A*.   [3] arborum *om. Aac.*

than four inches. In doing so take into account the size of the tree and the quality of the bark. When you have put in all the scions that the tree will stand, bind the tree with elm-bark or osiers. Next with well-worked clay mixed with chaff daub the whole of the wound and the space between the grafts so that the scions project two inches, and put moss on the top of the clay and bind it on so that rain may not penetrate. If the tree which you wish to engraft is small, cut it off near the ground so that it projects a foot and a half above the soil; then, after cutting it down, carefully smooth the wound and split the stock in the middle a little way with a sharp knife, so that there is a cleft three inches deep in it. Then insert a wedge in this cleft as far as it will allow and thrust down into it scions which have been pared away on both sides in such a way as to make the bark of the scion exactly meet the bark of the tree. When you have carefully fitted in the scions, pull out the wedge; then bind and daub the tree in the manner described above. Next heap up earth round the tree right up to the graft itself; this will give the best protection from wind and heat.

The third kind of grafting, being the most delicate, is not suited to every kind of tree. Generally speaking those trees admit of this kind of grafting which have moist, juicy and strong bark, like the fig-tree; for this yields a great abundance of milky juice and has a stout bark, and so a graft can be very successfully inserted by the following method. You should seek out young and healthy branches on the tree from which you wish to take your grafts, and you should look on them for a bud which has the appearance of being good and gives a sure promise of pro-

LUCIUS JUNIUS MODERATUS COLUMELLA

habebit: eam duobus digitis quadratis circumsignato, ut medio gemma sit et ita acuto scalpello circumcisam,[1] delibrato diligenter, ne gemmam laedas. Item quam arborem inserere voles, in ea nitidissimum ramum eligito, et eiusdem spatii corticem circumcidito, et a materie[2] delibrato; deinde in eam partem, quam nudaveras, gemmam, quam ex altera arbore sumpseras, aptato ita, ut emplastrum 9 circumcisae parti conveniat. Ubi haec feceris, circa gemmam bene vincito, ita ne laedas: deinde commissuras et vincula luto oblinito, spatio relicto, qua gemma libere germinet. Materiem, quam inseveris, si sobolem vel supra ramum habebit, omnia praecidito, ne quid sit quo possit succus avocari, aut cui magis quam insito serviat. Post unum et vicesimum diem solvito emplastrum. Hoc genere optime etiam olea inseritur. Quartum illud genus insitionis iam docuimus, cum de vitibus disputavimus: itaque supervacuum est hoc loco repetere[3] iam traditam rationem terebrationis.

XXVII. Sed cum antiqui negaverint posse omne genus surculorum in omnem arborem inseri, et illam quasi[4] finitionem, qua nos paulo ante usi sumus, veluti quandam legem sanxerint, eos tantum surculos posse coalescere, qui sint cortice ac libro et fructu consimiles iis arboribus, quibus inseruntur, existimavimus errorem huius opinionis discutiendum, tradendamque posteris rationem, qua possit omne

[1] circumcisam *Lundström* : circūcisę *S* : circūcisae *A*.
[2] a materie *S* : materiae *A* : materie *ac*.
[3] hoc loco repetere *S* : hoc corepetere *A*.
[4] quasi *S* : quas *Aac*.

---

[a] Ch. 8 § 3.

ducing a sprout. Make a mark round it enclosing two square inches with the " eye " in the middle and make an incision all round with a sharp knife and carefully remove the bark without damaging the bud. Also choose the healthiest branch on the tree which you wish to use for grafting and make an incision, enclosing the same dimensions as before, and remove the bark from the firm-wood; then apply the " eye " which you have taken from the other tree to the space which you have laid bare, so that the scutcheon fits into the part which you have cut round. Having done this, bind the bud well all round in such a way as not to damage it; then daub the edges and the ties round them with clay, leaving a space where the " eye " may bud without hindrance. If the firm-wood in which you have inserted the graft has any shoots and branches, above it cut everything away, so that there may be nowhere to which the sap may be drawn away and benefit another part rather than the graft. After the twenty-first day unbind the scutcheon. This kind of grafting is very successful with the olive-tree also. The fourth method of grafting we have already set forth when we treated of vines [a]; so it is superfluous to repeat here the method of " terebration " already described.

XXVII. Since the ancients denied that any kind of scion can be grafted on any kind of tree and established that limitation, which we quoted just now as a hard and fast rule, namely, that only those scions can unite which resemble the trees in which they are inserted, in their bark and rind and fruit, we have thought it advisable to disprove the error of this opinion and to hand down to posterity a method by which any kind of scion can be grafted upon any

Any kind of scion can be grafted on any kind of tree.

LUCIUS JUNIUS MODERATUS COLUMELLA

2 genus surculi omni generi arboris inseri. Quod ne longiori exordio legentes fatigemus, unum quasi exemplum subiciemus, qua similitudine quod quisque genus volet omni arbori poterit inserere.

3 Scrobem quoquoversus pedum quattuor ab arbore olivae tam longe fodito, ut extremi rami oleae possint eam contingere. In scrobem deinde fici arbusculam deponito, diligentiamque adhibeto, ut robusta et nitida fiat. Post triennium aut quinquennium, cum iam satis amplum incrementum ceperit,[1] ramum olivae qui videbitur nitidissimus, deflecte, et ad crus arboris ficulneae religa: atque ita amputatis ceteris ramulis et tantum cacumina, quae inserere voles, relinquito: tum arborem fici detruncato, plagamque
4 levato, et mediam[2] cuneo findito. Cacumina deinde olivae, sicuti matri inhaerent, utraque parte adradito, et ita fissurae fici aptato, cuneumque eximito, et diligenter colligato, ne qua vi revellantur. Sic interposito triennio coalescet ficus olivae: et tum demum quarto anno, cum bene coierint,[3] velut propaginis ramulos olivae ramos a matre resecabis. Hoc modo omne genus in omnem arborem inseritur.

XXVIII. Cytisum, quod Graeci aut zeas, aut carnicin, aut tripharin vocant,[4] quam plurimum habere expedit, quod gallinis, apibus, ovibus, capris, bubus quoque, et omni generi pecudum utilissimum est, quod ex eo cito pinguescit, et lactis plurimum praebet ovibus:[5] tum etiam quod octo mensibus

[1] coeperit *A*. [2] medium *Sc*.
[3] coierint *Schneider*: poterit *SAac*.
[4] quod ... vocant *om. S*. [5] ovis *S*: ovib, *A*.

kind of tree. That we may not weary our readers with too long a discourse, we will submit, as it were, a single example by following which anyone can graft any kind of scion upon any kind of tree.

Dig a trench measuring four feet each way at such a distance from an olive-tree that the ends of the branches of the olive-tree can reach it; then plant a small fig-tree in the trench, and be careful that it grows strong and healthy. After three or five years, when it has made enough growth, bend down the branch of the olive-tree which seems to be the healthiest, and bind it to the stock of the fig-tree. Then remove the rest of the small branches and leave only those tops which you wish to engraft; next cut the fig-tree and smooth off the wound and split it in the middle with a wedge. Then pare the tops of the olive, as they still adhere to the mother-tree, on both sides and fit them into the cleft in the fig-tree, and take away the wedge and carefully tie them to the fig-tree, so that no force may tear them away. After an interval of three years the fig-tree will coalesce with the olive-tree, and, finally, in the fourth year, when they have become properly united, you will cut off the olive-branches from the mother-tree as though they were the small branches of a layer. This is the way in which any kind of scion is grafted on any kind of tree.

XXVIII. You should have as much shrub-trefoil as possible, what the Greeks call *zea* or *carnicis* or *tripharis*, because it is most useful for chickens, bees, sheep, goats and oxen and cattle of every kind, which quickly grow fat upon it, while it makes ewes yield a very large quantity of milk; moreover also you could use it for eight months of the year as green

_{How to grow and make use of the shrub-trefoil.}

viridi eo pabulo uti, et postea arido possis. Praeterea in quolibet agro, quamvis macerrimo [1] celeriter comprehendit, omnemque iniuriam sine noxa patitur.

2 Mulieres quidem, si lactis inopia premuntur, cytisum aridum in aqua macerari [2] oportet: cum tota nocte permaduerit, postero die expressi [3] succi ternas heminas permisceri modico vino, atque ita potandum dari: [4] sic et ipsae [5] valebunt, et pueri abundantia lactis confirmabuntur. Satio autem cytisi vel autumno circa idus Octob. vel vere fieri potest.

3 Cum terram bene subegeris, in modum holerum [6] areas facito, ibique velut ocimum semen cytisi autumno serito: plantas deinde vere disponito, ita ut [7] inter se quoquoversus quattuor pedum spatio distent. Si semen non [8] habueris, cacumina cytisorum vere disponito, stercoratam terram circa aggerato. Si pluvia non incesserit,[9] rigato xv proximis diebus. Simulac novam frondem agere coeperit, sarrito. Post triennium deinde caedito, et

4 pecori praebeto. Equo abunde est viridis pondo xv, bovi pondo xx, ceterisque pecudibus pro portione virium. Potest autem etiam circa sepem agri satis commode ramis cytisus seri.[10] Aridum si dabis, exiguius dato, quoniam maiores vires habet, priusque aqua macerato, et exemptum [11] paleis permisceto.

---

[1] macerrimo *vett. edd.*: acerrimũ *S*: agessimũ *A*.
[2] aquã cerari *A*: aquam macerari *a*.
[3] expressit *SA*.
[4] dare *Sac*.
[5] ipsae *S*: ipsa *Aac*.
[6] olerum *Lundström*: herum *S*: herũ *A*.
[7] ita ut *Lundström*: haut *SA*: aut *c*: haud *a*.
[8] si semen non *a*: sistitiennon *SA*.
[9] ingesserit *SA*.
[10] agri—seri *add. Lundström*: *ex Varr. RR.* v.12.4.

fodder and afterwards as dry. Furthermore on any ground whatsoever, even if it be very lean, it quickly takes root and bears any ill-treatment without taking harm. Indeed if women suffer from lack of milk, 2 dry shrub-trefoil ought to be steeped in water and after it has soaked for a whole night, on the following day three *heminae* of the juice squeezed out of it ought to be mixed with a little wine and given them to drink; in this way they will themselves enjoy good health and their children will grow strong on the abundance of milk provided for them. Shrub-trefoil can be sown either in the autumn about October 15th or in the spring. When you have worked the 3 soil thoroughly, make beds like those for vegetables and in the autumn sow there the seed of the shrub-trefoil as you would sow basil; then in the spring set out the plants, so that they are distant four feet each way from one another. If you have not any seed, plant out tops of shrub-trefoil in the spring and heap well-manured soil round them. If rain has not come on, water them for the fifteen following days. As soon as a plant begins to put forth young foliage, hoe the ground. Then after three years cut down the plants and give them to the cattle. Fifteen pounds of shrub-trefoil when it is green is quite enough for a horse, and twenty pounds for an ox; and 4 it should be given to other animals according to their strength. Shrub-trefoil can also be very conveniently propagated by planting boughs round the fence of a field. If you give it as dry food, give it in smaller quantities, since it is stronger in this state, and steep it first in water and after taking it out of

---

[11] exemplum *SAac*.

LUCIUS JUNIUS MODERATUS COLUMELLA

Cytisum quod[1] aridum facere voles, circa mensem Septembrem cum semen eius grandescere incipiet, caedito; paucis deinde horis, dum flaccescat, in sole habeto. Deinde in umbra adsiccato, et ita condito.

XXIX. Salicem et genistam crescente luna vere circa calendas Martias serito. Salix humida loca desiderat, genista etiam sicca: utraque tamen circa vineam opportune seruntur, quoniam palmitibus idonea praebent vincula.

Harundo optime seritur radicibus,[2] quas alii bulbos, alii oculos[3] vocant. Simulatque terram bipalio repastinaveris,[4] radicem harundinis acuta falce praesectam impendenti pluvia disponito. Sunt qui harundines integras sternant, quoniam ex omnibus nodis strata[5] harundines emittat. Sed fere hoc genus evanidam[6] exilemque et humilem harundinem affert. Melior itaque satio est ea, quam prius demonstravimus. Placet autem omnibus annis, simulac harundinem cecideris, locum alte et aequaliter fodere, atque ita rigare.

XXX. Violam qui facturus est, terram stercoratam et repastinatam ne minus[7] alte pedem in pulvinos redigat. Atque ita plantas annotinas scrobiculis pedalibus factis ante calendas Martias dispositas habeat. Semen autem violae sicut holerum in areis duobus temporibus seritur, vere vel autumno. Colitur autem modo eo, quo et cetera holera, ut runcetur, ut sarriatur, ut interdum etiam rigetur.

[1] cytisum quod *a* : cytisūq. *SA*.
[2] radicibus *Gesnerus supplevit*.
[3] alioculos *SA*.
[4] bipalore pastinaveris *S* : bipalo repastinaveris *Aa*.
[5] strata *S* : rata *A* : sata *ac*.
[6] evanidum *SAac*.
[7] nimius *Aac*.

the water mix it with chaff. Cut shrub-trefoil which you wish to dry about the month of September, when the seed begins to grow large; then keep it in the sun for a few hours until it withers; then dry it in the shade, and so store it.

XXIX. Plant the willow and the broom when the moon is waxing about March 1st. The willow requires a damp situation, the broom one which is dry; both, however, can be conveniently planted round a vineyard, because they produce bands suitable for tying up vine-shoots.

The cultivation of willows, broom and reeds.

Reeds are best planted by means of roots which some people call bulbs, others " eyes ". As soon as you have trenched the ground for the second time with a double mattock, cut off the root of the reed with a sharp pruning-knife and set it when rain is threatening. Some people lay whole plants of reeds flat on the ground, because a plant when thus laid out puts forth reeds from all its knots, but this method generally produces a reed which is frail, thin and low-growing. That method of planting is better which we explained to you before; but it is a good plan every year, as soon as you have cut the reeds, to dig up the ground deeply and uniformly and thus to water it.

XXX. He who intends to grow the violet should form flower-beds of soil which has been well manured and trenched to the depth of at least a foot. Next he should have plants of a year's growth set in little trenches a foot deep made before March 1st. Seeds, on the other hand, of violet, like the other plants that are pot-herbs, are sown in beds at two seasons, spring or autumn. It is cultivated in the same manner as all the other pot-herbs by weeding, hoeing and also by watering from time to time.

How to grow violets and roses.

LUCIUS JUNIUS MODERATUS COLUMELLA

2 Rosam fruticibus[1] ac surculis disponi per sulcos pedales convenit per idem tempus, quo et viola. Sed omnibus annis fodiri ante calend. Martias et interputari oportet. Hoc modo culta multis annis perennat.

[1] fructibus *codd.*

ON TREES xxx. 2

The rose-tree should be planted at the same time as 2
the violet in furrows one foot deep in the form of
shrubs or cuttings; but it must be dug round every
year before March 1st and pruned here and there.
If cultivated in this way it lasts for many years.

INDEX OF PROPER NAMES

* indicates more than one occurrence.

A

ABDERA, I 31
Aborigines, I 47
Abydos, III 69
Accius, I 23
Achaean, I 47
Achaia, II 493; III 21
Achelous, III 29
Acrisia, III 25
Aeschrion, I 33
Africa, I 21, 31*, 181; II 217, 235; III 105
African, I 273; II 95, 321, 323*; III 139, 141, 143, 219
Agathocles, I 33
Ageladas, III 9
Ager Gallicus, I 255
Ajax, III 21
Alba, I 277
Alban, I 241, 273, 277*
Albanian, I 47
Albuelis, I 245
Alexandrian, III 159
Alexandrine, II 365
Alexis, III 33
Alfius, I 81
Allobroges, III 243
Allobrogian, I 243
Altina, II 185
Altinum, II 235
Ambivius, Marcus, III 197
Amerian, II 99
Amerine, I 451
Aminean, I 237, 239*, 241, 249, 271*, 277*, 279*, 281*; II 475, 477; III 229, 259, 261, 271, 279, 291
Amiternum, III 45
Ammoniac, II 173
Amphilochus, I 33
Amphipolis, I 33
Amphitrite, III 25
Amythaon, III 37
Anaxipolis, I 33
Andromeda, III 103
Androtion, I 33
Anician, II 99
Antigonus, I 33
Antipathies, On, III 169
Apelles, I 25
Apennines, II 125
Apian, I 243; III 291
Apollonius, I 33
Apulia, I 275; III 139
Apulian, II 235

413

INDEX OF PROPER NAMES

Aquarius, III 71*, 103
Aquila, I 267; III 95, 101*, 125*
Arabia, I 275
Arcadia, II 231
Arcadian, I 47; III 37
Arcelacan, I 249; 271, 341
Archer, III 379
Archytas, I 31
Arcturus, I 171; II 487*, 491*; III 43, 69, 81, 95, 103, 105, 115
Ardea, I 277
Argitis, I 245*, 249
Argo, III 83, 107
Aricia, III 19
Aries, III 87*
Aristaeus, I 27; II 431
Aristandrus, I 33
Aristomachus, II 477
Aristomenes, I 33
Aristotle, I 31; II 245, 431, 433
Armenian, III 42, 126
Arrow, III 83
Ascanius, II 193
Ascraean, III 45
Asia, I 21; II 125, 235; III 43
Asia Minor, II 407
Assyrian, III 15
Astronomers, III 69
Astropian, II 95
Athenagoras, I 33
Athenian, III 51, 175
Athens, I 31; II 123; III 45*
Atinian, II 45*; III 381

Atlantic, II 407; III 83
Atlas, I 141; III 11
Attic, III 41, 263*
Attica, I 33; II 123, 493
Atticus, Julius, I 33, 35, 261, 305*, 327, 329*, 331, 353, 357, 359, 371, 377, 381, 437*, 439*, 449*, 459
Augustus, III 49*
Autolycus, I 47
Autumn, III 9
Autumnus, I 343
Aventine, I 47

B

Bacchius, I 33
Bacchus, I 29, 233, 343; III 7, 9, 11, 25, 27, 29, 33, 45*, 81, 89
Baetica, I 17, 179, 309; II 7, 39, 73, 231, 235, 417; III 105, 145
Balance, III 107
Basilic, I 243, 249, 271, 277, 279, 281, 343, 347; III 107
Bear, III 393
Bias, I 33
Bithynia, I 33
Bituric, I 243, 249, 271, 277, 279, 281, 341, 347
Boeotia, I 31, 59
Bolus, II 273
Boreas, III 13, 33
Bowl, III 107
Bruttian, III 19, 229, 243
Brutus, I 25
Bryaxis, I 25

414

INDEX OF PROPER NAMES

Bucolics, I 279
Bull, III 91, 115, 121, 379

C

Cacus, I 47
Caecilian, III 23*, 145
Caecuban, I 277
Caelius, I 25
Caere, I 281
Caesar, Tiberius, III 161
Caesonius, I 53
Calabrian, II 235; III 139, 307
Calliope, III 27
Callisto, III 77
Callistruthian, II 95; III 43
Calvus, I 25
Campania, I 19, 59, 169, 239, 275; II 125
Campanian, II 365
Campus, I 85
Cancer, III 99
Canopus, III 21
Cappadocian, II 173; III 23*, 145
Capricorn, II 487; III 99, 125
Capua, III 17
Carpathian Sea, II 407
Carseoli, I 277
Carthaginian, I 33, 35, 37, 49, 325; II 407, 427, 449; III 141, 163, 195*, 271, 353
Casinum, II 403
Cassiopea, III 115
Cato, Marcus, I 35, 39, 43, 45, 47, 53, 57, 87, 111, 205, 223, 251, 253, 279, 377; II 121*; III 51, 321*
Caudine Forks, III 17
Caunian, III 43
Caurus, III 13
Cea, II 431
Celsus, Cornelius, I 35*, 85, 117, 123, 125, 151, 183, 231, 245, 247, 249, 251, 329, 353, 371, 377, 437; II 147, 165, 233, 245, 271, 387, 429*, 445*, 447, 469, 485, 493
Centaur, III 107, 115
Cepheus, III 99
Ceration, II 101
Ceraunian, I 233, 255, 281
Ceres, II 123
Cestine, II 99; III 289
Chaereas, I 33
Chalcidian, II 95, 325, 331; III 43
Chalcis, I 59
Chaldaean, III 69, 125
Chelidonian, III 43
Chian, III 43
Chios, I 33
Chiron, I 25; III 37
Chrestus, I 33
Christ's thorn, II 293; III 131
Cicero, I 23, 25, 275; III 51, 65, 175, 187
Cilicia, I 21, 169; III 101
Ciminian, II 403
Cimolian, II 173
Cincinnatus, Quinctius, I 13
Cinyrian, III 21
Circus, I 85, 275

415

INDEX OF PROPER NAMES

Cithaeron, III 25
Claudius, III 49
Clusian, I 139*
Cnossian, III 11
Colophon, I 33
Columella, Marcus, I 203; II 39, 235, 237; III 239, 241, 269, 277
Corduba, II 235
Corycian, II 437
Corycus, I 275
Corydon, III 33
Crab, II 483; III 71, 99*, 127
Crates, I 33
Crete, II 431
Crotus, III 11
Crown, III 99, 113*
Crustuminian, III 209
Culminian, II 71; III 307, 327
Cumae, III 17
Cupid, III 23
Curiatian, I 273
Cybele, III 25
Cyclades, I 17; II 493
Cydonitan, I 235
Cymê, I 33
Cyprian, III 145
Cyprus, III 23

D

Dadis, I 33
Daedalus, III 9
Damascene, III 43
Dardanian, III 37
Decimian, III 209
Delian, III 27

Delos, II 325
Delphic, III 25
Demeter, II 123
Democritus, I 25, 31, 309; II 273, 365, 485; III 131, 167, 169
Demosthenes, I 25
Denicales, I 223
Dentatus, Curius, I 13, 49
Deucalion, III 11
Dictaean Cave, II 429
Dicte, II 429
Dindyma, III 25
Diodorus, I 33
Dion, I 33
Dione, III 33
Dionysius, I 33*; II 217
Diophanes, I 33
Dog, Lesser, III 101
Dog-star, I 215, 305, 323, 373; II 39, 255, 303, 485, 487; III 9, 91, 105, 123, 125, 213, 243
Dolabellian, III 209
Dolphin, III 73, 97, 103, 125
Dracontion, I 249

E

Earth, I 3; III 21, 23, 25
Economicus III 51, 175, 177, 187
Egypt, I 125, 181, 309; II 367
Egyptian, I 273; II 273, 399; III 161, 219, 307, 325
Epicharmus, I 31; II 241
Epigenes, I 33
Epirus, II 125

INDEX OF PROPER NAMES

Erechtheus, II 431
Erigone, III 43
Etesian, III 103*
Etruria, II 125
Euboea, I 59; II 493
Eubulus, I 33
Eudoxus, I 25; II 489
Eugenian, I 241
Euhemerus, II 429, 431
Euphorion, I 33
Euphronius, I 33*
Euphyton, I 33
Eurus, III 13
Euthronius, II 431
Evagon, I 33
Evian, III 27

F

Fabricius, Gaius, I 13
Falernian, III 45
Fate, I 151
Faventia, I 255
Favonian, II 99
Favonius, I 211
Fereolan, I 249
Fidicula, III 91
Fire-star, III 33
Fish, Northern, III 107
Fish, Southern, III 105
Fishes, III 107
Fortune, III 35
Fragellan, I 249
Fulcan, II 95

G

Gades, II 235, 407; III 23, 145
Gaetulian, III 15
Galatian, I 149, 155

Gallic, I 447, 451*; II 45, 85; III 235, 279, 337
Gallinaria, II 323
Ganges, II 367
Gargara, II 193
Gaul, I 17, 171, 247, 461; II 67, 125, 425; III 43
Gaulish, II 235; III 371
Gauls, II 7, 41, 387, 407
Gaurus, I 59
Georgic, III 131
Georgics, II 269; III 5
German, I 275
Germany, I 275
Getae, II 233
Goat, III 91, 95, 125
Graecinus, Julius, I 35, 251, 255, 257, 259, 261, 305, 307, 359, 363, 437
Grecian, I 113; III 29
Greece, II 407
Greek, I 31, 35, 45, 47, 111, 163, 247, 249, 319, 451*; II 257, 259, 273, 315*, 389, 435, 437; III 5, 17, 195*, 261, 325
Greeks, the, II 123, 161, 167, 171*, 173, 249, 309, 321, 325*, 353, 437, 443, 469, 479, 483; III 131, 151, 169, 179, 195, 197, 405

H

Halcyon, III 83
Hamilcar, III 195
Harbinger, III 99
Hare, III 123
Hegesias, I 33
Hell, III 11

INDEX OF PROPER NAMES

Helle, III 19
Helvenacan, I 247; II 41
Helvolans, I 245, 341
Herculean, III 17
Hesiod, I 31
Hesperia, I 47
Hesperus, III 33
Hiberian, I 47
Hieron, I 31
Hipparchus, I 29; II 489*; III 125
Holy Mountain, II 193
Horconian, I 249
Horse, III 83, 87
Hyades, III 89, 91, 93, 95, 121, 123*
Hybla, II 493; III 21
Hyginus, I 35, 305; II 427*, 469, 473, 475, 477, 481, 493; III 119, 167
Hymettus, Mount, II 431; III 41
Hyperion, III 43

I

Iberian, III 33
Iberian Sea, II 407
India, I 275
Inerticulan, I 247
Irtiolan, I 249
Isauricus, II 405
Ischomachus, III 51*, 59
Italian, I 273, 461; II 45, 67, 407, 435, 437; III 89, 111, 113
Italy, I 13, 31, 79, 125, 169, 239, 251, 255, 261, 277*, 321; II 73, 235, 473; III 103, 105

J

Jewish, I 275
Jove, I 25; III 37*, 95*
Judea, I 275
Jupiter, II 429

K

Kids, III 69, 109, 113

L

Lateritan, II 99; III 209
Latin, I 35*; II 315*, 323; III 51, 177
Latium, I 17, 23; II 125
Latona, III 33
Lepine, III 17
Lernaean Crab, III 35
Lethaean, I 193
Lethe, III 11
Liber, III 225
Libera, III 225
Libra, II 487; III 89, 91
Libya, I 275; II 231
Libyan, I 233, 251; II 95; III 43
Licinian, II 71, 73; III 307, 327, 385
Licinius, I 49, 51; III 197
Liguria, I 275*
Ligurian, II 407; III 245
Lion, III 71, 77, 81, 101, 102, 379
Livian, II 95; III 43
Lucifer, III 33
Lucullus, Lucius, I 57; II 403
Lupercus, III 23
Lydian, II 95; III 43

INDEX OF PROPER NAMES

Lyre, III 73, 77*, 91, 93, 103*, 121, 127
Lyre, Little, III 119
Lysimachus, I 33
Lysippus, I 25

M

Maenalus, III 45
Maeonian, I 25
Maeotis, Lake, II 367
Mago, I 33, 35, 37, 309, 323*, 375; II 31, 125, 217, 485*, 495; III 195, 265, 285, 385
Mareotic, I 247
Mariscan, III 43
Maronea, I 33
Marrucine, III 17
Mars, III 23
Marsian, I 149; II 145; III 209
Massic, I 277
Matian, II 99; III 289
Matius, Gaius, III 281
Median, II 325, 329, 331
Megara, III 15
Melampus, I 25
Melian, II 325
Melissa, II 429
Menander, I 33
Mendes, III 161
Mendesium, II 273
Menestratus, I 33
Merican, I 249
Messala, I 25
Metellus, III 23
Meto(n), I 25; II 489
Mevania, I 275*
Milesian, II 235

Miletus, I 33
Minerva, I 25
Mnaseas, I 33; III 195
Muraena, Licinius, II 405
Murgentine, I 249
Muses, I 23, 113; III 9, 29
Mutina, II 235
Mysia, I 21, 275

N

Naevian, II 71, 99; III 209, 301
Naiad, III 33
Naples, III 17
Narycian, III 41
Nature, I 3
Nemeturican, III 235, 241, 243, 245
Nepa, III 11
Neptune, II 403; III 11
Nicander, II 431
Nomentan, I 241
Nomentum, I 255
Numa, II 401
Numantinus, II 405
Numidia, I 21, 125, 309; II 417
Numidian, II 323
Numisian, I 235, 249; III 277, 279
Nursia, III 45
Nysian, III 25

O

Ocean, III 9, 11
Olympian, I 25
Olympic, I 281
Orata, Sergius, II 405
Orator, I 23

INDEX OF PROPER NAMES

Orchis, II 71
Orcus, I 43
Ornithian, III 81, 93
Orpheus' lyre, III 13

P

Paestum, III 9
Palatine, I 47
Pales, III 7
Palladian, III 17
Pallas, III 39
Pamphylia, III 101
Pamphylian Sea, II 407
Pan, III 45
Paphian, III 23*, 39
Paphos, III 23
Parma, II 235
Parnassian, III 25
Parrhasius, I 25
Pausean, III 295*, 297, 299*, 301, 307, 385
Paxamus, III 195
Pegasus, III 83, 87
Pelasgian, I 47
Pelusian, II 99; III 15
Pergamus, I 33
Pergulan, I 249
Peripatetic, II 431
Persia, III 43
Persis, I 33
Phasis, II 367
Phidias, I 25
Philippus, Marcius, II 403
Philometor, Attalus, I 31
Phoebe, III 33
Phoebus, III 11, 27
Phradmon, III 9
Phrixus, III 19, 39
Phrygian, III 29

Phryxon, II 429
Picenum, I 255
Pierian, III 25
Plato, I 25, 339
Pleiads, II 481, 483*, 487*, 489; III 89, 91, 93*, 113, 115*, 119
Plentiphanes, I 33
Pollentia, II 235
Pollio, Naevius, I 25, 275
Polyclitus, I 25; III 9
Pompeian, I 249; III 17, 207
Pontus, II 367; III 69
Posia, II 71, 73
Praxiteles, I 25
Precian, I 245
Priapus, III 9, 15
Procyon, III 101
Promethean, III 11
Protogenes, I 25
Psithian, I 247
Punic, I 31, 53; III 27
Punicum, I 169
Pupinia, I 53
Pyrrhus, I 13
Pythagoras, I 25

Q

Quinquatria, III 85, 163
Quirites, I 17

R

Ram, II 481; III 87, 107
Regulus, Marcus Atilius, I 53*
Rhaetian, I 249
Rhodes, I 33

420

INDEX OF PROPER NAMES

Rhodian, I 233; II 95, 325, 331
Riphaean, III 13
Roman, I 13, 17, 33, 51, 129, 205, 275; III 45, 179, 195
Romulus, I 15; II 401
Rubigo, III 37

S

Sabaean, III 29
Sabatine, II 403
Sabine, I 13, 17, 451; II 73; III 19, 179
Sagittarius, III 81, 121, 125
Saserna, I 27, 29, 35, 81, 191, 193, 253, 309, 329, 377
Saturn, II 431
Saturnian, I 17
Scaevola, Quintus, I 57
Scaudian, II 99
Scirpulan, I 249
Scorpion, III 87*, 93, 115*, 119, 125, 379
Scrofa, Tremel(l)ius, I 27*, 31, 35, 105, 107, 143, 161, 193, 253, 305, 309; II 45
Scyros, II 493
Scythia, II 367
Sebethis, III 17
Seneca, I 255
Sergian, II 71, 73; III 327
Serpentarius, III 99
She-goat, III 107
Sicca, III 15
Sicilian, I 31, 47; II 437; III 21
Sicily, II 407, 493
Signia, III 17
Signian, I 71

Signine, II 99
Siler, III 19
Silvinus, Publius, I 3, 35, 105, 109, 227, 263, 271, 273, 281, 353, 361, 363*; II 3, 5, 119, 123, 231, 321, 421; III 3, 7, 45, 49, 175
Sircitulan, III 279
Sirius, III 33, 43
Snake, III 69
Socrates, I 31; III 51, 187
Sophortian, I 247
Spain, I 383; II 193
Spaniard, I 243; II 219
Spanish, II 173
Spartan, II 315
Spica Virginis, III 107
Spionian, I 249, 271, 343, 347
Stabiae, III 17
Stephanitan, I 235
Sticulan, I 249
Stolo, I 27, 377
Stygian, III 11
Surrentine, I 277
Surrentum, I 239
Syracusan, II 241
Syria, I 169
Syrian, II 99*, 261, 475; III 271, 339

T

Tages, III 37
Tanagran, II 325
Tarchon, III 37
Tarentine, I 21; II 95, 99, 235, 237, 257*; III 89, 393
Tarentum, I 31

421

INDEX OF PROPER NAMES

Tartessian, II 407; III 39
Tartessus, III 23
Taurus, III 119
Tethys, III 23
Thasian, I 247
Thasos, I 33
Thaumas, III 33
Thebes, I 59
Theophilus, I 33
Theophrastus, I 31
Thessaly, II 431
Tiber, II 403
Tibur, III 19
Tiburnian, II 95
Titan, III 9
Tmolus, I 275
Trebellius, Marcus, II 5
Triptolemus, I 27; II 123
Turkey-oak, II 425
Turnus, III 19
Tuscan, III 37*
Twin Stars, III 35, 95
Tyrian, III 33

U

Umbria, II 125
Utica, I 33

V

Varro, Marcus Terentius, I 13, 35, 253*, 279; II 7, 123, 217, 367, 373, 403*
Veline, II 403
Venuculan, I 235, 249; III 279
Venus, III 23, 25, 209
Venus-pear, II 99
Vergil, I 23, 35, 111, 119, 151, 163, 193, 221, 233, 297, 309, 323, 377; II 29, 34, 193, 429*, 431, 461, 485, 495; III 5, 69, 343 [Maro]; II 253, 431, 435, 449; III 45
Vergiliae, I 141*
Vesuvian, III 17
Vesuvius, I 239
Veterensis, Paridius, I 363
Vintager, III 83, 103
Virgo, III 103*, 107, 109, 113, 135
Visula, I 245*
Volsinian, II 403
Volusius, Publius, I 81
Vulcan, III 141, 151, 157
Vulturnus, II 39; III 107

W

Wagoner, III 113
Whale, III 107

X

Xenophon, I 31; III 51, 175, 177, 187

Z

Zephyr, III 13
Zodiac, II 489

MONTHS

January, I 155, 169, 175, 179*, 185; II 139, 345; III 71*, 73*, 79, 81, 85, 127*, 129, 135, 139, 143, 145, 149, 151, 321, 325, 375

February, I 135, 161, 165, 167, 169, 171, 173, 179*, 201, 207, 211, 319, 409; II 139, 391, 491; III 75, 77*, 79, 81*, 83*, 95, 129, 139*, 145, 155*, 157, 159, 165*, 329, 335, 375, 393, 395*

March, I 133, 155, 169, 171, 173, 177, 179, 181, 367; II 139, 345, 391, 401, 481; III 75, 83*, 85*, 89*, 93, 95, 129, 139*, 145*, 147, 151*, 153, 159*, 347, 361, 363, 381, 383, 393*, 397, 409

April, I 131, 135, 175; II 139, 345; III 89*, 91*, 95, 129, 139, 143, 145, 361, 395, 399

May, I 177; II 483; III 93*, 95*, 97, 99, 129, 135, 141, 149*

June, I 437; II 301, 483; III 97, 99*, 103, 129

July, I 131*, 169*; II 183; III 99*, 101*, 129

August, I 135, 171; III 101, 103*, 105, 129, 141, 151, 157, 167

September, I 131, 135, 157, 171, 177, 201, 203; II 487; III 93, 105*, 107*, 129, 135, 139, 151*, 167, 409

October, I 141, 143, 157, 167, 169, 173, 319, 367, 371, 387*, 409; II 345, 487; III 103, 111, 113*, 115*, 129, 141, 155, 291, 359

November, I 141, 169, 387, 437, 457; II 141*; III 111, 119*, 121*, 123*, 129, 135, 147, 153

December, I 167, 319, 375; II 141, 281, 487; III 79, 93, 123, 125*, 129, 143, 265, 375*

GENERAL INDEX

To keep this section of the Index within reasonable bounds a certain amount of compression was employed. In the Volumes in which marginal " boxes " or an outline of the subject matter of the author's work is furnished, many readers, no doubt, will not seek more assistance. Also, because of the somewhat lengthy account of some subjects discussed, reference to the beginning of such sketches would seem sufficient, for the reader will quite naturally continue to the end of the topic.

A

ablaqueatio (n), I 365, 375, 383, 389
acnua, II 7
acorn, II 399
actus, II 7
alder, III 29
alexanders, III 17, 141, 201, 203
all-heal, III, 15, 139, 147
almond, II 435; III 75, 127, 333, 337, 393, 397
amaranth, III 21
amethystos, I 247
ἀνδράφαξις, III 155
anise, III 35, 139, 155, 205, 207, 219, 289, 307, 325
ἀφροσκόρδου, III 141
apiary, II 433, 439, 449, 467

apiastrum, II 465
ἀποκατάστασις, I 271
ἀπόκρουσις, I 163
apple, I 221, 349; III 7, 19, 39, 43, 209, 271, 279, 289 (varieties), 293, 389
apple, tuber-, III 75, 127
apricot, III 42-3, 127
arbustum, I 355, 461
arbutus-berry, II 415
arepennis, II 7
argilla, I 305
arietillum, I 169
armenta, II 119
arum, III 27
ash(-tree), III, 81, 125, 381
asparagus, III 29, 41, 139, 155, 157, 201
asphalteion, II 171
asphodel, II 437

425

GENERAL INDEX

ass, II 231; III 37
assafoetida, III 339, 340
auger, Gallic, I 447; III 371

B

bailiff, III 51, 63, 181
balm, II 459, 465
balsam, III 33, 235, 325
barley, I 139, 145, 149, 155, 173, 187; II 393, 395, 399, 425, 427; III 75, 99, 113, 129, 199, 275
basil, III 35, 147, 149
basket, III 123, 225, 303
basse, II 401, 403, 413
bay(-tree), III 129, 249, 305
" beak " (of *falx*), I 427
bean, I 117, 139, 149, 159, 161, 163, 185, 187, 209; II 398–9 (Egyptian bean), 425; III 15, 117, 121
bear, III 379
bear's foot, II 437
beaver, III 379
bee, II 427, 429, 431, 435, 447, 449, 451, 459, 465, 469, 471, 473, 481, 503; III 123, 153, 211, 267 (-grape),
beech, III 389
beer, III 15
beet, III, 35, 139
beetle, II 449
" bend " (of *falx*), I 427
bes, II 11
birds, II 321
birds, amphibious, II 387
" black tails ", II 407

boar, wild, II 421, 427
braize, sea-, II 407
branding iron, III 77
brankursine, III 27
brine, III 199, 201, 203, 205, 207, 209, 215, 293, 295, 297, 299, 305, 331
broom, I 381, 453; III 39, 81, 123, 199, 201, 225, 283, 313, 409
brushwood, III 231
bryony, II 353 (Cretan); III 29, 37, 39, 201
bull, II 177
bulrush, III 325
butcher's broom, III 201
butterfly, II 449

C

cabbage, III 35, 139, 143, 145, 155, 201, 203
calavance (kidney-bean), I 139, 159, 189; III 41, III, 113, 207
calf, II 187*
callus, I 333, 413
canalis, III 311
candetum, II 7
candle, I 221
candosocci, II 41
canker-worm, III 37
canterius, I 381
cantherinum, I 153
caper, III 17, 139, 163, 201, 203
caraway, III 307
cardamum, III 325
cardoon, III 26, 27, 139, 147

GENERAL INDEX

cariosa(terra), I 131
carnice, III 405
carob, II 293
carrot, II 437; III 21
cassia, I 275; III 235
castratio(n), I 457; III 89
caterpillar, II 93, 485; III 35, 39, 169, 391
cattle, I 175, 179; II 121, 147, 183, 421
celadine, III 15
celery III, 20
centuria, II 7
cereal, II 51, 69
cevae, II 185
chaff, III 129
chalk, Cimolian, II 173
chamois, II 421
cheese, II 201, 215, 273, 285, 335, 337; III 41
cherry, III 75, 127
chervil, III 15, 139, 155
chestnut, I 449, 457, 459; III 393
chickling, III 129 *see* chickling-vetch
chickling-vetch, I 169, 173, 179, 189; III 113, 129
chick-pea, II 427
chicory, II 389, 393; III 207
cicely, III 24
ciconia, I 317
cinnamon, III 33
citron, III 42–3
clay, Carthaginian, II 449
clima, II 7
coccolobis, I 243
cockroach, II 449
colocassia, II 398–9

consiligo, II 145
convolvulus, III 379
coot, II 397
coriander, III 27, 139, 147, 205
cork-wood, II 423
cornel-berry, III 423
corn-flag, III 15, 33
corruda, III 155
costus, III 235
cow, II 179
cradle (grain), I 217
cremia, III 231
cress, II 395; III 27, 139
cripa, III 235
crocus, III 21
cuckoo-pint, III 27
cucumber, I 149; III 27, 41, 133, 159, 161, 163
culter, I 425
cumin, III 29, 307
custos, I 403
" cut-back ", I 403
cypress, I 429
cytisus, I 173

D

daffodil, III 33
damsons, III 423
darnel, II 351
date, III 235
decuria, I 99
deer, II 421, 427
defrutum, III 237
dens, I 125
dextans, II 11
diabetes, I 283
dill, III 17
distichum, I 155

427

GENERAL INDEX

dittander, III 139, 153, 205
dodrans, II 11
dog, II 305, 315
dog-fish, II 415
dragon-root, III 27
drone, II 495
duck, II 397
dung, ox-, II 489
duramentum, I 401

E

egg, II 341
elder, I 429; III 285
elecampane, III 17, 139, 151, 291, 293
elices, I 143
elm, III 7, 81, 369, 381
embryulkein, II 249
emmer(-wheat), I 137–41, 145, 187, 221; II 313, 395; III 113
emplastration, III 103, 397
endive, II 427; III 147
epityrum, III 299
eryngo, II 145
evergreen, III 87

F

faecinia, I 241
farrago, I 176
fauces, I 417
fennel, II 443; III 7, 17, 27, 123, 163, 201, 203, 207, 215, 223, 231, 259, 295, 297, 299, 301, 305, 307
fenugreek, I 139, 173, 177, 189; II 389; III 111, 115, 239, 251, 253, 325

fig, I 221, 349; II 93, 95, 415, 417, 491; III 43 (varieties), 103, 105, 117, 129, 215, 223, 343, 391
fir, III 199, 291
fish, II 321, 401, 403, 405, 407, 409, 411, 413, 415; III 337
flail (grain), I 217
flax, I 139, 167, 189; III 259, 263, 265, 285
flea, III 35, 167
flounder, II 405
focaneus, I 417
fowl, African, II 323
frankincense, I 275
frog, III 7
fruit-syrup, III 271
furniture, III 187, 191

G

galbanum, II 260–1
garden, III 7, 131, 135
garganey, II 397
garlic, II 143; III 14, 15, 35, 139, 141, 143
gecko, II 449
gilliflower, II 437
gilt-head, II 405, 407
gimlet, I 447
gleucine (oil), III 325
goat, II 277, 283
goby, II 417
goose, II 389
gourd, III 41, 133, 159, 161
grafting, I 437; II 101; III 75, 83, 85, 369, 397
granary, I 71, 185, 277

GENERAL INDEX

grape, I 233 (varieties); III 219, 273
grape-husk, II 399; III 219, 273, 373
grist-mill, I 77
"guardian", III 357
guinea-fowl, II 322–3, 385
gum, III 319
gypsum, III 239, 253, 277

H

hair-cloth, III 291
hake, II 413
halicastrum, I 139
haft's horn, II 351
"hatchet" (of *falx*), I 427
hay, I 211; III 93, 129
hazel, II 95; III 337, 393
hedge, III 39
hellebore, III 7
helvus, I 245
hemlock, III 7
hemp, I 139, 169, 189; III 115, 313
hen, II 323, 327, 331
hexastichum, I 153
hive, II 443, 485
holiday, I 221
honey, II 495; III 7, 41, 99, 103, 199, 211, 263, 265, 269, 271, 289, 297, 303, 335, 337
horehound, III 39, 257
hornbeam, III 125
hornet, II 429, 487
horse, II 189, 197, 199, 201, 207, 209, 211, 213
house-leek, III 201

hyacinth, tassel, III 14, 15
hyena, I 149
hyssop, III 259

I

iners (*vitis*), I 247
iris, III 251, 253, 325
irrigation, I 205, 209, 219; III 133, 135, 159
ivy, II 435
ἰχθυοτροφεῖα, II 321

J

jugerum, II 7
jujube-tree, II 435, 437; III 74–5, 126–7
jumenta, II 119
juniper, I 429; II 317 (prickly)

K

κάμπαι, see caterpillar
"keeper," I 403
kite's foot, III 201, 203
knife, vine dresser's, I 425
κυνόσβατον, III, 131

L

λαγοτροφεῖα, II 323
land, types of, I 109
land, measurements of, II 3
land, plants required for, II 23
land, trenching of, II 5
larkspur, II 437
laser, II 96, 173; III 203, 337, 339, 340, 395
laurel, I 429; III 87

GENERAL INDEX

leek, II 383; III 19, 21, 39, 139, 147, 149, 153, 169, 203, 205, 297, 335
lentil, I 139, 165, 167, 187, 189; II 363
lettuce, II 427; III 15, 21, 23, 139, 145, 207, 335
lily, II 437; III 15, 31, 235 (sword-lily)
lime-tree, linden, II 435; III 389 (lime-tree-wood)
linen, III 149, 291, 325
linseed, III 113
lixivum, III 269
lizard, II 447, 449
locust, II 383
lotus, Indian, II 398–9; "lotus", III 29
lovage, III 337
lucerne, I 139, 173, 175, 191
lupine, I 117, 139, 157, 159, 189, 193, 203; II 143; III 15, 95, 101, 111, 113, 117, 129
lye, III 241, 319

M

mackerel, II, 415
maigres, II 407
mallet-shoot, I 293; III 81, 349, 373
mallows, II 485; III 29
mandrake, III 7
manure, I 77, 137, 195; III 121, 137, 361
marble, "flower" of, III 237
marcus, I 247
marigold, III 15, 35

marjoram, II 435, 437, 443 (wild), 487, 493; III 21, 27, 33, 153, 203, 205, 207, 209, 293, 337
mast, III 119, 129
mastic, II 435; III 295, 297, 299, 301, 303, 305, 391
mead, III 161, 269, 293, 297
μελισσῶνες, II 323
melomeli, III 289
mergi, II 41
merle, sea-, II 413
mildew, III 379
milk, III 203, 205, 405, 407
millet, I 139, 155, 157, 169, 189, 195; II 363, 399; III 89, 111, 151
mint, III 17, 151, 205, 297, 335, 337
molluscs, II 405
moray, *see* murry
mortar, III 149, 325, 333, 335
moth, II 481, 489; *see* caterpillar
motherhood, I 95
mouse, shrew-, II 171
mucro (of *falx*), I 427
mulberry, III 43
mule, II 213, 225
mullet, II 401
murry, II 401, 405, 407, 409, 413
must, III 209, 231, 233, 235, 237, 239, 293, 295, 297, 299, 303, 325, 335
mustard, II, 437; III 17, 139, 147, 201, 333, 335

GENERAL INDEX

mygale, II 171
myrrh, I 275; III 21, 235
myrtle, II 373; III 87, 249, 261, 263, 265

N

narcissus, II 437; III 15
navew, I 171, 173, 209; III 45, 139, 141, 165, 169, 331, 333
Nelumbo, II 398–9
nettle, II 393
nitre, I 163; III 317, 333
νομοφύλακες, III 195
nursery (for the vine), III 345
nut(-tree), III 393
nymph, II 429

O

oak, II 293, 425, 435; III 125, 129, 285
oats, I 173, 177
ochre, II 455; III 347, 385, 389, 391
ointment, III 327
oistroi, II 483
olive, I 75; II 71 (varieties), 73, 75, 85, 441; III 87, 91, 93, 95, 119, 295, 299, 301, 307, 309, 343, 383
onion, III 17, 35, 139, 163, 165, 205, 207, 335
orach, III 41, 155
orchard, I 79; III 387
orchites, III 295, 299, 301
ordure (human), III 393
ὀρνιθοτροφία, II 327
ὀρνιθῶνες, II 321

oryx, II 421, 427
osier, I 449; III 123
ostigo, II 275
ox, I 123, 127; II 127, 129, 137, 141, 143, 145, 151, 153, 155, 157, 161, 167, 169, 175; III 117, 127, 327
oxyacanth, II 283, 293
oxymel, III 335
oyster, II 405; III 69

P

pace, II 7
palm, II 441; III 123, 325
palmation, III 127
pampinarium, I 287
panic grass, I 139, 155, 157, 169, 189, 195; III 399
"paring-edge" (of *falx*), I 427
parsley, III 21, 39, 141, 149, 201, 205, 297, 305, 335, 337
parsnip, III 15, 21, 139, 201
pastinum, I 329
pea, I 139, 159, 187; III 113
peach, III 75
pea-fowl, II 375
pear, I 221, 303, 349; II 97, 99; III 7, 29, 209 (varieties), 343
pelamid, II 415
pennyroyal, II 351; III 201, 203, 259, 335, 337
pepper, III 337
περιστερεῶνες, II 321
pestle, III 149, 333

431

GENERAL INDEX

πετροσέλινον, III 151
phagedaina, II, 479
pickling, III 201
pig, II 291, 299, 305; III 327, 329
pigeon, II 361
pilchard, II 415
pine, III 199, 249, 255, 333, 337, 393, 435
pitch, III 41, 111, 197, 215, 219, 223, 227, 229, 235, 241, 243, 245, 255, 265, 271, 273, 275, 277, 279, 281, 293, 301, 327, 331
plaice, II 405
planting-hole, III 85, 387
plaster, III 237
ploughing, I 135
plum, I 303; III 7, 42–3, 209
" point " (of *falx*), I 427
pollex, I 403
pomegranate, II 97; III 33, 139, 155, 271, 281, 285
pomiferous trees, II 87
pond, I 77
poplar, III 81, 129, 139, 285, 291, 381
poppy, III 15, 35, 155
porcae, I 133; II 7
pot-herb, II 427
poultry, II 323
praesidiarium, I 403
pratum (= *paratum*), I 205
preserves, III 197
privet, III 33
propagation (of vine), III 365
pruning (of vine), I 375, 409; III 73, 85, 89

pulla (*terra*), I 19, 121, 303
pullum, I 169
pulse, I 139; III 97, 127
pulveratio, I 435
puppy, I 221
purple-mollusc, II 405
purslane, III 41, 215

Q

quince, II 99; III 231, 271, 287, 295
quincunx, I 311

R

racemarius, I 333
radish, III 139, 141, 157, 165, 167, 169
radish, wild, II 175, 437; III 207
raisin, III 105, 265, 277, 293, 335, 337
raisin-wine, II 463, 477, 491
ram, II, 237
rape, III 111
rasis, III 235
rauca, III, 385
reed, III 297, 305, 313, 325, 343, 409
" reserve ", I 403
resex, I 403
resin, III 235
robus, I 137
rocket, III 15, 39, 139, 147, 335
roedeer, II 421
rose, II 437; III 9, 15, 29, 33, 81, 85, 253, 343, 411
rosemary, II 435, 437; III 249, 259, 261, 271

GENERAL INDEX

rostrum (of *falx*), I 427
royal, III 295, 385
rubellana, I 241
rue, III 17, 139, 151, 153, 203, 207, 301, 305, 335
rush, I 381; III 35, 235, 301, 325
" rustic " hen, II 385

S

saffron, I 275; III 33, 235
salt, II 131
salt, Ammoniac, II 173
salt, Cappadocian, II 173
salt, Spanish, II 173
samera, II 45
samphire, III 201, 215
satureia, II 435, 437
satyr, III 45
savory, II 435, 437, 443, 487, 493; III 27, 155, 165, 203, 205, 335, 337
scalprum, I 427
scamnum, I 125
scripulum, II 9
sea-acorn, II 405
sea-cabbage, III 21
securis (of *falx*), 427
sedge, I 381
sedum, I 149
seed, I 151
seedling, III 389
" septimontial " (sowing), I 161
septunx, II 11
seris, II 389
serpent, III 39
service(-apple), III, 221, 223

sesame, I 139, 167, 189; III 99, 101, 113, 219, 337
sextans, II 11
sextula, II 9
shad, II 399, 401, 415, 417
shark, II 415
sheep, II 233, 263; III 89, 95, 177
shrew-mouse, III 257, 379
shuttle, III 295, 385
sicilicus, II 9
silico, I 137, 149, 150
siliqua, I 177
silphium, I 167; II, 97, 173; III 203, 337, 340; *see* laser
σιλφιον, III 203
sinus, I 427
siri, I 73
skirwort, III 15, 139, 141, 335
slave, I 83, 89, 91, 97
σμυραῖον, III 151
snail, III 35
snake, II 447, 491; III 41
soapwort, III, 89
soil, I 121, 129
sole, II, 405, 413
solea, III 311
soot, III 39, 167, 317
sorb(-apple), II 415
sorrel, III 39
southern-wood, III 259
spado, I 293
sphondyli, II 405
spikenard (Celtic), III 235
" sponge ", III 155, 157
squill, III 39, 257
stadium, II 7
stake (and prop), III 77

433

GENERAL INDEX

staphylinos, II 437
starwort, II, 437
straw, III 101
stork, I 317
sturgeon, II 407
succory, III 15
" sucker ", I 411
suffrago, I 411
συζυγίαι ἐναντιοτήτων, I 111
" summer root," III 359
surmullet, II 413
swallow, III 13, 81, 83
" swimmer ", III 299, 305

T

tamarisk, II 435, 437, 487
tare, I 139, 173, 175, 179, 209; III 99, 111, 115, 117
taro, II 388–9
teal, II 397
tenant-farmer, I 79
terebinth, II 435
terebration, I 445
thamnum, see bryony
thorn, I 207; III 131 (Christ's thorn), 131 (dog's thorn)
threshing-floor, I 77, 215; III 135
" throat-shoot," I 415
thrush, II 371
" thumb," I 403, 405
thyme, II 435, 437, 487, 493; III 153, 205, 209, 259, 335, 337
tinnunculus, II 365
tisana, I 139
tree, cork, II 443

tree, strawberry, II 373, 425
trefoil, II 113, 389, 435; III 405, 407
trenching, I 311, 313, 353, III 75, 79, 355, 359
trench-fork, I 329, 333
trenching-spade, I 267, 355; III 345
triens, II 11, 13
tripharis, III 405
trifolium, III 29
τρυγητήρ, III 83
tuber-apple, III 75, 127
tudicala, III 311
turbot, II 405, 413
turnip, I 171, 209; III 45, 111, 139, 141, 165, 167, 331, 333; wild, II 437
turpentine, III 235
turtle-dove, II 369

U

ulcer, III 151
umerus, I 285
uncia, II 11
unio, III 209
urine, III 371

V

Valeriana, III 234
varia (terra), I 131, 245
vennuculum, I 139
veratrum, II 225
vermin, III 167, 169
vetch, I 139, 179, 209; II 363, 389, 427; III 75, 99, 111, 113, 115, 117, 129; *see* tare

434

GENERAL INDEX

villa fructuaria, I 65
villa rustica, I 65
villa urbana, I 65
vine, I 229; II 33, 39, 43, 45, 65, 67; III 343; vine, wild, II 355
vine-dresser's knife, I 425
vinegar, III 199, 201, 203, 207, 209, 215, 223, 259, 291, 297, 299, 303, 305, 333, 335, 337, 347
vineyard, II 27, 29; III 351
vintage, III 107, 109, 113, 225
violet, II 437; III 15, 29, 33, 343, 409

W

walnut, III 199, 337, 393
wax, III 319
wax-flower, II 459
weevil, I 73, 163, 167, 219
wheat, I 137, 139, 141, 145, 187, 191; II 313; III 75, 113; *see* emmer

willow, III 35, 37, 81, 123, 149, 287, 409
wine, I 75; III 229
wine, bee-, III 289
wine, faecinian, III 291
wine, raisin-, III 293, 297
withe, I 399; II 443, 501; III 369
wormwood, III 259, 379
wrasse, parrot, II 401, 407, 413, 415

X

χηνοτροφεῖα, II 323

Y

yew, III 7

Z

zea, III 405
zeus, II 407
Zizyphus, II 435, 437 (jujube); III 74–5, 126–7 (tuber-apples)

435